Radon Series on Computational and Applied Mathematics 4

Sergey Repin

A Posteriori Estimates for Partial Differential Equations

Walter de Gruyter · Berlin · New York

Author

Prof. Sergey I. Repin
V. A. Steklov Institute of Mathematics at St. Petersburg
Fontanka 27
191023 St. Petersburg
Russia
E-mail: repin@pdmi.ras.ru

Keywords

Partial differential equations, a posteriori error estimates, Poisson's equation, diffusion problems, elasticity, incompressible viscous fluids, nonlinear problems

Mathematics Subject Classification 2000

00-02, 35-02, 35J05, 35K05, 65M15, 65N15, 74Bxx, 76Dxx

∞ Printed on acid-free paper which falls within the guidelines of the ANSI to ensure permanence and durability.

ISBN 978-3-11-019153-0

Bibliographic information published by the Deutsche Nationalbibliothek

The Deutsche Nationalbibliothek lists this publication in the Deutsche Nationalbibliografie; detailed bibliographic data are available in the Internet at http://dnb.d-nb.de.

Printed in Germany
Cover design: Martin Zech, Bremen.
Typeset using the author's LaTeX files: Kay Dimler, Müncheberg.
Printing and binding: Hubert & Co. GmbH & Co. KG, Göttingen.

Contents

Preface

Pure mathematicians sometimes are satisfied with showing that the non-existence of a solution implies a logical contradiction, while engineers might consider a numerical result as the only reasonable goal. Such one sided views seem to reflect human limitations rather than objective values. In itself mathematics is an indivisible organism uniting theoretical contemplation and active application. R. COURANT [112]

Partial differential equations (PDE's) were introduced as mathematical models of various physical phenomena. In the 20th century, the theory of differential equations was mainly developed in the context of an *a priori conception* that can be expressed by the triad: *existence, regularity, and approximation.* In it, the accent is made on a priori mathematical analysis, and numerical experiment is often regarded as the very last (and in a sense technical) step, which is more related to practical applications than to theory.

A certain revision of views has started 20–30 years ago. It was stimulated by rapid development of numerical methods for PDE's. The experience accumulated in this area shows that a priori methods provide only one part of the information necessary for a comprehensive analysis of models based on PDE's. If differential equations are considered not as a self-contained branch of pure mathematics but as tools consigned to serve natural sciences, then the imperfection of purely a priori analysis is easy to observe. For example, almost all results of regularity theory and asymptotic analysis have a *qualitative* meaning and are addressed to the whole class (or a subclass) of boundary value problems considered. However, in the numerical experiment we always deal with an approximate solution of a particular problem the quality of which must be certified by a certain *quantitative* criterion. The latter task calls for further development of different mathematical methods focused on *a posteriori* analysis of approximate solutions.

The need for new mathematical approaches to the analysis of PDE's is motivated not only by "routine" arguments (such as getting accurate numerical approximations). There is a more fundamental problem: *validation of mathematical models.* Certainly, it can be solved only by joint efforts of mathematicians and specialists in a particular natural science and a posteriori error control methods able to guarantee the reliability of mathematical experiments must play an essential role in such research.

This book can be viewed as an introduction to a posteriori error estimation theory for PDE's, which is now in the process of formation and development. It includes an extended version of the lecture course "A posteriori estimates and adaptivity in continuum mechanics" that was prepared for the Special Radon Semester organized in 2005 by the Radon Institute of Computational and Applied Mathematics in Linz. That course was based on earlier lectures delivered in 2000–2001 for students of the St. Petersburg Polytechnical University (Russia) and in 2003 for students and scientific researchers of the University of Jyväskylä (Finland) and the University of Houston (USA). In 2006, I read a modified version of the course at the Helsinki University of Technology (Finland) and in 2007 at the University of Valenciennes (France). The work with lectures was also supported by the DAAD program of Germany and FIM (Switzerland) during long-term visits to the University of Saarbrücken (Germany) and the Swiss Federal Institute of Tecnology (ETH, Zurich).

For these years, the content and structure of the text varied. However, the main line of it remains the same: for each class of boundary value problems, a posteriori estimates are derived by purely functional methods, which are used in the theory of PDE's for analysis of the corresponding differential equations. In other words, the method suggested for deriving a posteriori error estimates (as well as methods, which study existence and regularity) exploits specific features of a particular mathematical problem, but does not attract properties of approximate solutions, mesh, and numerical method used (the latter information can be utilized later). As a result, we obtain estimates that contain no mesh-dependent constants and are valid for any approximation from the corresponding energy space. This new *functional approach* to the a posteriori error estimation developed in the last decade is the main subject of the book.

A posteriori estimates of the functional type came about from two sources. The variational statement of the problem (if it has variational form) generates the first derivation method, which is well exposed in many papers and in the book [244]. The second method is based on transformations of integral identities that define generalized solutions. Most results exposed in Chapters 3–9 are obtained with the help of this "nonvariational" method. The main idea of it is briefly as follows:

An integral identity (variational inequality) that defines a generalized solution is a source of guaranteed and computable bounds of the difference between this solution and any function from the corresponding energy space.

Chapter 2 contains a concise overview of various a posteriori error estimation methods developed in the 20th century, which are different from those considered in subsequent chapters. Its purpose is to give only a general presentation and to discuss several approaches to the error control problem. The reader interested in their detailed investigation is referred to relevant literature.

In Chapter 3, the basic ideas of the functional approach are explained with the paradigm of a simple elliptic problem. The next chapters are devoted to particular classes of problems: diffusion, linear elasticity, variational inequalities, etc. Their goal is not only to present new estimates, but to demonstrate the *method* used for deriving them. The latter problem is even more important because the understanding of basic ideas would allow the reader to derive a posteriori estimates for a concrete problem of interest.

The book rests upon a moderate background in functional analysis and the theory of PDE's. I hope that it will be useful for advanced specialists in the mathematics of computations, as well as for students specialized in applied mathematics, and for numerical analysts.

I wish to express my deep gratitude to Prof. O. A. Ladyzhenskaya for comments and advice she gave me in 2000–2003. During the last decade I has had fruitful discussions of problems related to the topic with many colleagues in Europe and USA; sincere thanks to all of them. Some of these discussions have resulted in joint works presented on the reference list.

I thank Dr. R. Plato and Dr. N. Lebedinskaya for their kind help in editing the book.

My special gratefulness is to the Radon Institute for Computational and Applied Mathematics (RICAM) in Linz, where the idea of this book appeared and was actively supported. Also, I am very grateful to the University of Jyväsylä and the Academy of Finland for a long-term support of my research.

Sergey Repin Saint Petersburg, 2008

1 Introduction

1.1 A priori and a posteriori methods of error estimation

Partial differential equations (PDE's) were divised as mathematical models of various physical phenomena. Over the 20th century, mathematical models based on PDE's have been intensively investigated as part of the new rapidly developing science *Mathematical Modeling*[1].

Originally, the analysis of models generated by differential equations was performed in the framework of an *a priori method* the main steps of which are as follows:

- proving existence and uniqueness of a solution;
- studying higher differentiability of a solution and deriving regularity estimates;
- establishing rate convergence estimates for sequences of approximations.

Proving the fact that the problem considered possesses a (unique) solution is the first principal step. The second step is focused on qualitative properties of exact solutions (the existence of higher derivatives, the continuity and smoothness of a solution, or the behavior in the vicinity of special points, the type of possible singularities). This a priori information can be further utilized in approximation methods. The third step started receiving serious attention in the 50–60s when first systematic results in the error control theory for PDE's were obtained. Classical convergence theory is perfectly presented, e.g., in P. Ciarlet [107] and G. Strang and G. Fix [344]. It is focused on *asymptotic estimates* of approximation errors. The ultimate aim of this theory is to prove that the difference between an exact solution u and an approximation u_k found in a finite-dimensional subspace of dimension k tends to zero as $k \to \infty$. In a sense, proving such a convergence can be viewed as a formal justification of the approximation method used. More exact estimates show the convergence rate, i.e., they establish the relation

$$\|u - u_k\|_V \leq c \left(\frac{1}{k}\right)^m, \tag{1.1.1}$$

where V is an appropriate space and c and m are some positive real numbers independent of k. The value of the constant c depends on the exact solution u and on the type of approximations used. These estimates give the most general information on the behavior of approximation errors and are often called *a priori* error estimates.

Usually, a priori error estimates establish the asymptotic behavior of an upper bound of the approximation error for the *whole set of solutions and their approximations of a certain type*. However, they are generally unable to evaluate efficiently the error

[1] In recent years, Mathematical Modeling has come under consideration as part of the Computational Science that "constitutes the third pillar of the scientific enterprize, peer alongside Theory and Physical Experiment" [267].

related to a particular approximate solution computed on a particular mesh. Moreover, they possess other features that make their practical exploitation rather difficult. First, they are valid *only for Galerkin approximations*. In practice, it is often difficult to guarantee that an approximate solution computed by a numerical procedure is indeed the exact solution of a respective finite-dimensional problem. Second, such estimates require *the extra regularity* of exact solution. However, in many practically important cases exact solutions do not have such regularity.

These drawbacks stimulated efforts focused on new error estimation methods able to characterize explicitly the accuracy of approximate solutions. In the late 70s and early 80s, it became clear that successful numerical methods for PDE's should be based on the so-called *mesh-adaptive* procedures that modify finite-dimensional spaces (and meshes) using the information comprised in the approximate solution computed at the previous step[2]. In fact, at that time it was understood that *a priori* methods in the theory of PDE's should be combined with *a posteriori* ones. In the a posteriori method, getting an approximate solution is not the final step but a starting point of a new analysis, which is aimed at (a) estimating the accuracy of the solution and (b) indicating the distribution of errors over the domain. Obviously, (a) gives a stopping criterion and (b) provides information needed for a correction of the finite-dimensional subspace.

A posteriori error estimates, together with other related topics (such as "mesh refinement" and "adaptivity"), have attracted much attention in the last decades. The above problems create the basis for a new wave of investigations in numerical analysis united by the common name "reliable modeling". Nowadays, *Fully Reliable Mathematical Modeling* based on advanced methods of computer simulation forms one of the most challenging scientific directions the development of which is of great importance for many applied sciences. Fully reliable modeling consists of

- efficient computation of a sequence of approximate solutions that converges to the desired exact solution and
- reliable verification of the accuracy of the approximation obtained.

The second problem is discussed in the present book.

1.2 Book structure

Throughout Chapter 1, we discuss the error control problem in general terms and present a short summary of mathematical notions and notation necessary for the understanding of the material.

[2]An interesting discussion on the reliability of different computer simulation methods in nonlinear mechanical problems is contained in [15]. It reflects early observations of the fact that approximations of strongly nonlinear problems (and not just them) may seriously depend on the structure of the finite-dimensional space (mesh) used.

Chapter 2 contains a concise overview of a posteriori error estimation methods. We discuss the very first error indicators suggested by Runge, Prager and Synge, Mikhlin, and Ostrowski. After that, we briefly present another group of a posteriori error estimates that follow from the monotonicity of operators. In the application to PDE's, the methods of this group are close to the theory of *positive* solutions and the *maximum* principle. Theoretically, they could give a posteriori error estimates of the strongest (pointwise) type. A short discussion of them presented in the chapter is intended to give preliminary knowledge (and to generate an interest in further study) rather than to expose a complete theory.

The subsequent material is devoted to error indication methods created in the 70s–90s for *finite element approximations*. We consider the classical *explicit residual* method, methods based on the post-processing of approximate solutions, and the *dual-weighted residual* method, which have gained high popularity in the computational community in the last decade. The purpose of this overview is to give only a general idea of various a posteriori error indicators, which are usually used in adaptive numerical methods. The reader interested in a detailed investigation of them is referred to relevant literature.

In Chapter 3, we start the study of functional a posteriori estimates. To express the main ideas in the most transparent form throughout the chapter we consider only one problem: Poisson's equation with Dirichlet boundary conditions. For this problem, we explain both derivation methods (variational and nonvariational) and show that they lead to exactly the same a posteriori error estimates. Further, we discuss properties, modifications, and practical implementation of the estimates.

Chapter 4 is devoted to diffusion problems. We begin with the linear stationary problem, which can be viewed as the first generalization of the problem treated in Chapter 3. A posteriori estimates are now derived for mixed, Neumann, and Robin boundary conditions. Further, we find error estimates for reaction-diffusion and convection-diffusion problems. In the last part of the chapter, we consider the case where the stationary diffusion equation involves both reaction and convection terms. A posteriori estimates for nonstationary diffusion equations are obtained in Chapter 9.

Chapter 5 deals with elliptic systems arising in linear elasticity theory. We obtain error estimates in terms of displacements and stresses and especially discuss the case of isotropic elasticity, which is typical of applications.

In Chapter 6, we present the error estimation method for models in the theory of viscous fluids. Among them we examine the stationary Stokes problem and its generalizations, the Oseen problem, and the stationary Navier–Stokes equation (for a sufficiently small velocity). Models related to generalized Newtonian fluids are discussed in Chapter 8.

The experience gained by studying Chapters 3–6 will help the reader to look at the theory from general positions. This is what Chapter 7 is intended for. Here, we evolve the variational and nonvariational methods of a posteriori error estimation in operator form. Using the material of this chapter it is easy to obtain a posteriori estimates for a particular class of problems that fits the general conditions imposed on the operators. In the last section of Chapter 7, we discuss mixed formulations of elliptic problems and show that the a posteriori estimates can also be derived by transforming the relations that define the saddle point of the corresponding Lagrangian.

The book is mainly focused on linear problems. A concise excursion to the theory of a posteriori error estimation for nonlinear problems is presented in Chapter 8. First, we deal with elliptic variational inequalities (of the first and second type) and show that proper transformations of variational inequalities also lead to computable estimates of approximation errors. It is of interest that the same estimates have earlier been derived by the variational method, which we also discuss in the chapter. Further, we consider generalizations of the variational and nonvariational methods to a wide class of nonlinear elliptic problems. A posteriori estimates of the functional type for various nonlinear problems have intensively been studying during the last decade. However, a consequent exposition of the corresponding results is beyond the scope of this book. Instead of this, in the last section of the chapter we give an overview of the results and a deeper discussion of the estimates derived for the α-Laplacian, for problems with nonlinear boundary conditions, and for generalized Newtonian fluids.

Chapter 9 is concerned with applications of error estimation theory to some other problems. It starts with equations of higher order and a class of problems associated with the operator curl. Certainly, the estimates presented follow from the general theory of Chapter 7. However, it seems pertinent to give a more transparent discussion of these (practically important) problems. A method that yields a posteriori functional estimates for evolutionary problems is given in the third section. Section 4 is devoted to a posteriori error estimates of the new type derived for optimal control problems with the help of functional error majorants. In the next section, we examine two error estimation methods for nonconforming approximations (they also follow from error estimation theory presented in Chapters 3–4). Another important topic concerns the influence of errors caused by indeterminacy of data, which always exists in real-life problems. In Section 6, we show that using new error estimation methods opens a way for measuring the errors that arise owing to data uncertainty. Finally, we discuss the problem of error estimation in terms of quantities other than the (global) energy norm. We show that with the help of functional error majorants one can obtain errors in local and weighted norms, in goal-oriented functionals, and in some other (nonenergy) norms.

In conclusion, it is worth adding a comment on practical applications of the a posteriori estimates presented in the book. In the last decade, serious efforts were concentrated on the study of this subject. The estimates were numerically tested for diffusion (elliptic and parabolic), stationary Stokes and Oseen problems, linear elasticity and thermoelasticity, variational inequalities, optimal control problems, and nonconforming approximations of elliptic problems. A posteriori estimates of the functional type were used together with standard software packages (such as MATLAB and ANSYS) and compared with a posteriori error estimates of other types. Summarizing the experience accumulated, we can say that new a posteriori estimates are more expensive (e.g., if we compare them with error indicators based on gradient averaging). However, they have two serious advantages: the error bounds are guaranteed and are valid for conforming approximations of all types. The latter property is especially important in practical computations where it is difficult to guarantee that the true solution is indeed regular, the approximate solution satisfies the Galerkin orthogonality condition, and it may be necessary to adapt quickly the error estimation method to approximations of new types. In addition, functional a posteriori estimates do not contain mesh-dependent constants and can be computed with the help of a unified numerical technology (and one computer code, which is a *checker* independent of a *solver*).

The reader will find references to publications including numerical results and their discussion in respective sections of the book. However, a systematic exposition of all these results would make the book unreasonably large. For this reason, it is concentrated on the explanation of key ideas, methods, and algorithms. Studying them would allow the thoughtful reader to implement easily the estimates in his/her own code and check how the error estimation technology works.

1.3 The error control problem

If we wish to use differential equations as mathematical models of real-life objects and processes, then the main question is how to solve them reliably. By the word "solve" we mean the ability to get detailed quantitative information on the exact solution. Regrettably, analytic methods yield solutions to a restricted amount of problems. Therefore, in the vast majority of cases "solving" is reduced to a numerical procedure resulting in an approximate solution. Thus, a mathematical experiment should be regarded as the crucial step that provides information, which is either difficult or impossible to get by other methods. The general principle of scientific objectivity suggests that

the mathematical experiment must obey the same strict authenticity rules as those commonly accepted in natural sciences.

In other words, mathematical modeling cannot be confident without answering the question:

What is the guaranteed accuracy of an approximate solution?

A priori rate convergence estimates supply a rather philosophical answer to this question, namely: "If the dimension of subspaces increases in a certain proportion (and if all subspaces are regular in some sense), then the upper bound of the error established for the whole class of approximations decreases with a certain rate". Such an answer cannot be accepted as a sufficient one. An a priori estimate may be unable to give a realistic estimate for a particular approximate solution. Moreover, there are several other questions that convincingly call for a different approach to the error control problem. For example:

- Which part of the difference between computed solutions and observable (physical) data is related to approximations (numerical method) and which is generated by the mathematical model used?
- How accurately do we know the coefficients, domain, boundary and initial conditions and how does the indeterminacy in this knowledge affects the solution?
- How sensitive is the solution with respect to changes in the mathematical model (e.g., with respect to adding (removing) lower terms to the equation, changing boundary conditions or using other modifications)?

These are difficult questions, and very often they are ignored in engineering and scientific computations. However, if they are not answered, then computer simulation is but a producer of more or less probable conjectures (which sometimes may lead to wrong conclusions).

Reliable answers to these and other questions cannot be found unless the following **main error control problem** is solved:

> *Given the data (coefficients,a domain, boundary conditions) of a boundary value problem having the exact solution u and a function v from the corresponding (energy) space V, compute the radii r_1 and r_2 of two balls $\mathcal{B}(v, r_1)$ and $\mathcal{B}(v, r_2)$ centered at v such that*

$$u \notin \mathcal{B}(v, r_1) \quad \text{and} \quad u \in \mathcal{B}(v, r_2). \tag{1.3.1}$$

We say that a method used to solve the above problem is *sharp* if one can find r_1 and r_2 such that $r_2 - r_1 \leq \epsilon$ for any given $\epsilon > 0$.

Obviously, this problem is solved if we have the estimate

$$\underline{\mathfrak{M}}\,(v, \mathcal{D}) \leq \|u - v\|_V \leq \overline{\mathfrak{M}}\,(v, \mathcal{D}), \qquad \forall v \in V, \tag{1.3.2}$$

where $\|\cdot\|_V$ stands for the norm of V, $\mathcal{D}$ denotes the set of known data and the functionals $\overline{\mathfrak{M}}$ (error majorant) and $\underline{\mathfrak{M}}$ (error minorant) are *directly computable*. This estimate establishes two-sided guaranteed error bounds for conforming approximations of all types.

Throughout the book, the functionals that give upper and lower bounds of approximation errors are denoted by $\overline{\mathfrak{M}}$ and $\underline{\mathfrak{M}}$, respectively. Subscripts are used to specify such a functional and relate it to a particular problem: for example, $\overline{\mathfrak{M}}_{ST}$ is the majorant for the Stokes problem.

Estimate (1.3.2) has a practical significance if the functionals $\underline{\mathfrak{M}}$ and $\overline{\mathfrak{M}}$ satisfy an additional *consistency* condition:

$$\underline{\mathfrak{M}}\,(v, \mathcal{D}) \to 0, \qquad \overline{\mathfrak{M}}\,(v, \mathcal{D}) \to 0 \quad \text{as } v \to u \text{ in } V, \tag{1.3.3}$$

which guarantees that the majorant (minorant) vanishes on any sequence converging (in the energy space) to the exact solution. The majorants discussed in the book satisfy this condition.

Another important requirement imposed on $\overline{\mathfrak{M}}$ and $\underline{\mathfrak{M}}$ is that (1.3.2) must hold *without extra regularity assumptions* on the exact solution u (which in a priori error estimates are usually required). From the practical point of view, such a property of an error estimate is very useful, because the regularity (smoothness) of exact solutions is unstable with respect to small variations in data (coefficients of a PDE and the geometry of a domain). In real life problems, these data are never known exactly, so that numerical analysts and engineers should use approximation and error control methods that do not exploit (explicitly or implicitly) higher regularity of exact solutions and are *stable with respect to small variations in problem data.*

In Chapters 3–9 we derive the functionals $\underline{\mathfrak{M}}$ and $\overline{\mathfrak{M}}$ for various classes of boundary-value problems. In addition to v, they also include the known data $\mathcal{D}$ and "free" function(s) y, which can be viewed as approximations of certain differential operator(s) applied to u. For example, in diffusion problems they are associated with ∇u or $A\nabla u$ (where A is the diffusion matrix) and in linear elasticity with $\mathbb{L}\varepsilon(u)$ (where $\mathbb{L}$ is the elasticity tensor and $\varepsilon(u)$ is the tensor of small strains). These differential complexes have a physical meaning and are often called "dual" variables. We consider them as arguments and rewrite (1.3.2) in the form

$$\underline{\mathfrak{M}}\,(v, y, \mathcal{D}) \leq \|u - v\|_V \leq \overline{\mathfrak{M}}\,(v, y, \mathcal{D}), \qquad \forall v \in V,\ y \in Y, \tag{1.3.4}$$

where Y is a certain set (usually it is a space) that contains admissible y.

Also we derive computable functionals that provide reliable bounds for the errors estimated in terms of combined (primal-dual) norms (e.g., in terms of the norm $\|(u - v)\|_V + \|(p - q)\|_Y$, where p denotes the exact dual function and q is an approximation of p).

Once the estimate (1.3.4) has been derived for a class of boundary value problems, the computer simulation methods for this class are *fully controllable*. Therefore, such estimates should be derived for each mathematical model used for the quantitative analysis. The aim of this book is not only to present $\overline{\mathfrak{M}}$ and $\underline{\mathfrak{M}}$ for particular classes of boundary value problems but to explain the basic principles that will help the reader to derive similar estimates for a particular mathematical model he/she studies.

Finally, we note that (1.3.4) generates the variational statement

$$\inf_{\substack{v \in V, \\ y \in Y}} \overline{\mathfrak{M}}\,(v, y, \mathcal{D}).$$

In principle, a sequence of approximations converging to the exact solution can be computed by a certain minimization procedure applied to $\overline{\mathfrak{M}}\,(v, y, \mathcal{D})$. The values of $\overline{\mathfrak{M}}$ indicate the quality of v and y as approximations of the exact solutions u and p, so that the efficiency of an approximation process is explicitly controlled. However, in practice such a minimization procedure may be rather expensive, and it is more efficient to find v and y by other (e.g., mixed) methods.

1.4 Mathematical background and notation

It is assumed that the reader is familiar with basic facts in functional analysis and the theory of differential equations and understands such notions as "generalized derivative", "weak solution", Lebesgue and Sobolev space. Those feel necessary to study the mathematical background are advised to use, e.g., [151, 197, 234, 211]. The content of these books is quite sufficient for the understanding of the material.

1.4.1 Vectors and tensors

By $\mathbb{R}^d$ and $\mathbb{M}^{d\times d}$ we denote the spaces of real d-dimensional vectors and $d \times d$ matrixes (tensors), respectively. The scalar product of vectors is denoted by $\cdot$, and for the product of tensors we use the symbol $:$, i.e.,

$$u \cdot v = u_i v_i, \qquad \tau : \sigma = \tau_{ij}\sigma_{ij},$$

where summation (from 1 to d) over repeated indices is implied. The norms of vectors and tensors are defined with the help of the products introduced above:

$$|a| := \sqrt{a \cdot a}, \qquad |\sigma| := \sqrt{\sigma : \sigma}.$$

Henceforth, the symbol $:=$ means "equals by definition". The multiplication of a matrix $A \in \mathbb{M}^{d\times d}$ and a vector $b \in \mathbb{R}^d$ is the vector, which we denote Ab. In the book, matrixes are usually denoted by capital Latin letters or by Greek letters (e.g., σ, τ, ε). A^{T} and A^{-1} denote the transposed and inverse matrixes, respectively.

Any tensor τ is decomposed into the *deviatoric* part τ^{D} and the *trace* $\operatorname{tr}\tau := \tau_{ii}$, so that $\tau := \tau^{\mathrm{D}} + \frac{1}{d}\mathbb{I}\operatorname{tr}\tau$, where $\mathbb{I}$ is the unit tensor. It is easy to check that

$$\tau : \mathbb{I} = \operatorname{tr}\tau, \quad \tau^{\mathrm{D}} : \mathbb{I} = 0, \tag{1.4.1}$$

$$|\tau|^2 = |\tau^{\mathrm{D}}|^2 + \frac{1}{d}(\operatorname{tr}\tau)^2, \tag{1.4.2}$$

so that we have an orthogonal decomposition of τ into two parts (which sometimes are called deviatorical and spherical).

In the book, we use various inequalities for scalars, vectors, and tensors. First, we recall the algebraic Young's inequality

$$2ab \leq \beta a^2 + \frac{1}{\beta}b^2, \tag{1.4.3}$$

which is valid for any $\beta > 0$. For a pair of vectors a and b we have a similar estimate

$$2a \cdot b \leq \beta|a|^2 + \frac{1}{\beta}|b|^2, \tag{1.4.4}$$

which implies the inequalities

$$|a+b|^2 \leq (1+\beta)|a|^2 + \frac{1+\beta}{\beta}|b|^2, \tag{1.4.5}$$

$$|a+b|^2 \geq \frac{1}{1+\beta}|a|^2 - \frac{1}{\beta}|b|^2. \tag{1.4.6}$$

Similarly, for a pair of tensors σ and τ we have

$$2\tau : \sigma \leq \beta|\tau|^2 + \frac{1}{\beta}|\sigma|^2, \tag{1.4.7}$$

$$|\tau+\sigma|^2 \leq (1+\beta)|\tau|^2 + \frac{1+\beta}{\beta}|\sigma|^2. \tag{1.4.8}$$

If H is a Hilbert space with scalar product $(\cdot,\cdot)$ and norm $\|\cdot\|$ associated with the product, then it is easy to extend (1.4.4)–(1.4.6) to the elements of H.

The inequality (1.4.1) is a particular form of the more general Young's inequality

$$ab \leq \frac{1}{p}(\beta a)^p + \frac{1}{p'}\left(\frac{b}{\beta}\right)^{p'}, \qquad \frac{1}{p}+\frac{1}{p'} = 1. \tag{1.4.9}$$

Differential operations for vectors and tensors are introduced with the help of a symbolic vector $\nabla = i_s \frac{\partial}{\partial x_s}$, where i_s $(s = 1,\dots,d)$ denote the init vectors of the Cartesian system. We recall that $i_s \cdot i_k = \delta_{sk}$, where $\delta_{sk} = 0$ if $s \neq k$ and $\delta_{kk} = 1$.

If ψ is a differentiable scalar-valued function then $\nabla\psi$ is the vector $\left\{\frac{\partial\psi}{\partial x_1}, \frac{\partial\psi}{\partial x_2}, \frac{\partial\psi}{\partial x_3}\right\}$, which is the *gradient* of ψ. For a vector-valued function $a = a_s i_s$ we have the following differential operations:

$$\nabla\cdot a = i_s\frac{\partial}{\partial x_s}\cdot a_k i_k = \frac{\partial a_s}{\partial x_s} = a_{s,s} := \operatorname{div} a,$$

$$\nabla\times a = \operatorname{curl} a := (a_{3,2}-a_{2,3}; a_{1,3}-a_{3,1}; a_{2,1}-a_{1,2}) \qquad (\text{for } d=3),$$

$$\nabla a = \nabla\otimes a = \frac{\partial a_k}{\partial x_s} i_s\otimes i_k = a_{k,s} i_s\otimes i_k.$$

Let ψ and a be a differentiable function and vector-valued function, respectively. It is easy to see that

$$\nabla\cdot(\psi a) = \frac{\partial}{\partial x_s} i_s\cdot(\psi a_k i_k) = \frac{\partial}{\partial x_s}(\delta_s^k \psi a_k) = \frac{\partial\psi}{\partial x_s} a_s + \frac{\partial a_s}{\partial x_s}\psi$$

and we arrive at the relation

$$\operatorname{div}\psi a = a\cdot\nabla\psi + \psi\operatorname{div} a. \tag{1.4.10}$$

Other differential relations, namely,

$$\begin{aligned}\nabla(\phi\psi) &= \phi\nabla\psi + \psi\nabla\phi,\\ \nabla(\psi\, a) &= \nabla\psi\otimes a + \psi\nabla a,\\ \nabla\times(\psi\, a) &= \psi(\text{curl } a) + \nabla\psi\times a,\\ \nabla\cdot(a\times b) &= b\cdot\text{curl } a - a\cdot\text{curl } b,\\ \nabla\cdot(a\otimes b) &= b\,\text{div } a + a\cdot\nabla b\end{aligned}$$

are also often used in the analysis of boundary value problems.

Differential operations of the second order are not always defined. Some others lead to a trivial result. We recall that

$$\begin{aligned}\nabla\times(\nabla\psi) = 0, \qquad \nabla\cdot(\nabla\times a) = 0,\\ \text{div}\,\nabla\psi = \nabla\cdot\nabla\psi = (\nabla\cdot\nabla)\psi = \Delta\psi,\end{aligned}$$

and

$$\nabla\times(\nabla\times a) = \nabla(\nabla\cdot a) - (\nabla\cdot\nabla)a = \text{grad}\,\text{div}\, a - \nabla^2 a.$$

A vector-valued function a is called *solenoidal* if it can be represented as curl b, where b of another vector-valued function. In this case, div $a = 0$, so that solenoidal fields are *divergence-free*.

Similar relations hold for tensors. For example, if τ is a tensor-valued function with differentiable components, then

$$\text{Div}\,(\tau a) = a\cdot\text{Div}\,\tau + \tau^{\mathrm{T}} : \nabla a. \tag{1.4.11}$$

By Div τ we denote the divergence of a tensor-valued function τ, which is a vector-valued function with components $\{\tau_{ij,j}\}$.

In subsequent chapters, we use several known integral relations. The first of them is the Ostrogradski formula:

$$\int_\Omega \text{div}\, a\, dx = \int_\Gamma a\cdot n\, ds, \tag{1.4.12}$$

where Γ is a closed surface of Ω.

By (1.4.10) and (1.4.12), we obtain

$$\int_\Omega \text{div}\,(\psi a)\, dx = \int_\Omega (\psi\,\text{div}\, a + \nabla\psi\cdot a)\, dx = \int_\Gamma \psi a\cdot n\, ds \tag{1.4.13}$$

If ψ vanishes on Γ, then from (1.4.13) it follows that

$$\int_\Omega (\psi\,\text{div}\, a + \nabla\psi\cdot a)\, dx = 0. \tag{1.4.14}$$

Another integral relation is

$$\int_\Omega \operatorname{curl} a \cdot v \, dx = \int_\Omega a \cdot \operatorname{curl} v \, dx - \int_\Gamma (a \times n) \cdot v \, ds. \tag{1.4.15}$$

In the above relations, we assume that the functions are sufficiently regular so that the corresponding volume and surface integrals exist.

Finally, we recall the Helmholtz's theorem, which is known as the fundamental theorem of vector calculus. It states that any sufficiently smooth, rapidly decaying vector-valued function can be resolved into curl-free and divergence-free (solenoidal) components, i.e.,

$$q = \nabla\psi + \operatorname{curl} a. \tag{1.4.16}$$

For a bounded domain Ω, Helmholtz decomposition states that q can be resolved into a solenoidal vector-valued function q_0 and $\nabla\psi$, where $\psi \in \overset{\circ}{H}{}^1(\Omega)$ (e.g., see [232]). Another version (which is often used in mathematical hydrodynamics) resolves q into q_0 and $\nabla\psi$ such that $\operatorname{div} q_0 = 0$, $q_0 \cdot n = 0$ on Γ, and ψ is a scalar-valued function having (generalized) derivatives of the first order (e.g., see [210] or [348]).

1.4.2 Spaces of functions

We denote a bounded connected domain in $\mathbb{R}^d$ by Ω and its boundary (which is assumed to be Lipschitz continuous) by Γ. Usually, ω stands for an open subset of Ω. The closure of sets is denoted by a bar and the Lebesgue measure of a set $\omega \in \mathbb{R}^k$ by $\operatorname{meas}_k \omega$. If ω is a domain in $\mathbb{R}^d$ and $\gamma \in \mathbb{R}^{d-1}$ is its boundary, then we also use simplified notation $|\omega|$ and $|\gamma|$ for the corresponding d and $d-1$ measures.

By $L^p(\omega)$ we denote the space of functions summable with power p with norm

$$\|w\|_{p,\omega} := \left(\int_\omega |w|^p dx \right)^{1/p}.$$

The vector-valued functions with components that are square summable in Ω form the Hilbert space $L^2(\Omega, \mathbb{R}^d)$. Analogously, $L^2(\Omega, \mathbb{M}^{d\times d})$ is the Hilbert space of tensor-valued functions (sometimes we use the special notation Σ for this space). If tensor-valued functions are assumed to be symmetric, then we write $\mathbb{M}_s^{d\times d}$ (and Σ_s instead of $L^2(\Omega, \mathbb{M}_s^{d\times d})$).

For $v \in L^2(\Omega, \mathbb{R}^d)$ and $\tau \in L^2(\Omega, \mathbb{M}^{d\times d})$, the norms are defined by the relations

$$\|v\|^2 := \int_\Omega |v|^2 \, dx \quad \text{and} \quad \|\sigma\|^2 := \int_\Omega |\sigma|^2 \, dx.$$

Since no confusion may arise, we denote the norm of $L^2(\Omega)$ and the norm of the space $L^2(\Omega, \mathbb{R}^d)$ by $\|\cdot\|$. The space of measurable essentially bounded functions is denoted by $L^\infty(\Omega)$. It is equipped with the norm

$$\|u\|_{\infty,\Omega} = \operatorname*{ess\,sup}_{x\in\Omega} |u(x)|.$$

By $\overset{\circ}{C}{}^{\infty}(\Omega)$ we denote the space of all infinitely differentiable functions with compact supports in Ω. The spaces of k-times differentiable scalar- and vector-valued functions are denoted by $C^k(\Omega)$ and $C^k(\Omega, \mathbb{R}^d)$, respectively; $\overset{\circ}{C}{}^k(\Omega)$ is the subspace of $C^k(\Omega)$ that contains functions vanishing at the boundary; $P^k(\Omega)$ denotes the set of polynomial functions defined in $\Omega \subset \mathbb{R}^d$, i.e., $v \in P^k(\Omega)$ if

$$v = \sum_{|\alpha| \leq m} a_\alpha x^\alpha, \quad m \leq k,$$

where $\alpha := (\alpha_1, \ldots, \alpha_d)$ is the so-called multi-index,

$$|\alpha| = \alpha_1 + \alpha_2 + \ldots + \alpha_d, \qquad a_\alpha = a_{\alpha_1,\ldots,\alpha_d},$$

and $x^\alpha = x^{\alpha_1} x^{\alpha_2} \ldots x^{\alpha_d}$.

Solenoidal vector-valued functions the components of which are square summable in Ω form the space $\mathrm{S}(\Omega)$.

For partial derivatives we keep the standard notation and write

$$\frac{\partial f}{\partial x_i} \quad \text{or} \quad f_{,i}.$$

Usually, we understood them in a generalized sense: a function $g = f_i$ is called the generalized derivative of $f \in L^1(\Omega)$ with respect to the x_i if it satisfies the relation

$$\int_\Omega f w_{,i}\, dx = -\int_\Omega g w\, dx, \qquad \forall w \in C_0^1(\Omega). \tag{1.4.17}$$

Generalized derivatives of higher orders are defined by similar integral relations (see S. Sobolev [336]).

By $\{g\}_S$ we denote the mean value of a function g on S, i.e.,

$$\{g\}_S := \frac{1}{|S|} \int_S g\, dx$$

and $\widetilde{g}_S := g - \{g\}_S$. The functions with zero mean form the space

$$\widetilde{L}^2(\Omega) := \left\{ q \in Q \,\middle|\, \{q\}_\Omega = 0 \right\}.$$

The space $H(\Omega, \mathrm{div})$ is a subspace of $L^2(\Omega, \mathbb{R}^d)$ that contains vector-valued functions with square-summable divergence, and $H(\Omega, \mathrm{Div})$ is a subspace of Σ that contains tensor-valued functions with square-summable divergence, i.e.,

$$H(\Omega, \mathrm{div}) := \{v \in L^2(\Omega, \mathbb{R}^d) \mid \mathrm{div}\, v \in L^2(\Omega)\},$$
$$H(\Omega, \mathrm{Div}) := \{\tau \in L^2(\Omega, \mathbb{M}^{d \times d}) \mid \mathrm{Div}\, \tau \in L^2(\Omega, \mathbb{R}^d)\}.$$

Both spaces $H(\Omega, \mathrm{div})$ and $H(\Omega, \mathrm{Div})$ are Hilbert spaces endowed with scalar products

$$(u, v)_{\mathrm{div}} := \int_\Omega (u \cdot v + \mathrm{div}\, u\, \mathrm{div}\, v)\, dx$$

and

$$(\sigma, \tau)_{\mathrm{Div}} := \int_\Omega (\sigma : \tau + \mathrm{Div}\, \sigma \cdot \mathrm{Div}\, \tau)\, dx,$$

respectively. The norms $\|\cdot\|_{\mathrm{div}}$ and $\|\cdot\|_{\mathrm{Div}}$ are associated with the above-defined scalar products.

Similarly, $H(\Omega, \mathrm{curl})$ is the Hilbert space of vector-valued functions having square-summable curl or, i.e.,

$$H(\Omega, \mathrm{curl}) := \{v \in L^2(\Omega, \mathbb{R}^d) \mid \mathrm{curl}\, v \in L^2(\Omega)\}.$$

This space can be defined as the closure of smooth functions with respect to the norm

$$\|w\|_{\mathrm{curl}} := \left(\|w\|_\Omega^2 + \|\mathrm{curl}\, w\|_\Omega^2\right)^{1/2}.$$

The Sobolev spaces $W^{m,p}(\Omega)$ (where m and p are positive integer numbers) contain functions summable with power p the generalized derivatives of which up to order m belong to L^p. For a function $f \in W^{m,p}(\Omega)$, the norm is defined as usual:

$$\|f\|_{m,p,\Omega} = \left(\int_\Omega \sum_{|\alpha| \le m} |D^\alpha f|^p\, dx\right)^{1/p}.$$

Here $\alpha = \{\alpha_1, \ldots, \alpha_d\}$ is the multi index and

$$D^\alpha v = \frac{\partial^{|\alpha|} v}{\partial x_1^{\alpha_1} \ldots \partial x_d^{\alpha_d}}$$

is the derivative of order $|\alpha|$. The Sobolev spaces with $p = 2$ are denoted by the letter H, i.e.,

$$H^m(\Omega) := \left\{v \in L^2(\Omega) \mid D^\alpha v \in L^2(\Omega), \quad \forall m : |\alpha| \le m\right\}.$$

These spaces belong to the class of Hilbert spaces. A subset of $H^m(\Omega)$ formed by the functions vanishing on Γ is denoted by $\overset{\circ}{H}{}^m(\Omega)$.

The functions in Sobolev spaces have counterparts on Γ (and on other manifolds of lower dimensions) that are associated with spaces of *traces*. Thus, there exist some bounded operators mapping the functions defined in Ω to functions defined on the boundary. For example, the operator $\gamma : H^1(\Omega) \to L^2(\Gamma)$ is called the *trace* operator if it satisfies the following conditions:

$$\gamma v = v|_\Gamma, \qquad \forall v \in C^1(\Omega), \tag{1.4.18}$$

$$\|\gamma v\|_{2,\Gamma} \le c_{T\Gamma} \|v\|_{1,2,\Omega}, \tag{1.4.19}$$

where $c_{T\Gamma}$ is a positive constant independent of v. From these relations, we observe that γv is a natural generalization of the trace defined for a continuous function (in the pointwise sense). The image of γ is a subset of $L^2(\Gamma)$, which is the space $H^{1/2}(\Gamma)$. Thus, $\gamma \in \mathcal{L}\big(H^1(\Omega), H^{1/2}(\Gamma)\big)$ and the space $\overset{\circ}{H}{}^1(\Omega)$ is the kernel of γ. The functions from other Sobolev spaces are also known to have traces in Sobolev spaces with fractional indices.

Also, for any $\phi \in H^{1/2}(\Gamma)$, one can define a continuation operator

$$\mu \in \mathcal{L}(H^{1/2}(\Gamma), H^1(\Omega))$$

such that

$$\mu\phi = w, \qquad w \in H^1(\Omega), \qquad \gamma w = \phi \qquad \text{on } \Gamma$$

and (e.g., see J.-L. Lions and E. Magenes [222])

$$\|\phi\|_{H^{1/2},\Gamma} \le c_\gamma \|w\|_{1,2,\Omega}, \qquad \|w\|_{1,2,\Omega} \le c_\mu \|\phi\|_{H^{1/2},\Gamma}. \tag{1.4.20}$$

Using the operator γ, we define subspaces of functions vanishing on Γ or on some part Γ_1 of Γ. Usually, such subspaces are marked by the zero subindex, e.g.,

$$V_0 := \{v \in V \mid \gamma v = 0 \text{ a.e. on } \Gamma_1\},$$

Henceforth, we understand the boundary values of functions in the sense of traces, so that the phrase "$u = \phi$ on Γ" means that the trace γu of a function u defined in Ω coincides with a given function ϕ defined on Γ (for the sake of simplicity, we usually omit γ). If for two functions u and v defined in Ω we say that $u = v$ on Γ, then we mean that $\gamma(u - v) = 0$ on Γ.

For $f \in L^2(\Omega)$, the functional

$$\langle f_{,i}, \varphi\rangle := -\int_\Omega f \frac{\partial \varphi}{\partial x_i}\, dx \tag{1.4.21}$$

is linear and continuous not only for functions in $\overset{\circ}{C}{}^\infty(\Omega)$ but also for all functions of the space $\overset{\circ}{H}{}^1(\Omega)$ (this fact follows from the density of smooth functions in $\overset{\circ}{H}{}^1(\Omega)$ and known theorems on the continuation of linear functionals). Such functionals can be viewed as generalized derivatives of square summable functions. They form the space $H^{-1}(\Omega)$ dual to $\overset{\circ}{H}{}^1(\Omega)$. It is easy to see that the quantity

$$| f_{,i} | := \sup_{\substack{\varphi \in \overset{\circ}{H}{}^1(\Omega) \\ \varphi \ne 0}} \frac{|\langle f_{,i}, \varphi\rangle|}{\|\nabla\varphi\|_\Omega} \tag{1.4.22}$$

is nonnegative and finite. It can be used to introduce the norm for $H^{-1}(\Omega)$.

1.4.3 Inequalities

In the subsequent chapters, we use several inequalities well known in functional analysis (e.g., see [336, 214]). For convenience of the reader, we collect and discuss them below.

First, we recall the inequality

$$|a \cdot b| \leq \left(\sum_{i=1}^{d} |a_i|^{\alpha} \right)^{1/\alpha} \left(\sum_{i=1}^{d} |b_i|^{\alpha'} \right)^{1/\alpha'}, \tag{1.4.23}$$

where $\frac{1}{\alpha'} + \frac{1}{\alpha} = 1$ and $a, b \in \mathbb{R}^d$. It is known as the discrete Hölder inequality. The Hölder inequality in functional form is as follows:

$$\int_\Omega uv\,dx \leq \|u\|_{\alpha,\Omega} \|v\|_{\alpha',\Omega}. \tag{1.4.24}$$

Let u and v be two functions in $L^{\alpha}(\Omega)$. Then

$$\begin{aligned}
\int_\Omega (u+v)^{\alpha}\,dx &= \int_\Omega u(u+v)^{\alpha-1}\,dx + \int_\Omega v(u+v)^{\alpha-1}\,dx \\
&\leq \|u\|_{\alpha,\Omega} \left(\int_\Omega (u+v)^{(\alpha-1)\alpha'}\,dx \right)^{1/\alpha'} + \|v\|_{\alpha,\Omega} \left(\int_\Omega (u+v)^{(\alpha-1)\alpha'}\,dx \right)^{1/\alpha'} \\
&= (\|u\|_{\alpha,\Omega} + \|v\|_{\alpha,\Omega}) \left(\int_\Omega (u+v)^{(\alpha-1)\alpha'}\,dx \right)^{(\alpha-1)/\alpha}
\end{aligned}$$

and we arrive at the Minkowski inequality

$$\left(\int_\Omega (u+v)^{\alpha} dx \right)^{1/\alpha} \leq \|u\|_{\alpha,\Omega} + \|v\|_{\alpha,\Omega}, \tag{1.4.25}$$

that states the triangle inequality for the norm. The Hölder and Minkowski inequalities also hold for spaces of vector- and tensor-valued functions.

For the functions in $\overset{\circ}{H}{}^1(\Omega)$, we have the Friedrichs inequality

$$\|w\|_\Omega \leq C_{F\Omega} \|\nabla w\|_\Omega, \quad \forall w \in \overset{\circ}{H}{}^1(\Omega), \tag{1.4.26}$$

where $C_{F\Omega}$ is a positive constant independent of w. It is not difficult to observe that the constant in (1.4.26) satisfies the relation

$$\frac{1}{C_{F\Omega}} = \lambda_\Omega := \inf_{\substack{w \in \overset{\circ}{H}{}^1(\Omega) \\ w \neq 0}} \frac{\|\nabla w\|}{\|w\|}. \tag{1.4.27}$$

Let $\Omega \subset \widehat{\Omega}$. For any $w \in \overset{\circ}{H}{}^1(\Omega)$, we can define $\widehat{w} = w$ in Ω and $\widehat{w}(x) = 0$ for any $x \in \widehat{\Omega} \setminus \Omega$. Obviously, $\widehat{w} \in \overset{\circ}{H}{}^1(\widehat{\Omega})$. Therefore,

$$\lambda_\Omega \geq \inf_{\substack{\widehat{w} \in \overset{\circ}{H}{}^1(\widehat{\Omega}) \\ \widehat{w} \neq 0}} \frac{\|\nabla \widehat{w}\|}{\|\widehat{w}\|} = \lambda_{C_{\widehat{\Omega}}} = \frac{1}{C_{F\widehat{\Omega}}}$$

and $C_{F\Omega} \le C_{F\widehat{\Omega}}$. Assume that

$$\Omega \subset \Pi := \{x \in \mathbb{R}^d \mid a_i < x < b_i, \quad b_i - a_i = l_i\}.$$

Then,

$$\lambda_\Omega \ge \lambda_\Pi = \pi \sqrt{\sum_i \frac{1}{l_i^2}},$$

and we obtain an explicit upper bound for $C_{F\Omega}$.

For $w \in H^1(\Omega)$, the Friedrichs inequality has a more general form

$$\|w\|_\Omega^2 \le c_{F\Omega}^2 \left(\|\nabla w\|_\Omega^2 + \int_\Gamma |w|^2 \, ds \right). \tag{1.4.28}$$

For $w \in H^1(\Omega)$, the Poincaré inequality reads

$$\|w\|_\Omega^2 \le C_{P\Omega}^2 \left(\|\nabla w\|_\Omega^2 + \left(\int_\Omega w \, dx \right)^2 \right). \tag{1.4.29}$$

From (1.4.29) it follows that

$$\|w\|_\Omega \le C_{P\Omega} \|\nabla w\|_\Omega, \qquad \forall w \in \widetilde{L}^2(\Omega). \tag{1.4.30}$$

If

$$\Omega = \Pi_l := \{x \in \mathbb{R}^d \mid x_i \in (0, l_i), \ l_i > 0\},$$

then the Poincaré inequality takes the form (e.g., see [217])

$$\|w\|_\Omega^2 \le \frac{1}{|\Pi_l|} \left(\int_{\Pi_l} w \, dx \right)^2 + \frac{d}{2} \int_{\Pi_l} \sum_{i=1}^d l_i^2 w_{,i}^2 \, dx. \tag{1.4.31}$$

In S. Mikhlin [234, 233], the reader will find more information concerning the constants in the Poincaré and Friedrichs inequalities.

In continuum mechanics, of importance is the following assertion known as the Korn's inequality. Let Ω be an open, bounded domain with Lipschitz continuous boundary. Then

$$\int_\Omega \left(|w|^2 + |\varepsilon(w)|^2 \right) dx \ge C_{K\Omega} \|w\|_{1,2,\Omega}^2, \qquad \forall w \in H^1(\Omega, \mathbb{R}^d), \tag{1.4.32}$$

where $C_{K\Omega}$ is a positive constant independent of w and $\varepsilon(w)$ denotes the symmetric part of the tensor ∇w, i.e.,

$$\varepsilon_{ij}(w) = \frac{1}{2} \left(\frac{\partial w_i}{\partial x_j} + \frac{\partial w_j}{\partial x_i} \right).$$

It is not difficult to verify that the left-hand side of (1.4.32) is bounded from above by the H^1-norm of w. Thus, it represents a norm equivalent to $\|\cdot\|_{1,2,\Omega}$. The kernel of $\varepsilon(w)$ is called the *space of rigid deflections* and is denoted by $\mathbf{R}(\Omega)$. If $w \in \mathbf{R}(\Omega)$, then it can be represented in the form $w = w_0 + \omega_0 x$, where w_0 is a vector independent of x and ω_0 is a skew-symmetric tensor with coefficients independent of x. It is easy to understand that the dimension of $\mathbf{R}(\Omega)$ is finite and equals $d + \frac{d(d-1)}{2}$.

For the functions in $\mathring{H}^1(\Omega)$, the Korn's inequality is easy to prove. Indeed,

$$\begin{aligned} |\varepsilon(w)|^2 &= \frac{1}{4}(w_{i,j} + w_{j,i})(w_{i,j} + w_{j,i}) \\ &= \frac{1}{4}(w_{i,j}w_{i,j} + w_{j,i}w_{j,i} + 2w_{i,j}w_{j,i}) = \frac{1}{2}(|\nabla w|^2 + w_{i,j}w_{j,i}), \end{aligned}$$

where the summation over repeated indices is implied. Therefore, for any $w \in \mathring{C}^2(\Omega)$ we have

$$\begin{aligned} \int_\Omega |\varepsilon(w)|^2 dx &= \frac{1}{2}\int_\Omega \left(|\nabla w|^2 + w_{i,j}w_{j,i}\right) dx = \frac{1}{2}\int_\Omega \left(|\nabla w|^2 - w_i w_{j,ij}\right) dx \\ &= \frac{1}{2}\int_\Omega \left(|\nabla w|^2 + w_{i,i}w_{j,j}\right) dx = \frac{1}{2}\int_\Omega \left(|\nabla w|^2 + |w_{i,i}|^2\right) dx \\ &\geq \frac{1}{2}\|\nabla w\|^2. \end{aligned}$$

Hence,

$$\|\nabla w\| \leq \sqrt{2}\|\varepsilon(w)\| \qquad \forall w \in \mathring{C}^2(\Omega). \tag{1.4.33}$$

Since $\mathring{C}^2(\Omega)$ is dense in $\mathring{H}^1(\Omega)$, this inequality is also valid for functions in $\mathring{H}^1(\Omega)$. The proofs of the Korn's inequality (1.4.32) are much more complicated (e.g., see [120]).

1.4.4 Convex functionals

Consider a Banach space V. A set $K \subset V$ is called *convex* if $\lambda_1 v_1 + \lambda_2 v_2 \in K$ for all $v_1, v_2 \in K$ and all $\lambda_1, \lambda_2 \in \mathbb{R}_+$ such that $\lambda_1 + \lambda_2 = 1$. Let K be a convex set. A functional $J : K \to \mathbb{R}$ is said to be *convex* if

$$J(\lambda_1 v_1 + \lambda_2 v_2) \leq \lambda_1 J(v_1) + \lambda_2 J(v_2) \tag{1.4.34}$$

for all $v_1, v_2 \in K$ and all $\lambda_1, \lambda_2 \in \mathbb{R}_+$ such that $\lambda_1 + \lambda_2 = 1$. A functional J is called *strictly convex* if

$$J(\lambda_1 v_1 + \lambda_2 v_2) < \lambda_1 J(v_1) + \lambda_2 J(v_2) \tag{1.4.35}$$

for all $v_1, v_2 \in K$ (such that $v_1 \neq v_2$) and $\lambda \in (0, 1)$. A functional J is called *concave* (resp., *strictly concave*) if the functional $(-J)$ is convex (resp., strictly convex).

The functional

$$\chi_K(v) = \begin{cases} 0 & \text{if } v \in K, \\ +\infty & \text{if } v \notin K \end{cases}$$

is called the *characteristic functional* of the set K. It is clear that it is convex if and only if the set K is convex.

If J_1 and J_2 are two convex functionals defined on a convex set K then the functionals $\alpha_1 J_1 + \alpha_2 J_2$ (for $\alpha_1, \alpha_2 \in \mathbb{R}_+$) and $\max\{J_1, J_2\}$ are also convex. It is worth noting that the latter fact remains valid for any amount of convex functionals, i.e., the upper bound taken over any set of convex functionals is a convex functional. Therefore, convex functionals are often represented as upper bounds of affine functionals.

By definition, the space V^* consists of all linear continuous functionals on V. It is called *topologically dual* to V. The value of $v^* \in V^*$ on $v \in V$ is denoted by $\langle v^*, v\rangle$. This product generates a *duality pairing* of the spaces V and V^*. If V is a Banach space, then V^* can also be normed by setting

$$\|v^*\|_* := \sup_{v \in V} \frac{\langle v^*, v\rangle}{\|v\|}. \tag{1.4.36}$$

Henceforth, we assume that the supremum (or infimum) of a quotient is taken with respect to all elements of V, except for the zero element 0_V.

Any affine functional defined on elements of V has the form $\langle v^*, v\rangle - \alpha$, where $v^* \in V^*$ and $\alpha \in \mathbb{R}$. A functional space is called *reflexive* if it coincides with the bidual space V^{**} (i.e., if there exists a one-to-one mapping of V to V^{**} and back that preserves the metric). All Hilbert spaces are reflexive. The same is true for the spaces L^p with $1 < p < +\infty$.

The theorem of F. Riesz asserts that for Hilbert spaces, any functional $v^* \in V^*$ can be written in the form of a scalar product introduced in such a space, i.e.,

$$(u, v) = \langle v^*, v\rangle, \qquad \forall v \in V, \tag{1.4.37}$$

where u is uniquely determined.

The functional $J^* : V^* \to \mathbb{R}$ defined by the relation

$$J^*(v^*) = \sup_{v \in V}\{\langle v^*, v\rangle - J(v)\} \tag{1.4.38}$$

is said to be *dual* (or *conjugate*) to J.

Remark 1.1. If J is a smooth function that increases at infinity faster than any linear function, then J^* is the Legendre transform of J. The dual functionals were studied by Young, Fenchel, Moreau, and Rockafellar (e.g., see [121, 132, 324]). The functional J^* is also called *polar* to J.

The functional

$$J^{**}(v) = \sup_{v^* \in V^*} \{\langle v^*, v\rangle - J^*(v^*)\} \tag{1.4.39}$$

is called the *second conjugate* to J (or *bipolar*).

If J is a convex functional attaining finite values, then J coincides with J^{**}.

To illustrate the definitions of conjugate functionals, consider functionals defined on the Euclidean space E^d. In this case, V and V^* consist of the same elements: d-dimensional vectors (denoted by ξ and ξ^*, respectively) and the quantity $\langle \xi^*, \xi\rangle$ is given by the scalar product $\xi^* \cdot \xi$.

Let $A = \{a_{ij}\}$ be a positive definite matrix. We have the following pair of mutually conjugate functionals:

$$J(\xi) = \frac{1}{2} A\xi \cdot \xi \quad \text{and} \quad J^*(\xi^*) = \frac{1}{2} A^{-1}\xi^* \cdot \xi^*. \tag{1.4.40}$$

Another example is given by the functionals

$$J(\xi) = \frac{1}{\alpha}|\xi|^{\alpha} \quad \text{and} \quad J^*(\xi^*) = \frac{1}{\alpha'}|\xi^*|^{\alpha'}, \tag{1.4.41}$$

where $\frac{1}{\alpha} + \frac{1}{\alpha'} = 1$. If φ is an odd convex function, then $(\varphi(\|u\|_V))^* = \varphi^*(\|u^*\|_{V^*})$.

Let a functional $J : V \to \mathbb{R}$ takes a finite value at $v_0 \in V$. The functional J is called subdifferentiable at v_0 if there exists an affine minorant l such that $J(v_0) = l(v_0)$. A minorant with this property is called the exact minorant at v_0. Obviously, any affine minorant exact at v_0 has the form

$$l(v) = \langle v^*, v - v_0\rangle + J(v_0), \quad l(v) \le J(v), \qquad \forall v \in V. \tag{1.4.42}$$

The element v^* is called a *subgradient* of J at v_0. The set of all subgradients of J at v_0 forms a *subdifferential*, which is usually denoted by $\partial J(v_0)$. It may be empty, may contain one element or infinitely many elements.

An important property of convex functionals follows directly from the fact that they have an exact affine minorant at any point (at which the functional attains a finite value). Assume that J is a convex functional and $v^* \in \partial J(v_0)$. Then there exists an affine minorant such that

$$\langle v^*, v\rangle - \alpha \le J(v), \qquad \forall v \in V,$$

and $\langle v^*, v_0\rangle - \alpha = J(v_0)$. Hence, we obtain

$$J(v) - J(v_0) \ge \langle v^*, v - v_0\rangle. \tag{1.4.43}$$

The inequality (1.4.43) represents the basic incremental relation for convex functionals. For proper convex functionals, there exists a simple criterion that enables one verify whether or not an element v^* belongs to the set $\partial J(v)$.

Proposition 1.2. *The following two statements are equivalent:*

$$J(v) + J^*(v^*) - \langle v^*, v\rangle = 0, \tag{1.4.44}$$

$$v^* \in \partial J(v), \tag{1.4.45}$$

$$v \in \partial J^*(v^*). \tag{1.4.46}$$

Proof. Assume that $v^* \in \partial J(v)$. In accordance with (1.4.43), we have

$$J(w) \geq J(v) + \langle v^*, w - v\rangle, \quad \forall w \in V.$$

Hence,

$$\langle v^*, v\rangle - J(v) \geq \langle v^*, w\rangle - J(w), \quad \forall w \in V$$

and, consequently,

$$\langle v^*, v\rangle - J(v) \geq \sup_{w\in V}\{\langle v^*, w\rangle - J(w)\} = J^*(v^*). \tag{1.4.47}$$

However, by the definition of J^*, we know that for any v and v^*

$$J^*(v^*) \geq \langle v^*, v\rangle - J(v). \tag{1.4.48}$$

We observe that (1.4.47) and (1.4.48) imply (1.4.44).

Assume that $v \in \partial J^*(v^*)$. Then $J^*(w^*) \geq J^*(v^*) + \langle w^* - v^*, v\rangle$, so that

$$\langle v^*, v\rangle - J^*(v^*) \geq \sup_{w^*\in V^*}\{\langle w^*, v\rangle - J^*(w^*)\} = J^{**}(v).$$

On the other hand,

$$\langle v^*, v\rangle - J^*(v^*) \geq J^{**}(v) = J(v),$$

and we again arrive at (1.4.44).

Assume that (1.4.44) holds. By the definition of J^*, we obtain

$$0 = J(v) + J^*(v^*) - \langle v^*, v\rangle \geq J(v) - J(w) - \langle v^*, v - w\rangle,$$

where w is an arbitrary element of V. Thus,

$$J(w) - J(v) \geq \langle v^*, w - v\rangle, \quad \forall w \in V,$$

which means that $J(v) + \langle v^*, v - w\rangle$ is an exact affine minorant of J (at v) and, consequently, (1.4.45) holds.

The proof of (1.4.46) is quite similar. □

Let J and J^* be a pair of conjugate functionals. Then

$$D_J(v, v^*) := J(v) + J^*(v^*) - \langle v^*, v\rangle$$

is called the *compound* functional. From Proposition 1.2 it follows that D_J is nonnegative and vanishes only if the arguments satisfy (1.4.45) and (1.4.46), which are also called the *duality relations* and very often represent the constitutive relations of a physical model. Compound functionals play an important role in the a posteriori error estimation of nonlinear problems (see Chapter 8). They serve as penalty functionals that penalize errors caused by dissatisfaction of the duality relations. For this reason, we denote the compound functionals by the letter D.

Note that the relation $D_J(v, v^*) \geq 0$ generates inequalities that can be viewed as generalizations of the Young's inequality (cf. (1.4.3)–(1.4.9)):

$$\langle v^*, v\rangle \leq J(v) + J^*(v^*). \tag{1.4.49}$$

In particular, if V and V^* coincide with $\mathbb{R}^d$ and $J(v) = \frac{|v|^\alpha}{\alpha}$, then $J^*(v^*) = \frac{|v^*|^{\alpha'}}{\alpha'}$ and (1.4.36) implies the estimate

$$v^* \cdot v \leq \frac{|v|^\alpha}{\alpha} + \frac{|v^*|^{\alpha'}}{\alpha'}, \qquad \forall\, v, v^* \in \mathbb{R}^d. \tag{1.4.50}$$

Finally, we recall some basic notions related to the differentiation of convex functionals. We say that J has a weak derivative $J'(v_0) \in V^*$ (at the point v_0) in the sense of Gâteaux if

$$\lim_{\lambda\to+0} \frac{J(v_0 + \lambda w) - J(v_0)}{\lambda} = \langle J'(v_0), w\rangle, \quad \forall w \in V. \tag{1.4.51}$$

Assume that J is differentiable in the above sense and $v^* \in \partial J(v_0)$. Then for any $v \in V$ we know that $J(v) - J(v_0) \geq \langle v^*, v - v_0\rangle$. Set $v = v_0 + \lambda w$, where $\lambda > 0$. Now, we have $J(v_0 + \lambda w) - J(v_0) \geq \lambda\langle v^*, w\rangle$. Therefore,

$$\langle J'(v_0), w\rangle = \lim_{\lambda\to+0} \frac{J(v_0 + \lambda w) - J(v_0)}{\lambda} \geq \langle v^*, w\rangle,$$

and $\langle J'(v_0) - v^*, w\rangle \geq 0$ for any $w \in V$. This inequality means that, in such a case, the Gâteaux derivative coincides with v^*.

2 Overview

2.1 Error indicator by Runge

First attempts to formulate accuracy criteria for approximate solutions date back to the very beginning of the 20th century when C. Runge suggested a heuristic error indication rule. Originally, it was applied to quadrature formulas, but later the rule was also adapted to integration procedures that he developed for ordinary differential equations. In essence, Runge's rule is based on comparing "coarse" and "refined" solutions.

Assume that an approximate solution u_h has been found on a mesh $\mathcal{T}_h$ with mesh size h and $u_{h_{\text{ref}}}$ is a solution on a refined mesh $\mathcal{T}_{h_{\text{ref}}}$. Let an asymptotic error estimate

$$u_h - u = \phi h^m + O(h^{m+p}), \quad m, p > 0, \tag{2.1.1}$$

be known, where ϕ is a certain (unknown) function. We have

$$u_{\text{ref}} - u = \phi h_{\text{ref}}^m + O(h_{\text{ref}}^{m+p}),$$

and, consequently,

$$u_h - u_{h_{\text{ref}}} \cong \phi(h^m - h_{\text{ref}}^m) + O(h^{m+p}) + O(h_{\text{ref}}^{m+p}).$$

From here, we conclude that, up to higher order terms,

$$u_h - u \cong h^m \frac{u_h - u_{h_{\text{ref}}}}{h^m - h_{\text{ref}}^m}. \tag{2.1.2}$$

If $h_{\text{ref}} = \kappa h$ with $\kappa \in (0, 1)$, then the above relation reads

$$u_h - u \cong \frac{1}{1 - \kappa^m} (u_h - u_{h_{\text{ref}}}). \tag{2.1.3}$$

Certainly, the estimate adequately presents the error only if h is small enough (so that the higher order terms are negligibly small) and if the asymptotic relation (2.1.1) holds. In general, these (rather strong) a priori requirements cannot be guaranteed, which means that (2.1.3) should be viewed as a conditional estimate.

Anyway, (2.1.3) partially justifies the following heuristic rule often used in engineering computations:

If $u_h - u_{h_{\text{ref}}}$ is small, then both approximations $u_{h_{\text{ref}}}$ and u_h are probably close to the exact solution u.

Indeed, (2.1.3) shows that for sufficiently small h, the error $u_h - u$ is proportional to $u_h - u_{h_{\text{ref}}}$. Using the modern terminology, we can say that

$$\mathcal{E}_R(x) := u_h(x) - u_{h_{\text{ref}}}(x)$$

is suggested as *an error indicator* and a certain measure of $\mathcal{E}_R$ as *a stopping criterion*. For this purpose, the quantities

$$\max_{x\in\Omega} |\mathcal{E}_R(x)| \quad \text{or} \quad \|\mathcal{E}_R(x)\|_{p,\Omega}, \quad p \geq 1$$

are often used. $\mathcal{E}_R$ implies a simple error verification method. In many cases, it gives quite acceptable (and even good) results. For these reasons, it would be logical to compare any new error indicator suggested for a particular class of problems with $\mathcal{E}_R$.

However, $\mathcal{E}_R$ is not fully reliable, because the fact that two subsequent elements of an approximation sequence are close to each other cannot guarantee that they are close to the exact solution. For example, $\mathcal{E}_R$ may be small if a refinement is improperly performed (i.e., if new degrees of freedom appended to $\mathcal{T}_{h_{\text{ref}}}$ do not correlate with the error). Also, it is often important to have computable and *guaranteed* upper and lower bounds for various norms (seminorms) of $u_h - u$, which cannot be derived by the above heuristic arguments.

In subsequent chapters, we will present quantities computed by u_h and $u_{h_{\text{ref}}}$ that *do provide guaranteed bounds of approximation errors* (the simplest form is exposed in the Section 3.6.2). However, their derivation is based on mathematical tools that did not exist at the time of C. Runge.

2.2 Prager–Synge estimate

In 1947, W. Prager and J. L. Synge [266] presented an a posteriori estimate valid for approximations of linear elliptic problems. Originally, the proof was motivated by an orthogonal decomposition of the energy space of a problem and purely geometric arguments (this approach is often called the *hypercircle method* [347]).

Let us discuss the idea with the paradigm of the problem

$$\Delta u + f = 0, \qquad \text{in } \Omega, \tag{2.2.1}$$

$$u = 0, \qquad \text{on } \Gamma. \tag{2.2.2}$$

By the Helmholtz decomposition, we know that $q \in L^2(\Omega, \mathbb{R}^d)$ is uniquely represented as $q_0 + \nabla\psi$, where $\psi \in \overset{\circ}{H}{}^1(\Omega)$, q_0 belongs to the set $\mathrm{S}(\Omega)$ of vector-valued solenoidal functions, and the orthogonality condition has the form

$$\int_\Omega q_0 \cdot \nabla w \, dx = 0, \qquad \forall w \in \overset{\circ}{H}{}^1(\Omega).$$

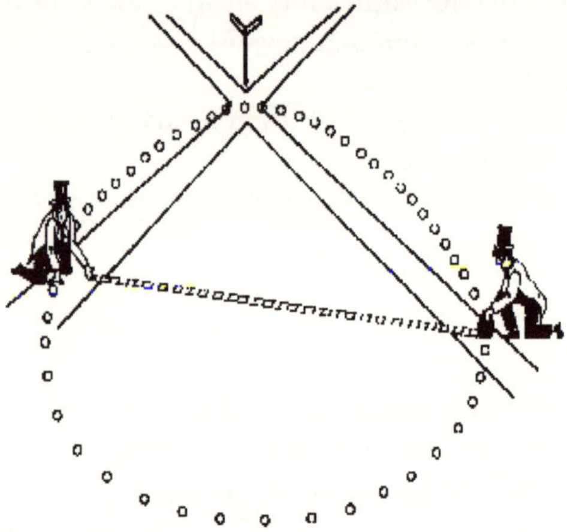

Figure 2.2.1: "Two blind men and their hypercircle." Geometric interpretation from [347]

Let

$$q \in Q_f := \left\{ \eta \in H(\Omega, \mathrm{div}) \,\middle|\, \int_\Omega \eta \cdot \nabla w \, dx = \int_\Omega f w \, dx \quad \forall w \in \mathring{H}^1(\Omega) \right\}.$$

The set Q_f contains vector-valued functions that satisfy (in a generalized sense) the relation $\mathrm{div}\, q + f = 0$.

Since $\nabla u - q \in Q_0 = \mathrm{S}(\Omega)$, we have the orthogonality relation

$$\int_\Omega \nabla(u - v) \cdot (\nabla u - q) \, dx = 0,$$

which implies the estimate

$$\boxed{\|\nabla(u - v)\|^2 + \|\nabla u - q\|^2 = \|\nabla v - q\|^2.} \tag{2.2.3}$$

Figure 2.2.1 presents a geometric interpretation of the hypercircle formula: if two blind men walking along two orthogonal roads are able to measure the distance between them, then they can construct a "hypercircle" and estimate the distance to the crosspoint.

From (2.2.3) it also follows that

$$\|\nabla(u - v)\| = \inf_{q \in Q_f} \|\nabla v - q\|. \tag{2.2.4}$$

This relation and its analogs for more complicated problems generate various a posteriori estimates that use equilibration of the dual variable (flux). We discuss some of them in Section 2.6.3.

2.3 Mikhlin estimate

S. Mikhlin [232] suggested to derive a posteriori estimates for linear elliptic problems with the help of *variational arguments*. The main idea of this approach is easy to explain with the paradigm of the problem (2.2.1)–(2.2.2).

First, we note that

$$\frac{1}{2}\|\nabla(u-v)\|^2 = J(v) - J(u), \tag{2.3.1}$$

where

$$J(u) := \inf_{v \in \overset{\circ}{H}{}^1(\Omega)} J(v), \quad J(v) := \frac{1}{2}\|\nabla v\|^2 - \int_\Omega f v\, dx.$$

This fact follows from the identity

$$J(v) - J(u) = \frac{1}{2}\|\nabla(u-v)\|^2 + \int_\Omega (\nabla u \cdot \nabla(v-u) - f(v-u))\, dx$$

and the relation

$$\int_\Omega \nabla u \cdot \nabla w\, dx = \int_\Omega f w\, dx, \qquad \forall w \in \overset{\circ}{H}{}^1(\Omega),$$

that holds for the minimizer u. Regrettably, the right-hand side of (2.3.1) is not computable because the value of $J(u)$ is not known. This difficulty can be bypassed if a computable lower bound of $J(u)$ is known. A natural quantity that serves this task is the value of the so-called *dual variational functional*

$$I^*(q) := -\frac{1}{2}\|q\|^2$$

whose maximum taken over the set of equilibrated fields coincides with $J(u)$, i.e.,

$$J(u) = I(p) := \sup_{q \in Q_f} I(q).$$

Thus, we find that for any $q \in Q_f$

$$\begin{aligned}
\frac{1}{2}\|\nabla(u-v)\|^2 &\le J(v) + \frac{1}{2}\|q\|^2 = \frac{1}{2}\|\nabla v\|^2 - \int_\Omega f v\, dx + \frac{1}{2}\|q\|^2 \\
&= \frac{1}{2}\|\nabla v\|^2 - \int_\Omega q \cdot \nabla v\, dx + \frac{1}{2}\|q\|^2 = \frac{1}{2}\|\nabla v - q\|^2.
\end{aligned} \tag{2.3.2}$$

Later (see, e.g, H. Gajewski, K. Gröger, and K. Zacharias [145], S. Mikhlin [234], P. Mosolov, and V. Myasnikov [239]), similar estimates were derived for some classes of convex variational problems. Estimates based on complementary principles were also obtained in A. M. Arthurth [14].

From the practical point of view estimates (2.2.3) and (2.3.2) have an essential drawback: *they are valid only for* $q \in Q_f$. However, the set Q_f is defined by the differential relation, which in general is difficult to exactly satisfy. In S. Mikhlin [232], a strategy close to the known *orthogonal projections method* (see also M. Vishik [362], H. Weil [369], S. Zaremba [375]) is discussed. It is based on the construction of an approximating sequence for the dual problem. However, practical realization of this approach within the framework of locally supported finite element approximations may be faced with serious technical difficulties. Certainly, for simple problems (e.g., homogeneous Laplace equation in a rectangular domain) it is possible to construct such approximations but for problems with strongly nonhomogeneous coefficients in complicated domains, vector-valued problems, problems with nonlinear and convective terms this task is much more difficult. For example, the so-called equilibrated finite element approximations for problems in solid mechanics (e.g., see C. Johnson and B. Mercier [188]) are constructed by means of macroelements.

Sometimes, numerical analysts use "almost equilibrated" approximations that may not exactly satisfy the differential relation. Substitution of an "almost equilibrated" function q may give a good error indicator but in this case the reliability of the upper bound in (2.3.2) is not guaranteed.

Also, it is worth remarking that an upper bound constructed by some $q \in Q_f$ is efficient only if q is close to the exact flux p. Since the latter is unknown, the only way to find a suitable q is to minimize $\|\nabla v - q\|$ over Q_f (or over a certain subspace of Q_f). Note that

$$\frac{1}{2}\|\nabla v - q\|^2 = \frac{1}{2}\|\nabla v\|^2 - \int_\Omega f v\, dx + \frac{1}{2}\|q\|^2,$$

where the first and second terms on the right-hand side do not depend on q. Therefore, the problem is reduced to

$$\inf_{q \in Q_f} \frac{1}{2}\|q\|^2,$$

which is the *dual* variational problem. Hence, in the Prager–Synge–Mikhlin method, getting sharp error bounds requires *solving the dual problem* with the help of conforming approximations of the set Q_f (which is defined by a differential relation). In the next chapter, we will see that the requirement $q \in Q_f$ is superfluous, and guaranteed upper bounds can be found using a much wider set for q.

2.4 Ostrowski estimates for contractive mappings

Many problems admit a fixed point formulation: Find $x_\circ$ in a complete metric space (X, d) such that

$$x_\circ = \mathrm{T} x_\circ, \tag{2.4.1}$$

where $T : X \to X$ is a continuous operator. Then approximations are usually constructed by the iteration procedure

$$x_i = Tx_{i-1}, \qquad i = 1, 2, \ldots, \tag{2.4.2}$$

where $x_0 \in X$ is a certain selected element. In this case, it is required

(a) to establish conditions that guarantee convergence of x_i to $x_\circ$ and

(b) to find computable estimates of the error $e_i = d(x_i, x_\circ)$.

Both problems can be effectively solved, provided that T possesses an additional property.

Definition 2.1. An operator $T : X \to X$ is called q-contractive on a set $S \subset X$ if there exists a positive real number q such that the inequality

$$d(Tx, Ty) \leq q\, d(x, y) \tag{2.4.3}$$

holds for any elements x and y of the set S.

The first step in the analysis of (2.4.2) is given by the following well-known theorem.

Theorem 2.2 (S. Banach). *Let* T *be a q-contractive mapping of a closed nonempty set $S \subset X$ to itself with $q < 1$. Then,* T *has a unique fixed point in S and the sequence x_i obtained by (2.4.2) converges to this point.*

Proof. It is easy to see that

$$d(x_{i+1}, x_i) = d(Tx_i, Tx_{i-1}) \leq q d(x_i, x_{i-1}) \leq \ldots \leq q^i d(x_1, x_0).$$

Therefore, for any $m > 1$ we have

$$\begin{aligned} d(x_{i+m}, x_i) &\leq d(x_{i+m}, x_{i+m-1}) + d(x_{i+m-1}, x_{i+m-2}) + \ldots + d(x_{i+1}, x_i) \\ &\leq q^i (q^{m-1} + q^{m-2} + \ldots + 1) d(x_1, x_0). \end{aligned} \tag{2.4.4}$$

Since

$$\sum_{k=0}^{m-1} q^k \leq \frac{1}{1-q},$$

(2.4.4) implies the estimate

$$d(x_{i+m}, x_i) \leq \frac{q^i}{1-q} d(x_1, x_0). \tag{2.4.5}$$

If $i \to \infty$, then the right-hand side of (2.4.5) tends to zero, so that $\{x_i\}$ is a Cauchy sequence. It has a limit $y \in X$. Then, $d(x_i, y) \to 0$ and

$$d(Tx_i, Ty) \leq q d(x_i, y) \to 0,$$

so that $\mathrm{d}(\mathrm{T}x_i, \mathrm{T}y) \to 0$ and $\mathrm{T}x_i \to \mathrm{T}y$. Pass to the limit in (2.4.2) as $i \to +\infty$. We observe that $\mathrm{T}y = y$. Hence, any limit of such a sequence is a fixed point.

Assume that two different fixed points $x_\circ^1$ and $x_\circ^2$ exist. Since

$$\mathrm{d}(x_\circ^1, x_\circ^2) = \mathrm{d}(\mathrm{T}x_\circ^1, \mathrm{T}x_\circ^2) \le q\mathrm{d}(x_\circ^1, x_\circ^2)$$

we arrive at a contradiction, which shows that $x_\circ$ is unique. □

Theorem 2.2 also implies an *a priori convergence estimate.*

Let $e_j = \mathrm{d}(x_j, x_\circ)$ denote the error at the jth step. Then

$$e_j = \mathrm{d}(\mathrm{T}x_{j-1}, \mathrm{T}x_\circ) \le q e_{j-1} \le q^j e_0$$

and

$$e_j \le q^j e_0.$$

This estimate gives a certain idea of how the error decreases. However, this a priori upper bound may be rather coarse.

A posteriori error bounds for the iteration method are given by the following theorem.

Theorem 2.3 (Ostrowski [257]). *For any x_j with $j \ge 1$, the following estimate holds:*

$$M_\ominus^j := \frac{\mathrm{d}(x_{j+1}, x_j)}{1+q} \le e_j \le M_\oplus^j := \frac{q\,\mathrm{d}(x_j, x_{j-1})}{1-q}. \qquad (2.4.6)$$

Proof. The right-hand side estimate in (2.4.6) follows from (2.4.5). For $i = 1$ it reads

$$\mathrm{d}(x_{1+m}, x_1) \le \frac{q}{1-q}\mathrm{d}(x_1, x_0).$$

Since $x_{1+m} \to x_\circ$ as $m \to +\infty$, we pass to the limit with respect to m and obtain

$$\mathrm{d}(x_\circ, x_1) \le \frac{q}{1-q}\mathrm{d}(x_1, x_0).$$

We may view x_{j-1} as the starting point of the sequence. Then, in the above relation $x_0 = x_{j-1}$ and $x_1 = x_j$, and we arrive at the following *upper bound* of the error:

$$\mathrm{d}(x_\circ, x_j) \le \frac{q}{1-q}\mathrm{d}(x_j, x_{j-1}).$$

The *lower bound* of the error follows from the relation

$$\mathrm{d}(x_j, x_{j-1}) \le \mathrm{d}(x_j, x_\circ) + \mathrm{d}(x_{j-1}, x_\circ) \le (1+q)\mathrm{d}(x_{j-1}, x_\circ),$$

which shows that

$$\mathrm{d}(x_{j-1}, x_\circ) \ge \frac{1}{1+q}\mathrm{d}(x_j, x_{j-1}).$$ □

We observe that two-sided estimates of e_j are easily computable, provided that q is estimated from above by a number less than 1.

Remark 2.4. Note that

$$\frac{M_{\oplus}^{j}}{M_{\ominus}^{j}} = \frac{1+q}{1-q} \frac{\mathrm{d}(x_j, x_{j-1})}{\mathrm{d}(x_{j+1}, x_j)} q \geq \frac{1+q}{1-q}.$$

Therefore, the efficiency of the upper and lower bounds in (2.4.6) crucially depends on the value of $1-q$.

Estimates (2.4.6) and their modifications can be applied to linear and nonlinear algebraic systems, integral equations, and other problems solved by iteration methods (e.g., see E. Zeidler [376]). The major difficulty in the application of the above a posteriori estimates is that in practice it may be difficult to find a sharp upper bound of q and to establish that it is indeed less than 1.

A posteriori methods for various iteration schemes have been investigated by many authors. Below we give some references that provide an idea of the results obtained and will help to find more pertinent information.

In G. Auchmuty [17], two sided p-norm error bounds for solutions of linear systems are presented. O. Scherzer, H. W. Engl, and K. Kunisch [328] studied a posteriori choice of the regularization parameter in a Tikhonov scheme for ill-posed problems (see also H. W. Engl and O. Scherzer [123]). A further study of this strategy is presented in Jin Qi-nian and Hou Zong-yi [268]. A posteriori estimates for linear ill-posed problems were investigated in A. Leonov [220]. More information on error estimates for iteration Newton-type methods can be found, e.g., in K. Braune [75], Jin Qi-nian [184], P. Meyer [231], F. Potra [263], K. Tsuruta and K. Ohmori [351], T. Yamamoto [373].

Boundary value problems are often reformulated in terms of integral equations. A posteriori error estimates for the respective numerical methods (e.g., boundary element methods) are usually derived with the help of the techniques that exploits special properties of the iteration operator. For the readers interested in iteration methods and a posteriori estimates for integral type formulations of linear and nonlinear boundary value problems we recommend the papers by C. Carstensen [89], Lin Qun and Shi Jan [269], M. Schultz and O. Steinbach [330], S. Shaw and J. R. Whiteman [333], and M. Schulz and W. L. Wendland [331], where they will also find more references related to the topic.

Finally, we note that the theory of contractive mappings can generate "numerically based existence theorems" that establish existence of localized solutions to nonlinear problems in the vicinity of a computed (constructed) approximate solution (see, e.g., J. G. Heywood, W. Nagata, and W. Xie [171], where such a method was applied to the Navier–Stokes equation).

2.5 Error estimates based on monotonicity

The theory of *monotone operators* gives another way of constructing explicitly computable error estimates. Such operators are defined on the so-called *ordered* (or *partially ordered*) spaces that introduce the relation $x \leq y$ for all (or almost all) elements x, y of the space. We recall that the operator T is called monotone if $x \leq y$ implies $\mathrm{T}x \leq \mathrm{T}y$ and antitone if $x \leq y$ implies $\mathrm{T}x \geq \mathrm{T}y$.

Consider an abstract fixed point problem: Find $x_\circ$ in a complete ordered (partially ordered) space X such that

$$x_\circ = \mathrm{T}x_\circ + f, \qquad f \in X. \tag{2.5.1}$$

Assume that $\mathrm{T} = \mathrm{T}_\oplus + \mathrm{T}_\ominus$, where $\mathrm{T}_\oplus$ is monotone and $\mathrm{T}_\ominus$ is antitone. Let $x_{\ominus 0}$, $x_{\ominus 1}$, $x_{\oplus 0}$, and $x_{\oplus 1}$ be such that $x_{\ominus 0} \leq x_{\ominus 1} \leq x_{\oplus 1} \leq x_{\oplus 0}$, and

$$x_{\ominus 1} = \mathrm{T}_\oplus x_{\ominus 0} + \mathrm{T}_\ominus x_{\oplus 0} + f, \qquad x_{\oplus 1} = \mathrm{T}_\oplus x_{\oplus 0} + \mathrm{T}_\ominus x_{\ominus 0} + f.$$

Then, we observe that

$$x_{\ominus 2} = \mathrm{T}_\oplus x_{\ominus 1} + \mathrm{T}_\ominus x_{\oplus 1} + f \geq \mathrm{T}_\oplus x_{\ominus 0} + \mathrm{T}_\ominus x_{\oplus 0} + f = x_{\ominus 1},$$
$$x_{\oplus 2} = \mathrm{T}_\oplus x_{\oplus 1} + \mathrm{T}_\ominus x_{\ominus 1} + f \leq \mathrm{T}_\oplus x_{\oplus 0} + \mathrm{T}_\ominus x_{\ominus 0} + f = x_{\oplus 1}.$$

By continuing the iterations we obtain elements such that

$$x_{\ominus k} \leq x_{\ominus (k+1)} \leq x_{\oplus (k+1)} \leq x_{\oplus k}. \tag{2.5.2}$$

If $x \to \mathrm{T}x + f$ maps a compact set $D \subset X$ to itself, then by the Schauder fixed point theorem $x_\circ$ exists and belongs to D. From (2.5.2) it follows that $x_\circ$ is bounded from below and above by the sequences $\{x_{\ominus k}\}$ and $\{x_{\oplus k}\}$, respectively.

Applications of this method are mainly oriented towards systems of linear simultaneous equations and integral equations (a detailed discussion of the monotonicity methods is presented in L. Collatz [110]). As an example, we consider the system of linear simultaneous equations

$$x = Ax + f,$$

which is supposed to have a unique solution $x_\circ$. Assume that

$$A = A_\oplus - A_\ominus, \qquad A_\ominus = \{a_{ij}^{\ominus}\} \in \mathbb{M}^{d \times d},$$
$$A_\oplus = \{a_{ij}^{\oplus}\} \in \mathbb{M}^{d \times d}, \quad a_{ij}^{\ominus} \geq 0, \quad a_{ij}^{\oplus} \geq 0.$$

We may partially order the space $\mathbb{R}^d$ by saying that $x \leq y$ if and only if $x_i \leq y_i$ for $i = 1, 2, \ldots, n$. Compute the vectors

$$x_{\ominus(k+1)} = A_\oplus x_{\ominus k} + A_\ominus x_{\oplus k} + f, \quad x_{\oplus(k+1)} = A_\oplus x_{\oplus k} + A_\ominus x_{\ominus k} + f.$$

If $x_{\ominus 0} \leq x_{\ominus 1} \leq x_{\circ} \leq x_{\oplus 1} \leq x_{\oplus 0}$, then for the components of $x_{\circ}$ we obtain two-sided estimates

$$x_{\ominus k}^{(i)} \leq x_{\ominus (k+1)}^{(i)} \leq x_{\circ}^{(i)} \leq x_{\oplus (k+1)}^{(i)} \leq x_{\oplus k}^{(i)}, \quad i = 1, 2, \ldots, n. \tag{2.5.3}$$

It should be noted that the convergence of $x_{\ominus k}^{(i)}$ and $x_{\oplus k}^{(i)}$ to $x_{\circ}$ (and the convergence rate) calls for a special investigation, which must use specific features of a particular problem.

The reader interested in estimates based on monotonicity can find more information and relevant references in, e.g., E. Geisler, A. Tal, and D. Garg [150] (where they are discussed in the context of ordinary nonlinear equations), J. Schröder [329], and M. Plum [262] (where monotonicity methods were used to perform computer-assisted existence proofs based on Schauder's fixed point theorem).

2.6 A posteriori error indicators for finite element approximations

It is commonly accepted that the mathematical concept of the finite element methods brings its origin in the paper by R. Courant [112] where the method was introduced as a variant of the B. Galerkin [147] and W. Ritz [323] methods using locally supported trial functions. In 60–70s, finite element methods have formed one of the main approaches to the numerical analysis of PDE's. The foundations of the finite element method are exposed in the books by Ph. Ciarlet [107], S. Brenner and R. L. Scott [76], C. Johnson [185], J. T. Oden and J. N. Reddy [254], G. Strang and G. Fix. [344], O. C. Zienkiewicz and K. Morgan [381], and many other publications. Achievements and some unsolved problems of the finite element method are discussed, e.g., in the papers by I. Babuška [20] and O. C. Zienkiewicz [379] (see also I. Babuška and J. E. Osborn [30]).

In the late 70s and early 80s, it became clear that the successive numerical methods for PDE's should be based on the so-called *mesh-adaptive* procedures that modify finite dimensional spaces with the help of the information contained in an approximate solution computed at the previous step. This new concept generated an interest to a posteriori error indicators that provide information for a proper "improvement" of finite-dimensional spaces. Nowadays, mesh-adaptive methods based on a posteriori estimates dominate in the numerical analysis of differential equations (e.g., see [261, 7, 8, 35, 40, 41, 100, 117, 124, 178, 187, 190, 322, 356, 383] and the papers cited therein). The majority of a posteriori methods for FEM are based either on the analysis of residuals or on special properties (e.g., additional regularity) of exact solutions.

A posteriori error estimation for finite element approximations is the main subject of the books by M. Ainsworth and J. T. Oden [8], I. Babuška and T. Strouboulis [35], W. Bangerth and R. Rannacher [40], K. Eriksson, D. Estep, P. Hansbo and C. Johnson [124], and R. Verfürth [356]. There, the reader will find detailed expositions of various approaches, results of numerical experiments, and a wide list of references. Below, we shortly discuss several a posteriori error estimation methods developed for finite element approximations. Certainly, the exposition is not complete. Its goal is to give a view of the mathematical ideas underlying the methods.

2.6.1 Explicit residual methods

From the mathematical viewpoint, the classical "residual method" is a method for finding *an upper bound of the residual functional evaluated in the topology of the image space of the respective operator.* It leads to the so-called "explicit residual a posteriori estimate", which yields an error bound for a Galerkin approximation as a sum of element-wise residuals and interelement jumps with weights given by constants in the so-called Clément's interpolation inequalities (see Ph. Clément [108]).

To present the main idea, we use the problem (2.2.1)–(2.2.2), the generalized solution of which meets the integral identity

$$\int_\Omega \nabla u \cdot \nabla w \, dx = \int_\Omega f w \, dx, \quad w \in V_0 := \overset{\circ}{H}{}^1(\Omega). \tag{2.6.1}$$

Let $v \in V_0$ be an approximate solution of this problem. Then,

$$\int_\Omega \nabla(u-v) \cdot \nabla w \, dx = \mathcal{F}_v(w), \quad w \in V_0, \tag{2.6.2}$$

where

$$\mathcal{F}_v(w) := \int_\Omega (f w - \nabla v \cdot \nabla w) \, dx$$

is a linear functional defined on V_0. It is easy to see that this functional is equal to zero if v coincides with u. In all other cases, the norm of this functional, defined by the relation

$$[\![\mathcal{F}_v]\!] := \sup_{\substack{w \in V_0, \\ w \neq 0}} \frac{|\mathcal{F}_v(w)|}{\|\nabla w\|}, \tag{2.6.3}$$

is positive. Therefore, it is natural to call $\mathcal{F}_v$ the *error functional.* It is easy to show that $[\![\mathcal{F}_v]\!]$ is indeed a measure of the deviation of v from u. By (2.6.2) and (2.6.3), we note that

$$\int_\Omega |\nabla(u-v)|^2 \, dx = \mathcal{F}_v(u-v) \leq [\![\mathcal{F}_v]\!] \, \|\nabla(u-v)\|.$$

Hence,

$$\|\nabla(u-v)\| \leq [\![\mathcal{F}_v]\!]. \tag{2.6.4}$$

However, $|\mathcal{F}_v(w)| \leq \|\nabla(u-v)\| \|\nabla w\|$, so that (2.6.3) implies the inequality opposite to (2.6.4). Thus, the norm of the norm of the deviation from the exact solution coincides with the norm of $\mathcal{F}_v$. Here arises the problem of how to compute $[\![\mathcal{F}_v]\!]$ practically for a given v. A straightforward computation of the norm based on (2.6.3) is hardly possible. A more promising way is to find computable upper bounds of $[\![\mathcal{F}_v]\!]$.

In the papers of I. Babuška and W. C. Rheinboldt [31, 32] and some other publications of them, a way was suggested for deriving such bounds, which was later called the *explicit residual method.*

The explicit residual method is applicable if v is a *Galerkin approximation* of the exact solution computed on a finite-dimensional space $V_h \in V_0$, i.e., if $v = u_h$, where

$$\int_\Omega \nabla u_h \cdot \nabla w_h \, dx = \int_\Omega f w_h \, dx, \quad w_h \in V_{0h}. \tag{2.6.5}$$

In this case,

$$\int_\Omega \nabla(u - u_h) \cdot \nabla w_h \, dx = 0, \quad w_h \in V_{0h}, \tag{2.6.6}$$

i.e., the error $\nabla(u - u_h)$ is orthogonal to ∇w_h for any $w_h \in V_{0h}$, and we find that

$$\mathcal{F}_{u_h}(w) = \int_\Omega \nabla(u - u_h) \cdot \nabla w \, dx = \int_\Omega \nabla(u - u_h) \cdot \nabla(w - \pi_h w) \, dx,$$

where $\pi_h : V_0 \to V_{0h}$ is a continuous mapping (typically π_h is defined by the Clément's interpolation operator [108]). Let Ω be divided into a collection of subdomains Ω_k, $k = 1, 2, \ldots, M$, and u_h be a smooth function in each subdomain. Then

$$\mathcal{F}_{u_h}(w) := \sum_{k=1}^{M} \int_{\Omega_k} \Delta(u_h - u)(w - \pi_h w) \, dx + \sum_{k,l=1}^{M} \int_{\Gamma_{kl}} \left[\frac{\partial(u - u_h)}{\partial \nu_{kl}} \right]_{\Gamma_{kl}} (w - \pi_h w) \, ds,$$

where Γ_{kl} is the common part of the boundaries of Ω_k and Ω_l, ν_{kl} is the unit normal vector to this boundary, and $[\,\xi\,]_{\Gamma_{kl}}$ denotes the jump of a quantity ξ at the boundary Γ_{kl}. If the Ω_k are simplexes, then for π_h the interpolation estimates are as follows:

$$\|w - \pi_h w\|_{\Omega_k} \le \gamma_{1k} \operatorname{diam}(\Omega_k) \|w\|_{1,2,\omega_{1k}}, \tag{2.6.7}$$

$$\|w - \pi_h w\|_{\Gamma_{kl}} \le \gamma_{2k} |\Gamma_{kl}|^{1/2} \|w\|_{1,2,\omega_{2k}}, \tag{2.6.8}$$

where ω_{1k} and ω_{2k} are certain domains (patches of neighbor elements) that contain Ω_k,

$$\operatorname{diam}(\Omega_k) = \sup_{x_1, x_2 \in \Omega_k} |x_1 - x_2|,$$

and γ_{1k} and γ_{2k} are the interpolation constants (which depend not only on Ω_k, but also on the form of all elements in the patches ω_{1k} and ω_{2k}, respectively). By these estimates we can represent $|\mathcal{F}_{u_h}|$ as the sum of local quantities

$$\eta_k^2 := \operatorname{diam}(\Omega_k)^2 \, \|\Delta u_h + f\|_{\Omega_k}^2 + \frac{1}{2} \sum_{l=1}^{M} |\Gamma_{kl}| \left\| \left[\frac{\partial u_h}{\partial \nu_{kl}} \right]_{\Gamma_{kl}} \right\|_{\Gamma_{kl}}^2,$$

which are related to the residual on Ω_k and the values of the jumps in the normal component of the gradient on the boundary. The respective estimate of the overall error is as follows:

$$\|\nabla(u - u_h)\|^2 \leq |\mathcal{F}_{u_h}|^2 \leq \sum_{k=1}^{M} c_k \eta_k^2, \tag{2.6.9}$$

where the constants c_k depend on γ_{1k} and γ_{2k} associated with the sampling considered. The quantities η_k are used as error indicators. Comparing their values on different elements, one can suggest an adequate mesh-adaptation procedure. In this case, the estimate is used in the form

$$\boxed{\|\nabla(u - u_h)\|^2 \leq C \sum_{k=1}^{M} \eta_k^2,} \tag{2.6.10}$$

where the different c_k are replaced by one common constant C. It should be noted that often the values of two terms in η_k are quite different. In this case, error indicators for mesh-adaptation can be constructed with the help of only the dominant part of η_k.

Difficulties arise if a guaranteed and sharp upper bound of $\|\nabla(u - u_h)\|$ is required. Indeed, we must find a large amount of local constants γ_{1k} and γ_{2k} in (2.6.7)–(2.6.8). In general, finding such a constant requires solving an infinite-dimensional problem on each patch. For example,

$$\gamma_{1k} = \operatorname{diam}(\Omega_k)^{-1} \sup_{\substack{w \in H^1(\omega_{1k}) \\ w \neq 0}} \frac{\|w - \pi_h w\|_{\Omega_k}}{\|w\|_{1,2,\omega_{1k}}}. \tag{2.6.11}$$

In general, this problem is of the same level as (2.6.3). In the literature, it is sometimes recommended to replace H^1 in (2.6.11) by a certain set of polynomial functions. This provides a possibility of computing an approximate value of γ_{1k} but, in such a case, the reliability of the upper bound of the error may be lost. Another unpleasant feature of the method is that γ_{1k} and γ_{2k} *depend on* $\mathcal{T}_h$ and, consequently, all of them must be recalculated if one sampling is replaced by another one.

Also, it is worth mentioning that the estimate (2.6.9) is derived by formal mathematical transformations that clearly overestimate $|\mathcal{F}_{u_h}|$. Obviously, in (2.6.10) such an overestimation may be much larger (especially for nonuniform meshes, where the constants γ_{1k} and γ_{2k} are quite different). Thus, such estimates may lead to a considerable overestimation of the total error. This fact was observed in some carefully performed tests (e.g., see C. Carstensen and S. A. Funken [93]). Nevertheless, the explicit residual method is widely exploited in practice, and the quantities η_k are often used not for the computation of a guaranteed upper bound of the total error, but serve as *error indicators* that give am idea of the distribution of local errors in Ω.

The amount of publications associated with residual based methods is huge. In addition to the books [8, 35, 124, 356], the reader will find discussions of the method, e.g., in M. Ainsworth and J. T. Oden [6, 7], C. Carstensen [87], C. Carstensen and S. A. Funken [92, 93] (in these papers the authors consider ways of computing bounds of interpolation constants in the residual based error estimator), C. Carstensen and R. Verfürth [100] (the authors show that the "edge component" dominates in the residual error estimator), K. Eriksson and C. Johnson [125], and C. Johnson and P. Hansbo [187].

In C. Carstensen and S. Sauter [99], a posteriori error estimates were derived for elliptic PDEs on domains with complicated structures. Estimates for problems with biharmonic operator are analyzed in A. Charbonneau, K. Dossou and R. Pierre [102]. Residual-type estimates for linear first order systems of PDE's were obtained in P. Houston, J. A. Mackenzie, E. Suli, and G. Warnecke [180].

A posteriori estimates taking into account the influence of the non-discretized part of the domain on the approximation error are considered in W. Dörfler and M. Rumpf [117]. Estimates in the L^∞-norm can be found, e.g., in S. W. Brady and A. R. Elcrat [66]. Adaptive methods for convection-diffusion problems are considered in C. Johnson [186] and R. Verfürth [360]. Papers by R. Verfürth [359] and K. Eriksson and C. Johnson [126] are devoted to parabolic type problems. A posteriori error estimates for anisotropic meshes are presented in K. G. Siebert [334] and G. Kunert [203, 204]. Also, we recommend papers by I. Babuška, R. Duran, and R. Rodriguez [25], G. F. Carey and D. L. Humphrey [84], J. T. Oden, L. Demkowicz, W. Rachowicz, and T. A. Westermann, [251], R. E. Ewing [129], D. W. Kelly, J. R. Gago, O. C. Zienkiewicz, and I. Babuška [195], B. I. Wohlmuth and R. H. W. Hoppe [372], where the reader will find discussions of various a posteriori estimators and more references.

Finally, we conclude this overview by saying several words about residual a posteriori estimates for nonlinear boundary value problems. In an abstract form (for nonlinear mappings) residual type estimates were considered in J. Pousin and J. Rappaz [264, 265] and R. Verfürth (e.g., see [355, 356]). The book [356] contains a systematic consideration of this question. Certain particular classes of nonlinear problems were analyzed by many authors. For example, nonlinear diffusion-convection problems were considered in J. Medina, M. Picasso and J. Rappaz [229] and for nonlinear parabolic problems in R. Verfürth [357]. A posteriori estimates for variational inequalities related to problems with obstacles were analyzed in M. Ainsworth, J. T. Oden, and C. Y. Lee [9], D. Braess [68], Z. Chen and R. H. Nochetto [103], R. H. W. Hoppe and R. Kornhuber [178], R. Kornhuber [198], and A. Vesser [361].

2.6.2 Implicit residual methods

In the so-called implicit residual methods, the error is represented as the solution of an auxiliary boundary value problem. For example, let $u_h \in V_{0h}$ be a conforming approximation of the problem (2.6.1). By (2.6.2) we have the relation

$$\int_\Omega \nabla(u - u_h) \cdot \nabla w \, dx = \mathcal{F}_{u_h}(w), \quad \forall w \in V_0 \tag{2.6.12}$$

which shows that $e = u - u_h$ is a solution of the problem

$$\Delta e + \mathrm{r}(u_h) = 0 \quad \text{in } \Omega, \tag{2.6.13}$$

$$e = 0 \quad \text{on } \Gamma, \tag{2.6.14}$$

where $\mathrm{r}(u_h) = f + \operatorname{div} \nabla u_h \in H^{-1}(\Omega)$ and (2.6.13) is understood in the sense of distributions. Formally, this idea can be extended to a wide class of linear problems. Indeed, if $\mathcal{A} : X \to Y$ is a linear operator for which we consider the problem

$$\mathcal{A}u + f = 0 \quad \text{in } \Omega, \tag{2.6.15}$$

$$u = 0 \quad \text{on } \Gamma, \tag{2.6.16}$$

then, for any $v \in X$,

$$\mathcal{A}e + \mathrm{r}(v) = 0 \quad \text{in } \Omega, \tag{2.6.17}$$

$$e = 0 \quad \text{on } \Gamma, \tag{2.6.18}$$

where $\mathrm{r}(v) = f + \mathcal{A}v$ is the residual.

We note that finding accurate approximate solutions of (2.6.13)–(2.6.14) (and of (2.6.17)–(2.6.18)) may be a difficult task (because, in general, the error functional is a distribution so that we have a boundary value problem with rather irregular right-hand side). Moreover, the accuracy of an approximate solution obviously affects the accuracy of error estimation, so that a new error estimation problem arises.

In practical applications, several methods are used in order to overcome the above mentioned difficulties. In particular, it is often suggested to split the residual functional into a number of functionals defined as solutions of local subproblems. For example, in R. E. Bank and A. Weiser [42]) error indicators are constructed with the help of local boundary value problems with data defined by residuals and interelement jumps (see also R. Duran and R. Rodriguez [119]). In the *equilibrated residual method* (e.g., see M. Ainsworth and J. T. Oden [8]), local boundary value problems are constructed on each element, using the residuals and suitable Neumann conditions on boundaries of the elements. Local Dirichlet and Neumann problems on patches are also used in E. Stein and S. Ohnimus [339] and R. Verfürth [358, 360]. Applications to nonlinear problems are discussed, e.g., in C. Carstensen, R. Klose, and A. Orlando [97].

Finally, we note that implicit type methods are often used for the indication of local errors. Concerning a posteriori methods developed to evaluate local errors of FEM approximations, we address the reader to the books by M. Ainsworth and T. Oden [8] and I. Babuška and T. Strouboulis [35]. Also we recommend papers by I. Babuška, F. Ihlenburg, A. Mathur, T. Strouboulis, S. K. Gangaraj, C. S. Upadhyay [27, 39, 39, 37], E. Stein and S. Ohnimus [338], R. Verfürth [358, 360], and M. Ainsworth, J. T. Oden and C. Y. Lee [9].

2.6.3 A posteriori estimates based on post-processing of approximate solutions

Post-processing methods exploit certain a priori known properties of exact solutions. In general terms, the situation that typically arises for finite element approximations is as follows. Consider a conforming approximation u_h (which belongs to a finite-dimensional subspace V_h) and the function Λu_h, where Λ is a certain linear operator (e.g., the operator ∇). Usually, Λu_h lies in a rather wide space U. For example, if approximations of (3.2.1)–(3.2.2) are constructed with the help of piece-wise affine continuous functions then $\nabla u_h \in L^2(\Omega)$. However, in many cases a priori estimates of the exact solution guarantee that

$$\Lambda u \in \mathbb{U},$$

where $\mathbb{U}$ is a subset of U. In particular, if $f \in L^2(\Omega)$, then the exact flux ∇u of the problem (3.2.1)–(3.2.2) is a vector-valued function in $\mathbb{U} = H(\Omega, \mathrm{div})$. Moreover, very often we know that $\nabla u \in H^1(\Omega, \mathbb{R}^d)$ (globally or locally). Another option is to set $\mathbb{U} = Q_f$. These observations suggest an idea to post-process ∇u_h and find a close vector-valued function that satisfies some of the above-menioned properties. Formally, the principal scheme is as follows.

Assume that we have a continuous mapping G such that

(a) $\mathrm{G}(\Lambda v_h) \in \mathbb{U}$,
(b) post-processing is a relatively inexpensive procedure,
(c) $\mathrm{G}(\Lambda v_h)$ is much closer to Λu than Λv_h.

If G satisfies (a)–(b), then the difference $\mathrm{G}(\Lambda v_h) - \Lambda v_h$ generates an efficient indicator of element-wise errors.

Regularization. Usually, elliptic problems with smooth coefficients have regular solutions in internal subdomains, which suggests an idea to project fluxes of approximate solutions to a set of more regular functions (e.g., see the paper by J. H. Bramble and A. H. Schatz [73], which is one of the earliest publications in this area). If the error caused by violations of a priori regularity properties dominates and a post-processing operator efficiently performs regularization of approximate solutions, then one may hope that the difference between the approximate solution and its regularized (smoothed) counterpart represents the major part of the error. Typically, regularization is applied to fluxes ∇u_h (or $A\nabla u_h$) and to stresses for problems related to continuum mechanics.

Assume that G performs efficient regularization of fluxes, i.e.,

$$\|\mathrm{G}\nabla u_h - \nabla u\| \leq \alpha \, \|\nabla u_h - \nabla u\| \,, \tag{2.6.19}$$

where $\alpha < 1$. In this case,

$$\|\nabla u_h - \nabla u\| \leq \|\mathrm{G}\nabla u_h - \nabla u_h\| + \alpha \, \|\nabla u_h - \nabla u\|$$

and

$$\|G\nabla u_h - \nabla u_h\| \leq \alpha \|\nabla u_h - \nabla u\| + \|\nabla u_h - \nabla u\| .$$

Hence,

$$(1 - \alpha)\|\nabla u_h - \nabla u\| \leq \|G\nabla u_h - \nabla u_h\| \leq (1 + \alpha)\|\nabla u_h - \nabla u\| . \quad (2.6.20)$$

If $\alpha \ll 1$, then (2.6.20) shows that the image of ∇u_h computed by a post-processing operator G is much closer to the exact solution than ∇u_h. Thus,

$$\|\nabla u_h - \nabla u\| \simeq \|G\nabla u_h - \nabla u_h\| , \quad (2.6.21)$$

and the function $|G\nabla u_h - \nabla u_h|(x)$ (which is easily computable) can be used as an error indicator. Certainly, the quality of such error indicator depends on the smallness of α.

The regularization of fluxes (or other solution components) leads to a variety of error indicators. Below we briefly describe some of them.

Local post-processing. In most cases, post-processing is performed by elementwise averaging procedures. Let O_i be a patch of finite elements (see Fig 2.6.1), i.e.,

$$\bar{O}_i = \bigcup_{j=1,2,\ldots,m_i} T_{ij} .$$

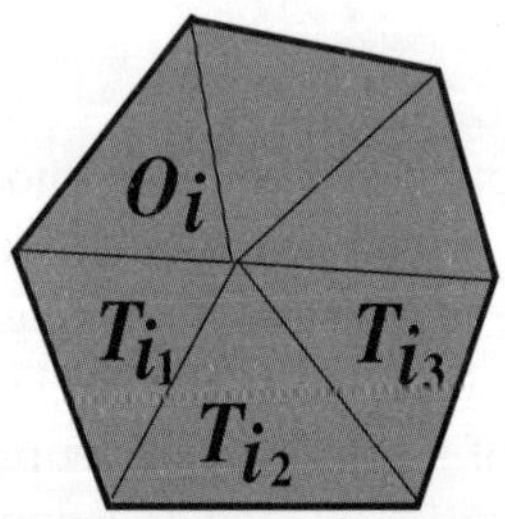

Figure 2.6.1 Patch O_i.

Define g_i as a vector-valued function in $P^k(O_i, \mathbb{R}^d)$ solving the minimization problem

$$\inf_{g \in P^k(O_i, \mathbb{R}^d)} \int_{O_i} |g - G\nabla u_h|^2 \, dx . \quad (2.6.22)$$

The minimizer g_i can be used to define the values of an averaged flux at some points. Further, these values are utilized by a prolongation procedure that defines an averaged function $G\nabla u_h : \Omega \to \mathbb{R}$. Consider the simplest case. Let u_h be a piecewise affine continuous function. Then,

$$\nabla u_h \in P^0(T_{ij}, \mathbb{R}^d) \quad \forall\, T_{ij} \in \mathcal{T}_h,$$

and $\nabla u_h \notin H^1(\Omega)$. Denote the values of ∇u_h on T_{ij} by $(\nabla u_h)_{ij}$. Set $k = 0$ and find $g_i \in P^0$ such that

$$\int_{O_i} |g_i - \nabla u_h|^2\, dx = \inf_{g \in P^0(O_i)} \int_{O_i} |g - \nabla u_h|^2\, dx$$
$$= \inf_{g \in P^0(O_i)} \left\{ |g|^2 |O_i| - 2g \cdot \sum_{j=1}^{M_i} (\nabla u_h)_{ij}\, |T_{ij}| + \sum_{j=1}^{M_i} |(\nabla u_h)_{ij}|^2 |T_{ij}| \right\}. \quad (2.6.23)$$

It is easy to see that g_i is given by a weighted sum of $(\nabla u_h)_{ij}$, namely,

$$g_i = \sum_{j=1}^{M_i} \frac{|T_{ij}|}{|O_i|} (\nabla u_h)_{ij}. \quad (2.6.24)$$

Now, we define the value of $G\nabla u_h(x_i)$ as g_i. Repeat this procedure for all nodes and define the vector-valued function $G\nabla(u_h)$ by piecewise affine prolongation of these values.

If the mesh is regular and all the quantities $|T_{ij}|$ are equal, then (2.6.24) reads

$$g_i = \sum_{j=1}^{M_i} \frac{1}{M_i} (\nabla u_h)_{ij}. \quad (2.6.25)$$

Various averaging formulas of this type are represented in the form

$$g_i = \sum_{j=1}^{M_i} \lambda_{ij} (\nabla u_h)_{ij}, \quad \sum_{j=1}^{M_i} \lambda_{ij} = 1, \quad (2.6.26)$$

where the quantities λ_{ij} are weight factors. For internal nodes, they may be taken in accordance with (2.6.24) or defined by the rule

$$\lambda_{ij} = \frac{|\gamma_{ij}|}{2\pi},$$

where $|\gamma_{ij}|$ is the radian measure of the angle of T_{ij} associated with the node i. However, if a node belongs to the boundary, then it is better to choose special weights. Their values depend on the mesh and on the type of the boundary. The reader can find a detailed consideration of this question in I. Hlaváček and M. Křižek [176].

Another way of defining g_i is to solve the problem

$$\inf_{g\in\mathbb{P}^k(O_i)} \sum_{s=1}^{m_i} |g(x_s) - \mathrm{G}\nabla u_h(x_s)|^2 , \tag{2.6.27}$$

where the points $x_s \in \bar{O}_i$ are the so-called *superconvergent* points. Evidently, in this case, the integral type averaging is replaced by a discrete one.

Global averaging. Local minimization problems on patches can be replaced by the following global problem: Find $\bar{g}_h \in \mathbb{U}_h$ such that

$$\|\bar{g}_h - \Lambda u_h\|^2_\Omega = \inf_{g_h\in\mathbb{U}_h} \|g_h - \Lambda u_h\|^2_\Omega , \tag{2.6.28}$$

where $\mathbb{U}_h$ is a certain finite-dimensional subspace of $\mathbb{U}$. The function $\bar{g}_h$ can be viewed as $\mathrm{G}\Lambda u_h$. Very often $\bar{g}_h$ is a better image of $\mathrm{G}\Lambda u$ than the functions obtained by local procedures. Moreover, mathematical justifications of the methods based on global averaging procedures can be performed under weaker assumptions what makes them applicable to a wider class of problems (e.g., see C. Carstensen and S. A. Funken [92] and C. Carstensen and S. Bartels [90] where it was shown that each averaging procedure leads to a certain a posteriori estimate).

Estimates based upon global averaging of gradients were also considered in B.-O. Heimsund, X.-C. Tai and J. Wang [170]. In J. Wang [367], it was suggested the so-called "least squares surface fitting" procedure that for problems with sufficiently smooth solutions lead to a recovered function with superconvergent properties. The analysis is based on the presentation

$$u - \mathcal{Q}_\tau u_h = (u - \mathcal{Q}_\tau u) + \mathcal{Q}_\tau(u - u_h),$$

where u is the exact solution of a linear elliptic problem, u_h is the Galerkin approximation computed on a mesh $\mathcal{T}_h$ and $\mathcal{Q}_\tau$ is the L^2-projection operator on the finite-dimensional space constructed on a mesh $\mathcal{T}_\tau$ with the help of piecewise polynomial functions of the order $r \geq 0$. The key fact is that

$$\|\mathcal{Q}_\tau u - \mathcal{Q}_\tau u_h\| \leq C h^{s-1+\alpha \min\{0,2-s\}} \|u - u_h\|_{H^1}, \tag{2.6.29}$$

where $\alpha \in (0,1)$ is a parameter that relates h and τ as $\tau = h^\alpha$, original problem is assumed to be H^s-regular with $1 \leq s \leq k+1$, and k is the degree of polynomials used in the Galerkin approximation. From (2.6.29) it follows that

$$\|u - \mathcal{Q}_\tau u_h\| \leqslant C\, h^{\beta(h,\mu,r,k)} \left(\|u_f\|_{r+1,\Omega_0} + \|u_f\|_{k+1,\Omega}\right), \tag{2.6.30}$$

provided that $u \in H^{k+1}(\Omega) \cap H^{r+1}(\Omega_0) \cap V_0$. In (2.6.30), the rate β depends on τ, h, r and k and is grater than 2 provided that u_f is regular enough, and the space V_τ is selected appropriately (i.e. it is sufficiently rich). Concrete values of the convergence rate for various k, r, and α are presented in ([367]).

We conclude this short overview of the methods using regularization of approximate solutions by several literature comments. One of the most well-known post-processing methods originates from the works of O. C. Zienkiewicz and J. Z. Zhu [382, 383] (in the literature it is often called ZZ error indicator). Simple and efficient error indicators based upon gradient averaging were suggested by many authors. Here, we refer to, e.g., M. Ainsworth, J. Z. Zhu, A. W. Craig and O. C. Zienkiewicz [10], I. Babuška and R. Rodriguez [33], B. Boroomand and O. C. Zienkiewicz [64], I. Babuška and R. Rodriguez [33], R. Duran, M. A. Muschietti and R. Rodriguez [118], R. Rodrigues [326], O. C. Zienkiewicz and J. Z. Zhu [384, 378], O. C. Zienkiewicz, B. Boroomand and J. Z. Zhu [380].

Mathematical justifications of the condition $\alpha \ll 1$ (which was used in (2.6.21)) are based on the so-called *superconvergence* phenomenon, which states that certain components of approximate solutions (Galerkin approximations) converge to the exact values with rates higher than the rate of the energy norm of the error. Probably, the earliest results on superconvergence were established in the papers by L. A. Oganesjan and L. A. Ruchovets [256] and M. Zlámal [385, 386]. Error indicators and adaptive methods based upon superconvergence phenomenon are discussed in, e.g., to J. Brandts [74], S.-S. Chow, G. F. Carey and R. D. Lazarov [106], R. E. Ewing, R. D. Lazarov and J. Wang [130], I. Hlaváček and M. Křižek [176], M. Křížek and P. Neittaanmäki [202, 242], R. Lazarov [217], R. Verfürth [356], J. Wang [367], J. Wang and X. Ye [368], N.-E. Wiberg, F. Abdulwahab and S. Ziukas [371], Z. Zhang and A. Naga [377], and in many other publications. In L. Wahlbin [366] readers can find a detailed exposition of the subject and other references. Surveys on superconvergence are presented, e.g., in M. Křížek and P. Neittaanmäki [201], J. R. Whiteman and G. Goodsell [370], and in the book [200].

Equilibration. Another group of methods exploits differential relations (usually they follow from conservation laws), which for exact solutions must be exactly satisfied. Consider again the problem (2.6.1)–(2.6.2) and set $\mathbb{U} = Q_f$. If G^{eq} is an equilibration operator that transforms ∇u_h into $q_h = \mathrm{G}^{eq}\nabla u_h \in Q_f$, then we can apply the estimate (2.2.4) and find that

$$\|\nabla(u - u_h)\| \leq \|\nabla u_h - q_h\|.$$

If the equilibration is sharp (i.e., $\operatorname{div} q_h + f = 0$), then this method provides a guaranteed upper bound of the error. This property is lost if q_h is equilibrated approximately but usually $\nabla u_h - q_h$ serves as a good error indicator provided that q_h is sufficiently close to the set of equilibrated fields.

Various approaches based on equilibration type post-processing of approximate solutions are presented in, e.g, M. Ainsworth and J. T. Oden [8], D. W. Kelly [194], P. Ladevéze and D. Leguillon [207] (the authors show that if the equilibrium equation is exactly satisfied, then the estimate is reduced to the error in constitutive relations; extensions of the method to some other other problems can be found in P. Ladevéze, J.-P. Pelle and Ph. Rougeot [208], P. Ladevéze and Ph. Rougeot [209], L. Gallimard, P. Ladevéze and J.-P. Pelle [148], P. Coorevits, J.-P. Dumeau and J.-P. Pelle [111]), E. Stein and S. Ohnimus [338], E. Stein, F. J. Bartold, S. Ohnimus, and M. Schmidt [337]).

Post-processing procedures based upon an equilibration of recovered stresses were considered in B. Boroomand and O. C. Zienkiewicz [63]. They result in stress fields that satisfy an equilibrium condition in a weak form. A method of equilibration is presented in P. Destuynder and B. Métivet [114]. This paper also contain numerical tests, in which a posteriori error estimates obtained by equilibration are compared with other estimates.

Recently a new equilibration method was suggested in publications of D. Braess and J. Schöberl [67, 69]. In it fluxes are projected to a subspace formed by Raviart–Thomas elements. After that a certain iteration procedure is performed on patches in which normal components of local fluxes are changed in order to satisfy the respective integral (balance) relation on each element.

2.6.4 A posteriori methods using adjoint problems

In the recent decades, it was developed a new approach to a posteriori error estimation based on the attraction of *adjoint boundary value problems.* Probably, first results in the error analysis for PDE's obtained with the help of adjoint problems are related to the works of J.-P. Aubin [16] and J. A. Nitsche [249] where it was suggested a way of deriving a priori rate convergence estimates in weaker norms. The idea to use adjoint problems in order to establish upper bounds for linear functionals of approximation errors is briefly discussed in the book by S. Mikhlin [232] (with a reference to a paper by M. Slobodyanskii [335]). Among early publications close to this approach, we also mention the work of T. Kato [192].

Nowadays, error estimates using adjoint problems are widely used in computer simulation. Concerning this subject, we first refer to the so-called *dual-weighted residual method.* One of the advantages of this method is that the attraction of the adjoint problem allows one to avoid difficulties with the evaluation of the interpolation constants.

Readers will find a detailed exposition of the method and applications to various problems in R. Rannacher [270] and in the book by W. Bangerth and R. Rannacher [40]. Also, we recommend the papers by R. Becker and R. Rannacher [48], C. Johnson and A. Szepessy [190], P. Houston, R. Rannacher, E. Süli [182], and R. Rannacher and F. T. Suttmeier [272, 273].

Adjoint problems are used in the so-called *goal-oriented* a posteriori error estimates that measure errors in terms of special "goal-oriented" quantities instead of global energy norms (e.g., see W. Bangerth and R. Rannacher [40], J. T. Oden and S. Prudhomme [253], J. Peraire and A. T. Patera [260], E. Stein, M. Rüter, and S. Ohnimus

[340], and the literature cited in these publications). The idea of this approach can be explained with the paradigm of the problem

$$\int_\Omega (A\nabla u \cdot \nabla w - fw)\,dx = 0 \qquad \forall w \in V_0 := \overset{\circ}{H}{}^1(\Omega), \tag{2.6.31}$$

where A is a positive definite real matrix and and $f \in L^2(\Omega)$ is a given function. Let $u_h \in V_0$ be an approximate solution computed on the mesh $\mathcal{T}_h$. Assume that it is required to estimate the quantity $\langle \ell, u - u_h\rangle$, where $\ell \in V_0^*$ is a given linear functional. Typically, ℓ is an integral type functional localized in a certain part of Ω. Define u_ℓ by the relation

$$\int_\Omega A^\star \nabla u_\ell \cdot \nabla w\,dx = \langle \ell, w\rangle \qquad \forall w \in V_0, \tag{2.6.32}$$

in which $A^\star$ is the matrix adjoint to A. From (2.6.31) and (2.6.32), it follows that

$$\begin{aligned}\langle \ell, u - u_h\rangle &= \int_\Omega A^\star \nabla u_\ell \cdot \nabla (u - u_h)\,dx \\ &= \int_\Omega (f \cdot u_\ell - A\nabla u_h \cdot \nabla u_\ell)\,dx = E(u_\ell, u_h). \end{aligned}\tag{2.6.33}$$

Hence, $\langle \ell, (u - v)\rangle$ can be easily estimated provided that u_ℓ is known. In practice, u_ℓ is replaced by an approximation $u_{\ell\tau}$ computed on an *adjoint mesh* $\mathcal{T}_\tau$ (which may not coincide with $\mathcal{T}_h$). If $u_{\ell\tau}$ is a sharp approximation of u_ℓ, then the quantity $E(u_{\ell\tau}, u_h)$ could be a good indicator of $\langle \ell, u - u_h\rangle$. To obtain another indicator, we rewrite (2.6.33) in the form

$$\langle \ell, u - u_h\rangle = E_1(u_h, u_{\ell\tau}) + E_2(u, u_h, u_\ell, u_{\ell\tau}), \tag{2.6.34}$$

where

$$E_1(u_h, u_{\ell\tau}) := \int_\Omega (f u_{\ell\tau} - A\nabla u_h \cdot \nabla u_{\ell\tau})\,dx$$

is a directly computable functional and

$$E_2(u, u_h, u_\ell, u_{\ell\tau}) := \int_\Omega A(\nabla u - \nabla u_h) \cdot (\nabla u_\ell - \nabla u_{\ell\tau})\,dx.$$

If $\mathcal{T}_\tau$ and $\mathcal{T}_h$ coincide, then $E_1(u_h, u_{\ell\tau}) = 0$ (because u_h is a Galerkin approximation). In this case,

$$|\langle \ell, u - u_h\rangle| \leq E_2(u, u_h, u_\ell, u_{\ell\tau}), \tag{2.6.35}$$

Estimate (2.6.34) serves as a source of various indicators. One of them is based on the idea to replace unknown functions ∇u and ∇u_ℓ by averaged gradients $G_h \nabla u_h$

and $G_\tau \nabla u_{\ell\tau}$, where G_h and G_τ are the respective averaging operators. It is proved that under the standard assumptions that guarantee superconvergence of the primal and adjoint approximations such a replacement leads to a higher order error (see [199, 244]). Then, the quantity

$$\widetilde{E}_2(u, u_h, u_\ell, u_{\ell\tau}) := \int_\Omega A(G_h \nabla u_h - \nabla u_h) \cdot (G_\tau \nabla u_{\ell\tau} - \nabla u_{\ell\tau})\, dx$$

can be used instead of E_2.

Finally, we note that the quantity $\langle \ell, u - v \rangle$ cannot completely characterize the error because it vanishes if $u - v$ is orthogonal to ℓ. Therefore, it is desirable to obtain estimates for various functionals ℓ_s, which amounts to solving several adjoint problems (e.g., see [244]). Recently, new forms of the above-discussed error indicators has been derived and tested (see P. Neittaanmäki, S. Repin, and P. Turchin [246]). They do not exploit superconvergence of the adjoint solution and, therefore, can be used if adjoint meshes are not very regular.

3 Poisson's equation

In this chapter, we begin studying estimates of the type (1.3.4). To present the main ideas in the most transparent form we, throughout the chapter consider only one elliptic problem:

$$\Delta u + f = 0 \quad \text{in } \Omega, \tag{3.0.1}$$

$$u = 0 \quad \text{on } \Gamma. \tag{3.0.2}$$

For this problem, we derive two-sided a posteriori estimates with the help of two different methods. The first method uses variational arguments, the second one is based on transformations of the corresponding integral identity. We discuss properties of the estimates, their practical implementation, and relationships between them and a posteriori estimates of other types.

3.1 The variational method

The variational method is based upon the variational statement of the problem (3.0.1): Find $u \in V_0$ such that

$$J(u) = \inf_{v \in V_0} J(v), \quad J(v) = \int_\Omega \left(\frac{1}{2} |\nabla v|^2 - f v \right) dx.$$

Henceforth, we call it *Problem $\mathcal{P}$*.

Note that

$$J(v) = \sup_{y \in Y} L(\nabla v, y), \quad L(\nabla v, y) = \int_\Omega \left(\nabla v \cdot y - \frac{1}{2} |y|^2 - f v \right) dx,$$

where $Y = L^2(\Omega, \mathbb{R}^d)$. Indeed, the value of the above supremum cannot exceed the quantity that we obtain if, for almost all $x \in \Omega$, the value of $y(x)$ is defined as the maximizer of the problem

$$\sup_{\xi \in \mathbb{R}^d} \left\{ \nabla v(x) \cdot \xi - \frac{1}{2} |\xi|^2 \right\}.$$

It is easy to observe, that (at any $x \in \Omega$) the solution of this problem is $\xi = \nabla v(x)$. Since $\nabla v(x) \in Y$, we conclude that

$$\sup_{y \in Y} L(\nabla v, y) = L(\nabla v, \nabla v) = J(v).$$

Then, the original (or *primal*) problem takes the *minimax* form:

$$(\mathcal{P}) \qquad \inf_{v\in V_0} \sup_{y\in Y} L(\nabla v, y). \tag{3.1.1}$$

If the order of inf and sup is changed, then we arrive at the so-called *dual problem*

$$(\mathcal{P}^*) \qquad \sup_{y\in Y} \inf_{v\in V_0} L(\nabla v, y). \tag{3.1.2}$$

Note that

$$\begin{aligned}\inf_{v\in V_0} \int_\Omega \left(\nabla v\cdot y - \frac{1}{2}|y|^2 - fv\right) dx &= -\frac{1}{2}\|y\|^2 + \inf_{v\in V_0}\int_\Omega (\nabla v\cdot y - fv)\, dx\\ &= \begin{cases} -\frac{1}{2}\|y\|^2 & \text{if } y\in Q_f,\\ -\infty & \text{if } y\notin Q_f,\end{cases}\end{aligned}$$

where the set Q_f is defined in Section 2.2. Hence, the dual problem has the form: Find $p\in Q_f$ such that

$$I^*(p) = \sup_{y\in Q_f} I^*(y), \tag{3.1.3}$$

where

$$I^*(q) = -\frac{1}{2}\|q\|^2.$$

How are the problems $(\mathcal{P})$ and $(\mathcal{P}^*)$ related to each other? To answer this question, we first establish a relation that holds regardless of the structure of $L(x, y)$.

Lemma 3.1. *Let $L(x, y)$ be a functional defined on the elements of two nonempty sets X and Y. Then*

$$\sup_{y\in Y}\inf_{x\in X} L(x, y) \le \inf_{x\in X}\sup_{y\in Y} L(x, y). \tag{3.1.4}$$

Proof. It is easy to see that

$$L(x, y) \ge \inf_{\xi\in X} L(\xi, y), \quad \forall x\in X,\ y\in Y.$$

Pass to the supremum over $y\in Y$. We obtain

$$\sup_{y\in Y} L(x, y) \ge \sup_{y\subset Y}\inf_{\xi\in X} L(\xi, y), \quad \forall x\in X.$$

The left-hand side depends on x, whereas the right-hand side is a number. Thus, we take the infimum over $x\in X$ and conclude that

$$\inf_{x\in X}\sup_{y\in Y} L(x, y) \ge \sup_{y\in Y}\inf_{\xi\in X} L(\xi, y).$$

□

Therefore, we always have $\sup \mathcal{P}^* \le \inf \mathcal{P}$. However, in our case we have a stronger relation, namely, $\sup \mathcal{P}^* = \inf \mathcal{P}$. To prove this fact, we note that

$$\int_\Omega \nabla u \cdot \nabla v \, dx = \int_\Omega f v \, dx, \quad \forall v \in V_0.$$

Therefore $\nabla u \in Q_f$ and

$$I^*(p) \ge I^*(\nabla u) = -\frac{1}{2}\|\nabla u\|^2 = \int_\Omega \left(\frac{1}{2}|\nabla u|^2 - |\nabla u|^2\right) dx$$
$$= \int_\Omega \left(\frac{1}{2}|\nabla u|^2 - f u\right) dx = J(u).$$

Thus, we conclude that $I^*(p) = J(u)$ and, consequently, u and $p = \nabla u$ are the solutions of the primal and dual problems, respectively.

Recall the estimate by Mikhlin (2.3.1)

$$\frac{1}{2}\|\nabla(u - v)\|^2 = J(v) - I^*(p) \le J(v) - I^*(q), \qquad \forall q \in Q_f,$$

from which we obtain (see Section 2.3)

$$\|\nabla(v - u)\| \le \|\nabla v - q\|, \qquad \forall q \in Q_f.$$

Take arbitrary $y \in L^2(\Omega)$. Then,

$$\|\nabla(v - u)\| \le \|\nabla v - y\| + \inf_{q \in Q_f} \|y - q\|. \tag{3.1.5}$$

Lemma 3.2. *For any* $y \in L^2(\Omega)$,

$$\inf_{q \in Q_f} \|y - q\| \le |\!|\!| \operatorname{div} y + f |\!|\!| \tag{3.1.6}$$

and for any $y \in H(\Omega, \operatorname{div})$,

$$\inf_{q \in Q_f} \|y - q\| \le C_\Omega \|\operatorname{div} y + f\|. \tag{3.1.7}$$

Proof. Consider an auxiliary problem

$$\Delta w_f + f + \operatorname{div} y = 0 \qquad \text{in } \Omega$$
$$w_f = 0 \qquad \text{on } \Gamma,$$

where $y \in L^2(\Omega)$ and, therefore, $f + \operatorname{div} y \in H^{-1}$. The corresponding solution w_f exists, unique, and satisfies the relation

$$\int_\Omega \nabla w_f \cdot \nabla w \, dx = \int_\Omega (f w - y \cdot \nabla w) \, dx. \tag{3.1.8}$$

From (3.1.8), we find that

$$\|\nabla w_f\| \leq [\![\operatorname{div} y + f]\!] := \sup_{\substack{w \in V_0 \\ w \neq 0}} \frac{\int_\Omega (y \cdot \nabla w - f w)\, dx}{\|\nabla w\|}. \tag{3.1.9}$$

Also, (3.1.8) has the form

$$\int_\Omega (\nabla w_f + y) \cdot \nabla w \, dx = \int_\Omega f w \, dx, \quad \forall w \in V_0,$$

which means that $\bar{q} := \nabla w_f + y \in Q_f$. Therefore,

$$\inf_{q \in Q_f} \|y - q\| \leq \|y - \bar{q}\| = \|\nabla w_f\| \leq [\![\operatorname{div} y + f]\!].$$

If y has a square summable divergence, then

$$\begin{aligned} [\![\operatorname{div} y + f]\!] &= \sup_{\substack{w \in V_0 \\ w \neq 0}} \frac{\int_\Omega (\operatorname{div} y + f) w \, dx}{\|\nabla w\|} \\ &\leq \sup_{\substack{w \in V_0 \\ w \neq 0}} \frac{\|\operatorname{div} y + f\| \|w\|}{\|\nabla w\|} \leq C_{F\Omega} \|\operatorname{div} y + f\|, \end{aligned} \tag{3.1.10}$$

where $C_{F\Omega}$ is the constant in the Friedrichs inequality for the domain Ω. □

From (3.1.5) and (3.1.6) it follows that

$$\|\nabla(v - u)\| \leq \|\nabla v - y\| + [\![\operatorname{div} y + f]\!], \qquad \forall y \in L^2(\Omega). \tag{3.1.11}$$

However, the right-hand side of this estimate includes the norm $[\![\cdot]\!]$, which is defined as the supremum over a functional space and, therefore, is not explicitly computable. A computable estimate follows from (3.1.10) and (3.1.11).

Theorem 3.3 ([276, 277, 282]). *For any $v \in u_0 + V_0$, the upper bound of the error is given by the estimate*

$$\boxed{\|\nabla(v - u)\| \leq \|\nabla v - y\| + C_{F\Omega} \|\operatorname{div} y + f\|,} \tag{3.1.12}$$

where y is an arbitrary function in $H(\Omega, \operatorname{div})$.

Remark 3.4. From (3.1.12) it follows that the quantity

$$\|\nabla v - y\| + C \|\operatorname{div} y + f\|$$

provides a guaranteed upper bound of the error for any constant $C \in [C_{F\Omega}, +\infty)$. If we set $C = +\infty$, then we obtain the hypercircle estimate (2.2.4).

Henceforth, we denote the right-hand side of (3.1.12) by $\overline{\mathfrak{M}}_{\Delta}(v, y)$ and call it the majorant of the deviation from exact solution or the *error majorant*. The majorant $\overline{\mathfrak{M}}_{\Delta}(v, y)$ also depends on the external data $\mathcal{D}$ (represented by Ω and f) but for the sake of simplicity we do not write them as explicit arguments.

Estimate (3.1.12) is the simplest one among the class of functional a posteriori estimates. However, it possesses all principal features typical of all of them (see Section 3.3). A consequent exposition of the variational approach to a posteriori error estimation is presented in the author's papers [276, 277, 278, 279, 282, 286] and in the book by P. Neittaanmäki and S. Repin [244].

Remark 3.5. Let v be a Galerkin approximation u_h computed on a finite element partition $\mathcal{T}_h$. Set $y = \nabla u_h$. Then (3.1.11) implies the estimate

$$\|\nabla(u - u_h)\| \leq | \Delta u_h + f |.$$

If the right-hand side is estimated from above using the Galerkin orthogonality property and the $H^1 \to L^2$ projection estimates on the patches, then we arrive at the explicit residual estimate considered in Section 2.6.1.

A lower bound of the error is given in the theorem below.

Theorem 3.6. *For any* $v \in V_0$,

$$\|\nabla(u - v)\|^2 \geq \underline{\mathfrak{M}}^2_{\Delta}(v, w), \tag{3.1.13}$$

where

$$\underline{\mathfrak{M}}^2_{\Delta}(v, w) := 2\mathcal{F}_v(w) - \|\nabla w\|^2, \tag{3.1.14}$$

w is an arbitrary function in V_0, and $\mathcal{F}_v(w)$ is the residual functional (cf. 2.6.2).

Proof. From the relation

$$2(J(v) - J(u)) = \|\nabla(u - v)\|^2,$$

it follows that

$$\|\nabla(u - v)\|^2 \geq 2(J(v) - J(v + w)),$$

where w is an arbitrary function in V_0. Therefore,

$$\|\nabla(u - v)\|^2 \geq \int_\Omega \left(-|\nabla w|^2 - 2\nabla v \cdot \nabla w\right) dx + 2\int_\Omega f \cdot w \, dx,$$

and we arrive at (3.1.13). □

3.2 The method of integral identities

The modern theory of partial differential equations considers integral relations as one of the major mathematical objects. Integral identities define generalized solutions of differential equations and provide the basis for the analysis of their properties (e.g., see [151, 217, 214, 222]). In this section, we show that two-sided guaranteed bounds of the error can be derived by transformations of an integral identity. Originally, this *modus operandi* was suggested in [287, 283]. It should come as no surprise that the corresponding estimates coincide with those derived by the variational method. As before, we explain the method with the paradigm of the problem (3.0.1) whose generalized solution is defined by the integral identity

$$\int_\Omega \nabla u \cdot \nabla w \, dx = \int_\Omega f w \, dx, \qquad \forall w \in V_0. \tag{3.2.1}$$

Upper bound of the error. Let $v \in V_0$ be a function viewed as an approximate solution. Insert it in (3.2.1). We have

$$\int_\Omega \nabla(u - v) \cdot \nabla w \, dx = \int_\Omega (f w - \nabla v \cdot \nabla w) \, dx. \tag{3.2.2}$$

Note that for a vector-valued function $y \in H(\Omega, \text{div})$ we have (cf. (1.4.14))

$$\int_\Omega (w \operatorname{div} y + \nabla w \cdot y) \, dx = 0.$$

In view of this relation,

$$\int_\Omega \nabla(u-v) \cdot \nabla w \, dx = \int_\Omega \Big((y - \nabla v) \cdot \nabla w + (\operatorname{div} y + f) w\Big) \, dx. \tag{3.2.3}$$

Set $w = u - v$, then we obtain

$$\|\nabla(u - v)\|^2 \le \|\nabla v - y\| \, \|\nabla(u - v)\| + \|f + \operatorname{div} y\| \, \|u - v\|.$$

Hence,

$$\|\nabla(u - v)\| \le \|\nabla v - y\| + C_{F\Omega} \|f + \operatorname{div} y\|$$

and the estimate (3.1.12) is derived by another method.

Remark 3.7. If $\Delta v \in L^2(\Omega)$, then from (3.2.2) we deduce the estimate

$$\|\nabla(u - v)\| \le C_{F\Omega} \|\Delta v + f\|. \tag{3.2.4}$$

Formally, it can be used for sufficiently regular approximations. However, it violates the consistency condition (1.3.3). If a sequence $\{v_k\}$ of approximate solutions converges to u in $\overset{\circ}{H}{}^1(\Omega)$, then the left-hand side of (3.2.4) tends to zero but this may be not true for the right-hand one.

Sometimes, it is required to find approximate solutions of problems, the right-hand sides of which are defined by linear functionals of a more general type (associated with generalized derivatives of L^2-functions). Let u be defined by the integral identity

$$\int_\Omega \nabla u \cdot \nabla w \, dx = \langle \ell, w \rangle, \qquad \forall w \in V_0, \tag{3.2.5}$$

where

$$\langle \ell, w \rangle = \int_\Omega (fw + \tau \cdot \nabla w) \, dx$$

and $\tau \in L^2(\Omega, \mathbb{R}^d)$ is a given vector-valued function. In this case,

$$\int_\Omega \nabla(u - v) \cdot \nabla w \, dx = \int_\Omega ((f + \operatorname{div} y)w + (y - \nabla v) \cdot \nabla w + \tau \cdot \nabla w) \, dx.$$

This relation implies the estimate

$$\|\nabla(u - v)\| \leq \|y + \tau - \nabla v\| + C_{F\Omega}\|f + \operatorname{div} y\|. \tag{3.2.6}$$

By (3.2.5) we know that $\bar{y} = \nabla u - \tau \in Q_f$. It is easy to see that for $y = \bar{y}$ the right-hand side of (3.2.6) is equal to the error.

Let τ be represented as the sum of a divergence-free function and gradient of a scalar-valued function, i.e., $\tau = \tau_0 + \nabla\vartheta$, where τ_0 belongs to the space

$$\mathrm{S}(\Omega) := \left\{\eta \in L^2(\Omega, \mathbb{R}^d) \mid \operatorname{div} \eta = 0, \text{ a.e. in } \Omega\right\},$$

and $\vartheta \in V_0$. Then, the estimate (3.2.6) takes the form

$$\|\nabla(u - v)\| \leq \|\nabla(v - \vartheta) - y\| + C_{F\Omega}\|\operatorname{div} y + f\|. \tag{3.2.7}$$

Remark 3.8. From the computational point of view, it is convenient to square both parts of (3.1.12) and rewrite the upper bound in the form of a quadratic functional, namely,

$$\begin{aligned}\|\nabla(u - v)\|^2 &\leq \overline{\mathfrak{M}}^{\,2}_{\beta,\Delta}(v, y) \\ &:= (1 + \beta)\|\nabla v - y\|^2 + \left(1 + \frac{1}{\beta}\right) C^2_{F\Omega}\|\operatorname{div} y + f\|^2.\end{aligned} \tag{3.2.8}$$

Here, β is a positive constant that comes from Young's inequality (1.4.3).

Lower bound of the error. Lower bounds of the error can also be derived by non-variational arguments. First, we note that

$$\|\nabla(u-v)\| = \sup_{w\in V_0} \frac{\int_\Omega (fw - \nabla v\cdot\nabla w)\,dx}{\|\nabla w\|}$$

from which we conclude that a lower bound of the error is given by the quantity

$$\underline{\mathfrak{M}}_{\Delta,(V_{0h})}(v) := \sup_{w_h\in V_{0h}} \frac{\int_\Omega (fw_h - \nabla v\cdot\nabla w_h)\,dx}{\|\nabla w_h\|}, \tag{3.2.9}$$

where V_{0h} is a finite-dimensional subspace of V_0. Finding $\underline{\mathfrak{M}}_{\Delta,(V_{0h})}(v)$ requires solving a finite-dimensional maximization problem (so that it is indeed computable). The more trial functions are contained in V_{0h} the sharper estimate will be computed. However, (3.2.9) exploits a quotient type functional, which may lead to certain difficulties in maximization procedures.

Another lower bound is obtained with the help of a quadratic functional. We have

$$\begin{aligned}\sup_{w\in V_0}\int_\Omega (\nabla(u-v)\cdot\nabla w)\,dx - \frac{1}{2}\|\nabla w\|^2 &\le \sup_{\tau\in L^2}\left\{\int_\Omega (\nabla(u-v)\cdot\tau - \frac{1}{2}|\tau|^2)\,dx\right\}\\ &= \frac{1}{2}\|\nabla(u-v)\|^2.\end{aligned}$$

On the other hand

$$\sup_{w\in V_0}\int_\Omega (\nabla(u-v)\cdot\nabla w - \frac{1}{2}|\nabla w|^2)\,dx \ge \frac{1}{2}\|\nabla(u-v)\|^2.$$

Thus, we conclude that

$$\begin{aligned}\|\nabla(u-v)\|^2 &= \sup_{w\in V_0}\int_\Omega (2\nabla(u-v)\cdot\nabla w - |\nabla w|^2)\,dx\\ &\ge \sup_{w\in V_0}\left\{-\|\nabla w\|^2 - 2\int_\Omega (\nabla v\cdot\nabla w - fw)\,dx\right\} = \sup_{w\in V_0}\underline{\mathfrak{M}}^2_\Delta(v,w)\end{aligned}$$

and we arrive at (3.1.13).

3.3 Properties of a posteriori estimates

Structure of the majorant First, we note that both terms on the right-hand side of (3.1.11) and (3.1.12) have a clear meaning: they represent measures of the errors in the basic relations

$$p = \nabla u, \tag{3.3.1}$$

$$\operatorname{div} p + f = 0, \tag{3.3.2}$$

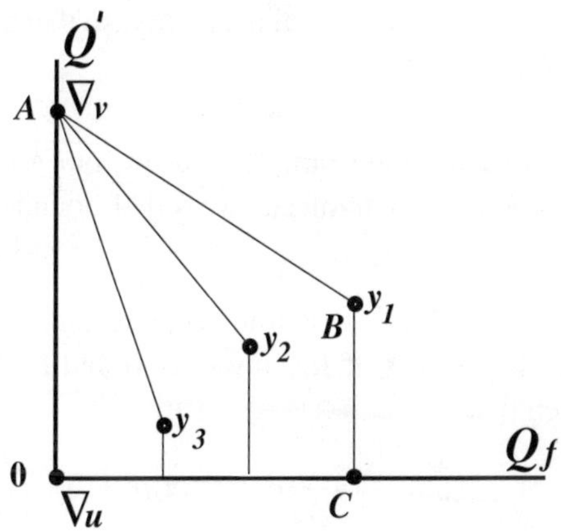

Figure 3.3.1 Geometrical interpretation of $\overline{\mathfrak{M}}_{\Delta}$.

which jointly form the equation. In (3.1.11), the equation $\operatorname{div} p + f = 0$ is understood in a weak sense, whereas in (3.1.12) the respective penalty is given in terms of the L^2-norm.

A geometrical interpretation of the majorant is shown in Fig. 3.3.1. Here Q' denotes the set of vector-valued functions representable as gradients of $\overset{\circ}{H}{}^1$-functions. It is clear that $\nabla u \in Q' \cap Q_f$. Moreover, for any $\eta \in Q_f$,

$$\int_\Omega (\eta - \nabla u) \cdot \nabla v \, dx = 0, \qquad \forall v \in Q',$$

so that the intersection is orthogonal. We observe that the error (associated with the interval OA) is estimated from above by the length of two-chained curve ABC. The length of AB is given by the first term of $\overline{\mathfrak{M}}_{\Delta}$, and the length of BC is estimated by the second one. This estimate is valid for any y, but the closer y lies to the exact flux $p = \nabla u$ the sharper is the estimate. Indeed, it is easy to see that the length of the two-chained curve associated with y_2 provides a better approximation of $|OA|$ than $|AB| + |BC|$. For the curve associated with y_3, the approximation is better than for y_2. Also, it is clear that the length of the curve cannot be smaller than $|OA|$, so that the estimate *always yields a guaranteed upper bound of the error.*

It should be outlined that in (3.1.11), the terms $\|\nabla v - y\|$ and $\|f + \operatorname{div} y\|$ have *sharp* multipliers. If the multiplier of the first term is less than 1 and/or the multiplier of the second one is less than $C_{F\Omega}$, then such a sum cannot be a guaranteed upper bound of the error. Indeed, apply $\overline{\mathfrak{M}}_{\Delta}(v, y)$ to the case where $v = 0$ and $y = 0$. Then, we arrive at the energy estimate for the generalized solution

$$\|\nabla u\| \leq C_{F\Omega} \|f\|.$$

Therefore, no constant less than $C_{F\Omega}$ can be used as a multiplier of the second term.

Next, set $y = \nabla u$. Then (3.1.12) holds as the equality, so that any constant less than 1 cannot be a multiplier of the first term.

Asymptotic properties of two-sided bounds. Now, our goal is to show that $\underline{\mathfrak{M}}_{\Delta}$ and $\overline{\mathfrak{M}}_{\Delta}$ allow one to compute guaranteed two-sided bounds of the error with *any desired accuracy.*

Definition 3.9. A sequence $\{X_k\}_{k=1}^{\infty}$ of finite-dimensional subspaces of a Banach space X is called limit dense in X if for any $\epsilon > 0$ and any $\xi \in X$, one can find a natural number k_ϵ such that

$$\inf_{\xi_m \in X_m} \|\xi_m - \xi\|_X \leq \epsilon, \qquad \forall m > k_\epsilon. \tag{3.3.3}$$

Proposition 3.10. *Let the spaces $\{Y_k\}_{k=1}^{\infty}$ be limit dense in $H(\Omega, \operatorname{div})$. Then*

$$\lim_{m\to\infty} \inf_{y_m \in Y_m} \overline{\mathfrak{M}}_{\Delta}(v, y_m) = \|\nabla(v-u)\|\,, \tag{3.3.4}$$

$$\lim_{m\to\infty} \inf_{\substack{y_m \in Y_m \\ \beta\in\mathbb{R}_+}} \overline{\mathfrak{M}}^{\,2}_{\beta,\Delta}(v, y_m) = \|\nabla(v-u)\|^2\,. \tag{3.3.5}$$

Proof. The proof of (3.3.4) is straightforward. Take an arbitrary small $\epsilon > 0$ and find a respective $k_\epsilon > 0$ such that $\|p - p_m\|_{\operatorname{div}} \leq \epsilon$ for some $p_m \in Y_m$ if $m > k_\epsilon$. Then,

$$\begin{aligned}\overline{\mathfrak{M}}_{\Delta}(u, y_m) \leq \overline{\mathfrak{M}}_{\Delta}(u, p_m) &\leq \|\nabla(u-v)\| + \|p - p_m\| + C_{F\Omega}\|\operatorname{div}(p - p_m)\| \\ &\leq \|\nabla(u-v)\| + \max\{1, C_{F\Omega}\}\,\epsilon.\end{aligned}$$

Analogously, for (3.2.8) we have

$$\begin{aligned}\inf_{\substack{y_m \in Y_m \\ \beta\in\mathbb{R}_+}} \overline{\mathfrak{M}}^{\,2}_{\beta,\Delta}(v, y_m) &\leq \overline{\mathfrak{M}}^{\,2}_{\epsilon,\Delta}(v, p_m) \\ &= (1+\epsilon)\|\nabla v - p_m\|^2 + \left(1 + \frac{1}{\epsilon}\right) C^2_{F\Omega}\|f + \operatorname{div} p_m\|^2.\end{aligned} \tag{3.3.6}$$

Since

$$\|\nabla v - p_m\| \leq \|p_m - p\| + \|\nabla(v-u)\| \leq \epsilon + \|\nabla(v-u)\|$$

we obtain

$$\|\nabla v - p_m\|^2 \leq \|\nabla(v-u)\|^2 + \mu_1\epsilon, \quad \mu_1 = \epsilon + 2\|\nabla(v-u)\|.$$

Also,

$$\|f + \operatorname{div} p_m\| \leq \|\operatorname{div}(p_m - p)\| \leq \|p_m - p\|_{\operatorname{div}} < \epsilon.$$

Therefore,

$$\overline{\mathfrak{M}}^{\,2}_{\epsilon,\Delta}(v,p_m) \leq (1+\epsilon)\,\|\nabla(v-u)\|^2 + (\mu_1 + C^2_{F\Omega})(\epsilon+\epsilon^2)$$
$$= \|\nabla(v-u)\|^2 + o(\epsilon). \tag{3.3.7}$$

Now, (3.3.6) and (3.3.7) imply (3.3.5). □

For any $\beta > 0$, the functional $\overline{\mathfrak{M}}^{\,2}_{\beta,\Delta}(v,y)$ is a quadratic functional with respect to y. Therefore, the quantities (approximate upper bounds of the error)

$$M^2_{\oplus k} = \inf_{\substack{y_k\in Y_k\\ \beta\in\mathbb{R}_+}} \overline{\mathfrak{M}}^{\,2}_{\beta,\Delta}(v,y_k) \tag{3.3.8}$$

can be found by well-known methods. In view of Proposition 3.10, they form a sequence of computable upper bounds such that

$$M^2_{\oplus k} \to \|\nabla(v-u)\|^2 \quad \text{as} \quad k\to\infty. \tag{3.3.9}$$

Lower bounds given by (3.1.13) posscss similar properties.

Proposition 3.11. *If the spaces $\{V_k\}_{k=1}^{\infty}$ are limit dense in V_0, then*

$$\lim_{m\to\infty} M^2_{\ominus m} = \|\nabla(v-u)\|^2, \tag{3.3.10}$$

where

$$M^2_{\ominus m} := \sup_{w_m\in V_m} \underline{\mathfrak{M}}^{\,2}_{\Delta}(v,w_m).$$

Proof. Take an arbitrary small $\epsilon > 0$ and find k_ϵ such that $\|\nabla(u-v-w_m)\| \leq \epsilon$ for $m > k_\epsilon$. Then,

$$\underline{\mathfrak{M}}^{\,2}_{\Delta}(v,w_m) = -\|\nabla w_m\|^2 - 2\int_\Omega (\nabla v\cdot\nabla w_m - f w_m)\,dx$$
$$= -\|\nabla w_m\|^2 + 2\int_\Omega \nabla(u-v)\cdot\nabla w_m\,dx$$
$$= \|\nabla(u-v)\|^2 - \|\nabla(w_m-(u-v))\|^2.$$

Hence,

$$\|\nabla(u-v)\|^2 \geq M^2_{\ominus m} \geq \|\nabla(u-v)\|^2 - \varepsilon^2. \tag{3.3.11}$$

This estimate shows that

$$M^2_{\ominus m} \to \|\nabla(u-v)\|^2 \quad \text{as } m\to\infty. \quad \square \tag{3.3.12}$$

Thus,

$$M_{\ominus m}^2 \leq \|\nabla(v-u)\|^2 \leq M_{\oplus k}^2, \tag{3.3.13}$$

i.e., the error is bounded from below and above by two sequences of *computable numbers*. The relation (3.3.13) means that conforming approximations of the problem (3.0.1) are *fully controllable* (i.e., in principle one can estimate the quality of any conforming approximation with any desirable accuracy).

Definition 3.12. Assume that $M_{\oplus k}$ and $M_{\ominus m}$ have been computed. Then we have the quantity

$$I_{\text{eff}}^{(mk)} := \frac{M_{\oplus k}}{M_{\ominus m}} \geq 1, \tag{3.3.14}$$

which provides an idea of the quality of the error estimation. We call $I_{\text{eff}}^{(mk)}$ the computable efficiency index.

We note that unlike the efficiency indexes comparing error estimates with the norm of the true error (which is known only in specially selected test problems), the quantity $I_{\text{eff}}^{(mk)}$ is indeed computable. From (3.3.10) and (3.3.12) it follows that

$$I_{\text{eff}}^{(mk)} \to 1 \qquad \text{as} \quad m,k \to +\infty. \tag{3.3.15}$$

The estimate (3.3.15) shows that the error bounds $M_{\ominus m}$ and $M_{\oplus k}$ are asymptotically exact.

Properties of the minimizer. It is easy to prove that the exact lower bound of $\overline{\mathfrak{M}}_{\Delta}(v,y)$ (and of $\overline{\mathfrak{M}}_{\beta,\Delta}(v,y)$) with respect to y is attained on a certain element of $H(\Omega,\text{div})$. Indeed, for any $v \in V_0$ (and any $\beta > 0$) the majorant is convex, continuous, and coercive on $H(\Omega,\text{div})$. By known results in the calculus of variations (e.g., see [121]), we conclude that a minimizer $\bar{y}(v)$ exists. Since $\overline{\mathfrak{M}}_{\beta,\Delta}(v,y)$ is a quadratic functional, the corresponding minimizer $\bar{y}(v,\beta)$ is unique (in this case, it depends on β).

Lemma 3.13. *Let $\bar{y} \in H(\Omega,\text{div})$ be such that*

$$\overline{\mathfrak{M}}_{\Delta}(v,\bar{y}) = \inf_{y \in H(\Omega,\text{div})} \overline{\mathfrak{M}}_{\Delta}(v,y). \tag{3.3.16}$$

There exists $\bar{w} \in V_0$ such that $\bar{y} = \nabla \bar{w}$.

Proof. For any $y_0 \in \mathrm{S}(\Omega)$ we have

$$\|\nabla v - \bar{y}\| + C_{F\Omega}\|\text{div}\,\bar{y} + f\| \leq \|\nabla v - y_0 - \bar{y}\| + C_{F\Omega}\|\text{div}\,\bar{y} + f\|.$$

From the above we conclude that for any y_0,

$$\int_\Omega \bar{y} \cdot y_0 \, dx + \frac{1}{2}\|y_0\|^2 \geq 0.$$

This inequality holds if and only if

$$\int_\Omega \bar{y} \cdot y_0 \, dx = 0, \qquad \forall y_0 \in \mathrm{S}(\Omega). \tag{3.3.17}$$

Recall that $\bar{y} \in L^2(\Omega, \mathbb{R}^d)$ admits the decomposition $\bar{y} = \nabla\bar{w} + \tau_0$, where $\bar{w} \in V_0$ and τ_0 is a solenoidal field. Set $y_0 = \tau_0$. From (3.3.17), it follows that $\|\tau_0\| = 0$. Thus, $\bar{y} = \nabla\bar{w}$. □

The minimizer of $\overline{\mathfrak{M}}_{\beta,\Delta}(v, y)$ has a similar property. We leave proving this fact for the reader.

3.4 Two-sided bounds in combined norms

In the so-called *mixed* formulations, the solution of a boundary value problem is defined as a pair of functions. For (3.0.1) it is the saddle point (u, p) of the Lagrangian L (cf. (3.1.1)–(3.1.2)). The majorant $\overline{\mathfrak{M}}_{\Delta}(v, q)$ also considers v and q as independent functions. Therefore, it is natural to measure the respective error in terms of combined (primal-dual) norms of the product space

$$W := V_0 \times H(\Omega, \mathrm{div}),$$

for which we introduce the norm

$$\|(v, y)\|_W := \quad \|\nabla v\| + \|y\| + \|\mathrm{div}\, y\| = \|\nabla v\| + \|y\|_{\mathrm{div}}.$$

Two other equivalent norms are as follows:

$$\|(v, y)\|_W^{(1)} := \|\nabla v\| + \|y\| + C_{F\Omega}\|\mathrm{div}\, y\|,$$
$$\|(v, y)\|_W^{(2)} := \left(\|\nabla v\|^2 + \|y\|^2 + \|\mathrm{div}\, y\|^2\right)^{1/2}.$$

It is easy to see that

$$\gamma_1\|(v, y)\|_W \leq \|(v, y)\|_W^{(1)} \leq \gamma_2\|(v, y)\|_W, \tag{3.4.1}$$
$$\frac{1}{\sqrt{3}}\|(v, y)\|_W \leq \|(v, y)\|_W^{(2)} \leq \|(v, y)\|_W, \tag{3.4.2}$$

where $\gamma_1 = \min\{1, C_{F\Omega}\}$ and $\gamma_2 = \max\{1, C_{F\Omega}\}$.

Let us show that the majorant $\overline{\mathfrak{M}}_{\triangle}(v, y)$ is *equivalent to the error* in the combined norm $\|(u - v, p - y)\|^{(1)}_W$. Since

$$\|p - y\| = \|\nabla u - y\| \leq \|\nabla(u - v)\| + \|\nabla v - y\|$$

and $\|\operatorname{div}(p - y)\| = \|\operatorname{div} y + f\|$, we find that

$$\begin{aligned}\|(u - v, p - y)\|^{(1)}_W &:= \|\nabla(u - v)\| + \|p - y\| + C_{F\Omega}\|\operatorname{div} y + f\| \\ &\leq 2\|\nabla(u - v)\| + \|\nabla v - y\| + C_{F\Omega}\|\operatorname{div} y + f\| \\ &\leq 3\overline{\mathfrak{M}}_{\triangle}(v, y).\end{aligned}$$

On the other hand,

$$\overline{\mathfrak{M}}_{\triangle}(v, y) \leq \|\nabla(v - u)\| + \|p - y\| + C_{F\Omega}\|\operatorname{div} y + f\|. \tag{3.4.3}$$

Thus, we note that the following two-sided estimate holds:

$$\overline{\mathfrak{M}}_{\triangle}(v, y) \leq \|(u - v, p - y)\|^{(1)}_W \leq 3\overline{\mathfrak{M}}_{\triangle}(v, y). \tag{3.4.4}$$

By (3.4.4) we conclude that $\overline{\mathfrak{M}}_{\triangle}$ is an *efficient* and *reliable* measure of the error in the combined norm $\|(u - v, p - y)\|^{(1)}_W$.

In view of (3.4.1) and (3.4.2), the majorant is also equivalent to two other combined norms, namely,

$$\frac{1}{\gamma_2}\overline{\mathfrak{M}}_{\triangle}(v, y) \leq \|(u - v, p - y)\|_W \leq \frac{3}{\gamma_1}\overline{\mathfrak{M}}_{\triangle}(v, y), \tag{3.4.5}$$

$$\frac{1}{\sqrt{3}\gamma_2}\overline{\mathfrak{M}}_{\triangle}(v, y) \leq \|(u - v, p - y)\|^{(2)}_W \leq \frac{3}{\gamma_1}\overline{\mathfrak{M}}_{\triangle}(v, y). \tag{3.4.6}$$

Also, we can define lower and upper bounds for the norm $\|(u - v, p - y)\|_W$ with the help of the functionals

$$\overline{\mathbb{M}}_{\Delta}(v, y) = 3\|\nabla v - y\| + (1 + 2C_{F\Omega})\,\|\operatorname{div} y + f\|$$

and

$$\underline{\mathbb{M}}_{\Delta}(v, y) := \|\nabla v - y\| + \|\operatorname{div} y + f\|,$$

which consist of the same terms as those in $\overline{\mathfrak{M}}_{\triangle}$ but with different weights. We have

$$\begin{aligned}\|(u - v, p - y)\|_W &:= \|\nabla(u - v)\| + \|p - y\| + \|\operatorname{div} y + f\| \\ &\leq 2\|\nabla(u - v)\| + \|\nabla v - y\| + \|\operatorname{div} y + f\| \\ &\leq 3\|\nabla v - y\| + (1 + 2C_{F\Omega})\|\operatorname{div} y + f\| =: \overline{\mathbb{M}}_{\Delta}(v, y).\end{aligned}$$

Hence, we find that

$$\underline{\mathbb{M}}_{\Delta}(v, y) \leq \|(u - v, p - y)\|_W \leq \overline{\mathbb{M}}_{\Delta}(v, y). \tag{3.4.7}$$

Similarly,

$$\gamma_1 \underline{\mathbb{M}}_\Delta(v,y) \le \|(u-v,p-y)\|_W^{(1)} \le \gamma_2 \overline{\mathbb{M}}_\Delta(v,y), \tag{3.4.8}$$

$$\frac{1}{\sqrt{3}} \underline{\mathbb{M}}_\Delta(v,y) \le \|(u-v,p-y)\|_W^{(2)} \le \overline{\mathbb{M}}_\Delta(v,y). \tag{3.4.9}$$

Finally, we note that

$$\overline{\mathfrak{M}}_\Delta(v,p) = \|\nabla(u-v)\|, \tag{3.4.10}$$

$$\overline{\mathfrak{M}}_\Delta(u,y) = \|y - \nabla u\| + C_{F\Omega}\|\mathrm{div}\,(y-p)\|. \tag{3.4.11}$$

Therefore,

$$\|(u-v,p-y)\|_W^{(1)} := \overline{\mathfrak{M}}_\Delta(v,p) + \overline{\mathfrak{M}}_\Delta(u,y). \tag{3.4.12}$$

3.5 Modifications of estimates

3.5.1 Galerkin approximations

Let V_{0h} be a finite-dimensional subspace of V_0. The Galerkin approximation u_h satisfies the orthogonality relation (cf. (2.6.6))

$$\int_\Omega \nabla(u-u_h)\cdot\nabla v_h\,dx = 0, \qquad \forall v_h \in V_{0h}.$$

Therefore,

$$\|\nabla(u-u_h)\|^2 = \int_\Omega (\nabla u_h - y - \nabla v_h)\cdot\nabla(u-u_h)\,dx + \int_\Omega (\mathrm{div}\,y + f)(u-u_h)\,dx$$

and we find that

$$\|\nabla(u-u_h)\| \le \|y - \nabla u_h - \nabla v_h\| + C_{F\Omega}\|f + \mathrm{div}\,y\|. \tag{3.5.1}$$

Set $v_h = u_h - w_h$, where w_h is an arbitrary function in V_{0h}. Then (3.5.1) has the form

$$\|\nabla(u-u_h)\| \le \|y - \nabla w_h\| + C_{F\Omega}\|f + \mathrm{div}\,y\|. \tag{3.5.2}$$

From (3.5.2) it follows that

$$\|\nabla(u-u_h)\| \le C_{F\Omega}\|f + \mathrm{div}\,y\| + \inf_{w_h \in V_{0h}} \|y - \nabla w_h\|. \tag{3.5.3}$$

Analogously, we obtain

$$\|\nabla(u-u_h)\| \le |\!| f + \mathrm{div}\,y |\!| + \inf_{w_h \in V_{0h}} \|y - \nabla w_h\|. \tag{3.5.4}$$

Note that the projection error estimate (Céa lemma) follows from (3.5.3) (and (3.5.4)) if we set $y = \nabla u$. In this case, the first term vanishes and we arrive at the well-known projection estimate

$$\|\nabla(u-u_h)\| \le \inf_{w_h \in V_{0h}} \|\nabla(u-w_h)\|.$$

3.5.2 Advanced forms of error bounds

In view of (3.2.3), we have

$$\|\nabla(u-v)\|^2 = \int_\Omega (y-\nabla v)\cdot\nabla(u-v)\,dx + \int_\Omega \mathrm{r}(y)(u-v)\,dx, \tag{3.5.5}$$

where

$$\mathrm{r}(y) := f + \operatorname{div} y.$$

If y is properly selected (e.g., with the help of a post-processing procedure that generates a function similar to p), then $y-\nabla v \approx \nabla(u-v)$ and, therefore, the quantity $\|y-\nabla v\|\|\nabla(u-v)\|$ does not essentially overestimate the first term on the right-hand side of (3.5.5). However, the quantity $\|\mathrm{r}(y)\|\|u-v\|$ may essentially exceed the integral $\int_\Omega \mathrm{r}(y)(u-v)\,dx$. Thus, the second term of (3.1.12) may be larger than the first one.

One can try to improve the estimate as follows. Take a function ϑ such that

$$\vartheta \in V_0^+ := \left\{\vartheta \in V_0 \mid \Delta\vartheta \in L^2(\Omega)\right\}.$$

Since $u-v$ vanishes at the boundary we observe that

$$\int_\Omega (v-u)\Delta\vartheta\,dx = \int_\Omega \nabla\vartheta\cdot\nabla(u-v)\,dx = \int_\Omega (f\vartheta - \nabla v\cdot\nabla\vartheta)\,dx = \mathcal{F}_v(\vartheta).$$

Note that $\mathcal{F}_v(\vartheta)$ is easily computable and $\mathcal{F}_u(\vartheta) = 0$ for any $\vartheta = 0$. Now, we arrive at the identity

$$\|\nabla(u-v)\|^2 = \int_\Omega ((\mathrm{r}(y)+\Delta\vartheta)(u-v)\,dx + (y-\nabla v)\cdot\nabla(u-v))\,dx + \mathcal{F}_v(\vartheta),$$

which implies the estimate

$$\|\nabla(u-v)\|^2 \le M(v,\vartheta,y)\|\nabla(u-v)\| + \mathcal{F}_v(\vartheta), \tag{3.5.6}$$

where $M(v,\vartheta,y) := C_{F\Omega}\|\mathrm{r}(y)+\Delta\vartheta\| + \|y-\nabla v\|$. Hence,

$$2\|\nabla(u-v)\| \le M(v,\vartheta,y) + \sqrt{M^2(v,\vartheta,y) + 4\mathcal{F}_v(\vartheta)}. \tag{3.5.7}$$

By (1.4.3) and (3.5.6), we can obtain another upper bound

$$\|\nabla(u-v)\|^2 \le \frac{\gamma}{2}M^2(v,\vartheta,y) + \frac{1}{2\gamma}\|\nabla(u-v)\|^2 + \mathcal{F}_v(\vartheta), \tag{3.5.8}$$

which yields (for $\gamma > 1/2$) the estimate

$$\|\nabla(u-v)\|^2 \le \frac{\gamma^2}{2\gamma-1}M^2(v,\vartheta,y) + \frac{2\gamma}{2\gamma-1}\mathcal{F}_v(\vartheta). \tag{3.5.9}$$

Since $\min_{\gamma>1/2}\frac{\gamma^2}{2\gamma-1} = 1$, we observe that (3.5.9) converts to (3.1.12) if $\vartheta = 0$.

We note that (3.5.9) has an advantage with respect to (3.1.12) only if $\|\mathrm{r}(y)\|$ is sufficiently large and $\Delta\vartheta$ compensates a considerable part of it. For this purpose, we need to find a proper function ϑ. A straightforward way is to take a collection of linearly independent functions $\vartheta_i \in V_0^+$, $i = 1, 2, \ldots, k$, and find α_i such that

$$\|\mathrm{r}(y) - \sum_{i=1}^{k} \alpha_i \Delta\vartheta_i\| \to \min .$$

Then, we set $\vartheta = \sum_{i=1}^{k} \alpha_i \vartheta_i$. If $\|\mathrm{r}(y) + \Delta\vartheta\|$ is essentially smaller than $\|\mathrm{r}(y)\|$, then (3.5.7) with $\vartheta = \vartheta_k$ supplies a sharper error bound than the basic estimate (3.1.12). Certainly, the efficiency of this method depends on the system of functions $\{\vartheta_i\}$.

Theoretically, the best choice of ϑ is the function ϑ_y that satisfies the equation $\Delta\vartheta_y + \mathrm{r}(y) = 0$ with homogeneous boundary conditions. In this case, we obtain the estimate

$$\|\nabla(u-v)\|^2 \le \frac{\gamma^2}{2\gamma-1}\|y - \nabla v\|^2 + \frac{2\gamma}{2\gamma-1}\mathcal{F}_v(\vartheta_y). \tag{3.5.10}$$

In practice, instead of the unknown function ϑ_y (which is the exact solution of a boundary value problem) a certain approximation of it can be used (see 3.6.4).

Another form of the error majorant follows from the relation

$$\|\nabla(u-v)\|^2 = \mathcal{F}_v(\vartheta) + \int_\Omega (\mathrm{r}(y)(u-v) + (y - \nabla v - \nabla\vartheta)\cdot\nabla(u-v))\,dx, \tag{3.5.11}$$

where ϑ is a function from the space V_0 (which is wider than V_0^+ and admits simpler approximations). Then, we arrive at (3.5.7) with $M(v, \vartheta, y)$ replaced by

$$\hat{M}(v, \vartheta, y) := C_{F\Omega}\|\mathrm{r}(y)\| + \|y - \nabla\vartheta - \nabla v\|.$$

If $y \in Q_f$, then we have the estimate

$$2\|\nabla(u-v)\| \le \|y - \nabla\vartheta - \nabla v\| + \sqrt{\|y - \nabla\vartheta - \nabla v\|^2 + 4\mathcal{F}_v(\vartheta)}, \tag{3.5.12}$$

which does not contain $C_{F\Omega}$. From (3.5.11) it also follows that

$$2\|\nabla(u-v)\| \le \hat{M}(\varphi, \tau_0, y) + \sqrt{\hat{M}^2(\varphi, \tau_0, y) + 4\mathcal{F}_v(\varphi) - 4\mathcal{F}_v(v)}, \tag{3.5.13}$$

where

$$\hat{M}(\varphi, \tau_0, y) := C_{F\Omega}\|\mathrm{r}(y)\| + \|y - \nabla\varphi - \tau_0\|$$

and ϕ and τ_0 are arbitrary functions in V_0 and $\mathrm{S}(\Omega)$, respectively.

By (3.5.11) and (1.4.3), we also obtain

$$\begin{aligned}\|\nabla(u-v)\|^2 &\geq \mathcal{F}_v(\vartheta) - \hat{M}(v,\vartheta,y)\|\nabla(u-v)\| \\ &\geq \mathcal{F}_v(\vartheta) - \frac{\gamma}{2}\hat{M}^2(v,\vartheta,y) - \frac{1}{2\gamma}\|\nabla(u-v)\|^2.\end{aligned}$$

Hence,

$$\frac{2\gamma+1}{2\gamma}\|\nabla(u-v)\|^2 \geq \mathcal{F}_v(\vartheta) - \frac{\gamma}{2}\hat{M}^2(v,\vartheta,y). \tag{3.5.14}$$

Note that if $\vartheta = u - v$ and $y = \nabla u$ then $\hat{M}(v,\vartheta,y) = 0$ and

$$\mathcal{F}_v(\vartheta) = \int_\Omega (f(u-v) - \nabla v \cdot \nabla(u-v))\,dx = \|\nabla(u-v)\|^2.$$

It is easy to see that in this case the right-hand side of (3.5.14) coincides with the left-hand one if $\gamma \to +\infty$.

Remark 3.14. In (3.5.12) and (3.5.14), the function ϑ should be selected such that $\|y - \nabla\vartheta - \nabla v\|$ is minimal. For example, one can take $\vartheta = \vartheta_{h'}$, where

$$\int_\Omega \nabla\vartheta_{h'} \cdot \nabla w_{h'}\,dx = \int_\Omega (y - \nabla v)\cdot\nabla w_{h'}\,dx, \qquad \forall w_{h'} \in V_{0h'},$$

where $V_{0h'}$ is a certain finite dimensional subspace of V_0. The function ϕ in (3.5.13) can be taken as $\vartheta_{h'} + v$.

3.5.3 Decomposition of the domain

Assume that Ω is decomposed into a set $\mathcal{T}$ of subdomains Ω_i (in particular, Ω_i may coincide with finite elements) with Lipschitz continuous boundaries, i.e.,

$$\overline{\Omega} = \bigcup_{i=1,\dots,N} \overline{\Omega}_i, \qquad \text{and} \qquad \Omega_i \cap \Omega_j = \emptyset \text{ if } i \neq j.$$

It is not difficult to see that

$$\begin{aligned}|\!|\!|\mathrm{r}(y)|\!|\!| &:= \sup_{w\in V_0}\frac{\int_\Omega (y\cdot\nabla w - fw)\,dx}{\|\nabla w\|} = \sup_{w\in V_0}\frac{\sum_{i=1}^N \int_{\Omega_i}(y\cdot\nabla w - fw)\,dx}{\|\nabla w\|} \\ &= \sup_{w\in V_0}\frac{\sum_{i=1}^N \int_{\Omega_i}\mathrm{r}(y)\,w\,dx}{\|\nabla w\|} \leq \sup_{w\in V_0}\frac{\sum_{i=1}^N \int_{\Omega_i}(\mathrm{r}(y) - \{\mathrm{r}(y)\}_{\Omega_i})w\,dx}{\|\nabla w\|} \\ &\qquad + \sup_{w\in V_0}\frac{\sum_{i=1}^N \{\mathrm{r}(y)\}_{\Omega_i}\int_{\Omega_i} w\,dx}{\|\nabla w\|}.\end{aligned} \tag{3.5.15}$$

For the first term on the right-hand side we have (cf. (1.4.23))

$$\sum_{i=1}^{N} \int_{\Omega_i} \widetilde{\mathrm{r}(y)}_{\Omega_i} w\,dx \leq \sum_{i=1}^{N} \|\widetilde{\mathrm{r}(y)}_{\Omega_i}\|_{\Omega_i} C_{P\Omega_i} \|\nabla w\|_{\Omega_i}$$

$$\leq \|\nabla w\| \sqrt{\sum_{i=1}^{N} \|\widetilde{\mathrm{r}(y)}_{\Omega_i}\|^2_{\Omega_i} C^2_{P\Omega_i}} \tag{3.5.16}$$

and for the second one

$$\sum_{i=1}^{N} \{\mathrm{r}(y)\}_{\Omega_i} \int_{\Omega_i} w\,dx \leq \sum_{i=1}^{N} \{\mathrm{r}(y)\}_{\Omega_i} |\Omega_i|^{1/2} \|w\|_{\Omega_i} \leq \sqrt{\sum_{i=1}^{N} \{\mathrm{r}(y)\}^2_{\Omega_i} |\Omega_i|}\; \|w\|_{\Omega}$$

$$\leq C_{F\Omega} \|\nabla w\| \sqrt{\sum_{i=1}^{N} \{\mathrm{r}(y)\}^2_{\Omega_i} |\Omega_i|}. \tag{3.5.17}$$

From (3.5.15)–(3.5.17) we deduce the estimate

$$[\![\mathrm{r}(y)]\!] \leq \sqrt{\sum_{i=1}^{N} \|\widetilde{\mathrm{r}(y)}_{\Omega_i}\|^2_{\Omega_i} C^2_{iP}} + C_{F\Omega} \sqrt{\sum_{i=1}^{N} \{\mathrm{r}(y)\}^2_{\Omega_i} |\Omega_i|}. \tag{3.5.18}$$

Let $y \in \widetilde{Q}(\mathcal{T})$, where

$$\widetilde{Q}(\mathcal{T}) := \{y \in H(\Omega, \mathrm{div}) \mid \{\mathrm{div}\, y + f\}_{\Omega_i} = 0 \quad \forall\, \Omega_i \in \mathcal{T}\}.$$

Then the second term of (3.5.18) vanishes and we obtain

$$[\![\mathrm{r}(y)]\!] \leq \sqrt{\sum_{i=1}^{N} \|\mathrm{r}(y)\|^2_{\Omega_i} C^2_{P\Omega_i}}. \tag{3.5.19}$$

This relation infers the estimate

$$\|\nabla(u - v)\| \leq \|\nabla v - y\| + \sqrt{\sum_{i=1}^{N} \|\mathrm{r}(y)\|^2_{\Omega_i} C^2_{P\Omega_i}}, \tag{3.5.20}$$

which, instead of $C_{F\Omega}$, involves constants in the Poincaré inequalities associated with the subdomains Ω_i.

Consider a special but important case, where $\mathcal{T}$ is a regular simplicial decomposition $\mathcal{T}_h$ consisting of simplexes T_i and

$$\mu_1 h \leq \mathrm{diam} T_i \leq \mu_2 h, \qquad \forall i = 1, 2, \ldots, N.$$

Let $C_{\max P} = \max_i\{C_{P\Omega_i}\}$. For regular triangulations the constant $C_{\max P}$ is of the same order as all other constants $C_{P\Omega_i}$, so that without big overestimation we can replace all these constants by $C_{\max P}$. Then, we arrive at the upper bound

$$\|\nabla(u-v)\| \le \|\nabla v - y\| + C_{\max P}\|\operatorname{div} y + f\|, \tag{3.5.21}$$

where $y \in \widetilde{Q}(\mathcal{T})$.

Remark 3.15. We note that $C_{\max P}$ can be expressed throughout the constant $\widehat{C}_{\max P}$ for a similar simplex the diameter of which is equal to one and the mesh parameter h.

3.5.4 Estimates with partially equilibrated fluxes

Assume that we have a vector-valued function $y_{\bar{f}}$ such that

$$\operatorname{div} y_{\bar{f}} + \bar{f} = 0,$$

where $\bar{f}$ is close to f in L^2-norm.

Set $y = y_{\bar{f}} + \tau_0$, where $\tau_0 \in \mathrm{S}(\Omega)$. Then $\operatorname{div} y + \bar{f} = 0$. We use (3.1.12) and arrive at the estimate

$$\|\nabla(u-v)\| \le \|\nabla v - \tau_0 - y_{\bar{f}}\| + C_{F\Omega}\|f - \bar{f}\|. \tag{3.5.22}$$

In particular, we can set $\tau_0 = \operatorname{curl}\eta$, where η is an arbitrary vector-valued function in $H(\Omega, \operatorname{curl})$. If the value of $\|f - \bar{f}\|$ is significantly smaller than the tolerance level accepted for approximations, then finding a sharp upper bound is reduced to the problem

$$\min_{\eta \in H(\Omega,\operatorname{curl})} \|\tau - \operatorname{curl}\eta\|,$$

where $\tau = \nabla v - y_{\bar{f}}$ is the given vector-valued function. Using a suitable finite-dimensional subspace for η (which is constructed with the help of conforming finite element approximations of $H(\Omega, \operatorname{curl})$), we find an upper bound by solving a quadratic minimization problem.

Partially equilibrated flux $\bar{y}$ can be used in (3.5.13). We have

$$2\|\nabla(u-v)\| \le \hat{M}(\varphi, y_{\bar{f}}) + \sqrt{\hat{M}^2(\varphi, y_{\bar{f}}) + 4\mathcal{F}_v(\varphi) - 4\mathcal{F}_v(v)}, \tag{3.5.23}$$

where

$$\hat{M}(\varphi, y_{\bar{f}}) := C_{F\Omega}\|f - \bar{f}\| + \|y_{\bar{f}} - \nabla\varphi\|.$$

Here ϕ is any function in V_0. This freedom can be used to minimize right-hand side of (3.5.23).

3.6 How can one use functional a posteriori estimates in practical computations?

3.6.1 Post-processing of fluxes

Let $V_{0h} \subset V_0$ be a finite-dimensional space. For example, V_{0h} may contain piecewise affine finite element approximations generated by the triangulation $\mathcal{T}_h$. Assume that $v_h \in V_{0h}$ is an approximate solution computed. In particular, v_h may coincide with the *Galerkin approximation* u_h defined by the relation (2.6.5). Also, it may be any other approximation, which differs from u_h owing to the presence of a roundoff, integration, or other errors. Using v_h, we find a rough approximation of the flux

$$p_h := \nabla v_h \in L^2(\Omega, \mathbb{R}^d). \tag{3.6.1}$$

Generally, p_h does not belong to $H(\Omega, \mathrm{div})$ and we cannot directly substitute $y = p_h$ in (3.1.12). For this reason, it is necessary to regularize p_h by a post-processing operator $\mathrm{G}_h : L^2(\Omega, \mathbb{R}^d) \to H(\Omega, \mathrm{div})$. After that, we obtain a vector-valued function $\mathrm{G}_h p_h$, which yields an easily computable estimate

$$\|\nabla(u - u_h)\| \leq \|\nabla u_h - \mathrm{G}_h p_h\| + C_{F\Omega} \|\mathrm{div}\, \mathrm{G}_h p_h + f\| . \tag{3.6.2}$$

The quality of the upper bound given by (3.6.2) depends on properties of the post-processing operator used. In Chapter 2, we have discussed thr main classes of post-processing (gradient averaging) operators. Any of them can be used in (3.6.2).

Raviart–Thomas elements of the lowest order (which are described and studied in, e.g., F. Brezzi and M. Fortin [79] and J. E. Roberts and J.-M. Thomas [325]) suggest one more post-processing operator, which we denote G_{RT}. Consider a patch formed by two elements having a common edge E_{ln} (see Fig. 3.6.1). If u_h is constructed by P^1-approximations, then $(\nabla u_h)|_{T_i}$ and $(\nabla u_h)|_{T_j}$ are constant vectors. Define the normal flux on E_{ln} as follows:

$$(y \cdot n_{ln})\ |_{E_{ln}} = \kappa_{ln}(\nabla u_h)|_{T_i} + (1 - \kappa_{ln})(\nabla u_h)|_{T_j},$$

where $\kappa_{ln} \in (0, 1)$. In the simplest case, we set $\kappa_{ln} = 1/2$. Another option (which takes into account sizes of elements) is

$$\kappa_{ln} = \frac{|T_i|}{|T_i| + |T_j|}.$$

For the boundary faces, we use the only one existing flux. Thus, we define three normal fluxes on three sides of each element. The field inside is obtained by the standard RT^0-extension of normal fluxes. As a result, we have a function $\mathrm{G}_{\mathrm{RT}} p_h \in H(\Omega, \mathrm{div})$.

We note that $\mathcal{E}_{\mathrm{RT}} := \nabla u_h - G_{\mathrm{RT}} p_h$ is an error indicator generated by the procedure. If the value of the term $\|\mathrm{div}\, \mathrm{G}_{\mathrm{RT}} p_h + f\|$ is too large (in comparison with the term $\|\nabla u_h - \mathrm{G}_{\mathrm{RT}}\, p_h\|$), then we can apply (3.5.7) or (3.5.9) in order to reduce it with the

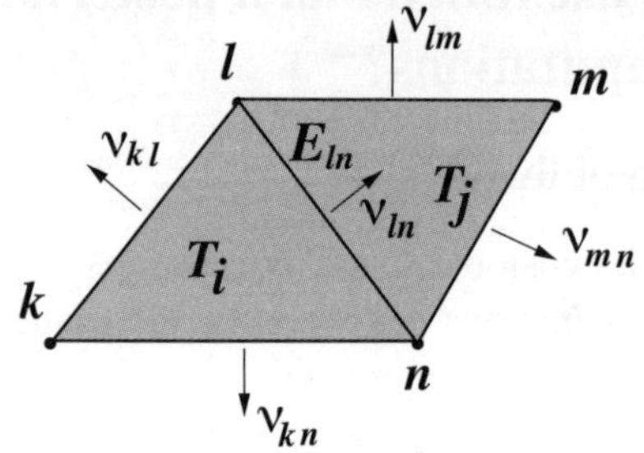

Figure 3.6.1 Patch related to E_{ln}.

help of ϑ. However, in general, substituting a post-processed gradient does not give a very accurate upper bound. Numerical experiments have shown that if G is constructed with the help of simple patch-averaging on the same mesh, then the upper bound given by the right-hand side of (3.6.2) is rather coarse. More sophisticated post-processing procedures usually lead to better estimates.

3.6.2 Runge type estimate

Let $u_{h_1}, u_{h_2}, \ldots, u_{h_k}, \ldots$ be a sequence of approximations on meshes $\mathcal{T}_{h_k}$. Compute $p_{h_k} := \nabla u_{h_k}$ and average it by an averaging operator G_{h_k} acting on $\mathcal{T}_{h_k}$. Then the accuracy of the approximation $u_{h_{k-1}}$ can be measured by the estimate

$$\|\nabla(u - u_{h_{k-1}})\| \le \|\nabla u_{h_{k-1}} - \mathrm{G}_{h_k} p_{h_k}\| + C_{F\Omega} \left\| \operatorname{div} \mathrm{G}_{h_k} p_{h_k} + f \right\|. \quad (3.6.3)$$

This estimate involves approximate solutions computed on two consequent meshes $\mathcal{T}_{h_{k-1}}$ and $\mathcal{T}_{h_k}$. Thus, it follows the same strategy as the Runge indicator. However, the estimate (3.6.3) is mathematically justified and provides a guaranteed upper bound for any pair of consequent meshes.

3.6.3 Minimization of the majorant

Minimization of the majorant with respect to y. Another strategy is to find y by minimizing the majorant on a certain subspace $Y_\tau \subset H(\Omega, \operatorname{div})$. In general, Y_τ may be constructed using a mesh $\mathcal{T}_\tau$ that differs from $\mathcal{T}_h$. Then

$$\|\nabla(u - u_h)\| \le \inf_{y_\tau \in Y_\tau} \{\|\nabla u_h - y_\tau\| + C_{F\Omega} \|\operatorname{div} y_\tau + f\|\}.$$

The wider is Y_τ, the sharper upper bound is obtained. A detailed discussion of the minimization methods and numerical results can be found in [134, 137, 244, 278, 303, 307] and some other publications cited therein.

If we intend to define y_τ by minimization of the majorant, then it is preferable to represent the problem in the quadratic form:

$$\min_{\beta>0} \min_{y_\tau \in Y_\tau} \overline{\mathfrak{M}}^2_{\beta,\Delta}(v, y),$$

where

$$\overline{\mathfrak{M}}^2_{\beta,\Delta}(v, y) = (1+\beta)\|\nabla v - y\|^2 + \left(1 + \frac{1}{\beta}\right) C^2_{F\Omega}\|\operatorname{div} y + f\|^2.$$

Practical computations can be performed by the following minimization algorithm:

Step 1. Set $k = 0$, $y_{\tau,0} = \mathrm{G}_\tau(\nabla u_h)$, where G_τ is a post-processing operator defined for $\mathcal{T}_\tau$. In particular, if $\mathcal{T}_h = \mathcal{T}_\tau$, then any standard gradient averaging operator (see Section 2.6.3) on $\mathcal{T}_h$ can be used.

Step 2. Find β_k such that

$$\overline{\mathfrak{M}}^2_{\beta_k,\Delta}(v, y_{\tau,k}) = \min_{\beta>0} \overline{\mathfrak{M}}^2_{\beta,\Delta}(v, y_{\tau,k}).$$

Here, we have a simple minimization problem, which is solved analytically.

Step 3. Stop if the quantity $\overline{M}_k = \overline{\mathfrak{M}}^2_{\beta_k,\Delta}(v, y_{\tau,k})$ is less than the desired accuracy level (in this case, we guarantee that a sufficiently accurate approximate solution has already been constructed).

Otherwise go to Step 4.

Step 4. Define $y_{\tau,k+1}$ by the relation

$$\overline{\mathfrak{M}}^2_{\beta_k,\Delta}(v, y_{\tau,k+1}) = \min_{y_\tau \in Y_\tau} \overline{\mathfrak{M}}^2_{\beta_k,\Delta}(v, y_\tau).$$

Step 5. Set $k = k + 1$ and go to Step 2.

This algorithm generates a sequence $\overline{M}_0, \overline{M}_1, \ldots, \overline{M}_k$ of monotonically decreasing upper bounds of the error. We terminate it if either the desired accuracy is confirmed or if the difference $\overline{M}_{k+1} - \overline{M}_k$ is considered insignificant (or if we have exceeded the time limit). In any case, the value $\overline{M}_k$ obtained at the very last step provides a guaranteed upper bound of the error.

Remark 3.16. From Lemma 3.13, it follows that

$$\|\nabla(u - v)\| = \inf_{y \in \mathfrak{Y}_k} \overline{\mathfrak{M}}_\Delta(v, y),$$

where

$$\mathfrak{Y}_k := \left\{ y \in H(\Omega, \operatorname{div}) \mid \int_\Omega y \cdot y_{0i}\, dx = 0, \quad \text{for } y_{01}, y_{02}, \ldots, y_{0k},\ y_{0i} \in \mathrm{S}(\Omega) \right\}.$$

If $\Phi(y) \geq 0$ is a penalty functional vanishing on $\mathfrak{Y}_k$, then

$$\inf_{y \in H(\Omega,\operatorname{div})} \overline{\mathfrak{M}}_\Delta(v, y) = \inf_{y \in H(\Omega,\operatorname{div})} \left\{ \overline{\mathfrak{M}}_\Delta(v, y) + \Phi(y) \right\}.$$

Analogously, the term Φ can be used with the squared majorant (3.2.8), which gives the estimate

$$\|\nabla(u-v)\|^2 \leq (1+\beta)\|\nabla v - y\|^2 + \left(1+\frac{1}{\beta}\right) C_{F\Omega}^2 \|\operatorname{div} y + f\|^2 + \Phi(y).$$

If the majorant is numerically minimized with respect to y, then such penalized forms may have certain advantages because they include a "stabilization term" Φ that penalizes deviations from the subspace, in which the exact minimizer lies.

On the construction of $\mathcal{T}_\tau$. One way to construct a finite element subspace of $H(\Omega, \operatorname{div})$ is to use standard piecewise affine approximations of vector-valued functions. It is well motivated if $\mathcal{T}_\tau$ coincides with the mesh $\mathcal{T}_h$ and v is a finite element approximation computed on this mesh.

Another natural class of conforming approximations of the space $H(\Omega, \operatorname{div})$ is represented by RT elements. In Section 3.6.1 we discussed the corresponding operator G_{RT} constructed by averaging of fluxes on the edges. If $\mathcal{T}_\tau$ and $\mathcal{T}_h$ coincide, then another regularization operator (which is used on the second step of the above-described minimization algorithm) can be constructed as follows. First, we define $y = \mathrm{G}_{\mathrm{RT}}\nabla u_h$.

Now, we describe a simple minimization procedure that can be used to obtain an almost equilibrated flux without big computational expenditures. It operates with the quantities $\gamma_{ln} = y \cdot \nu_{ln}$, which completely define a piecewise affine vector-valued function y. By (1.4.12), we have

$$\int_{T_i} \operatorname{div} y \, dx = \gamma_{lk}|E_{lk}| + \gamma_{ln}|E_{ln}| + \gamma_{kn}|E_{kn}|.$$

In view of this relation,

$$\mu_i + \gamma_{ln} d_i = (\operatorname{div} y)_{T_i}, \quad \mu_i = \frac{\gamma_{kl}|E_{kl}| + \gamma_{kn}|E_{kn}|}{|T_i|}, \; d_i = \frac{|E_{ln}|}{|T_i|}.$$

Using analogous relation for T_j, we obtain

$$\mu_j - \gamma_{ln} d_j = (\operatorname{div} y)_{T_j}, \quad \mu_j = \frac{\gamma_{lm}|E_{lm}| + \gamma_{mn}|E_{mn}|}{|T_j|}, \; d_j = \frac{|E_{ln}|}{|T_j|}.$$

Our goal is to select γ_{ln} in such a way that

$$\int_{T_i} \left((\operatorname{div} y)_{T_i} + f\right)^2 dx + \int_{T_j} \left((\operatorname{div} y)_{T_j} + f\right)^2 dx \to \min.$$

Since $(\operatorname{div} y)_{T_i}$ and $(\operatorname{div} y)_{T_j}$ are constant on T_i and T_j, respectively, we find the corresponding value of γ_{ln} by the relation

$$\gamma_{ln} = \frac{\mu_j|T_i| - \mu_i|T_j| + |T_i||T_j|(\{f\}_{T_j} - \{f\}_{T_i})}{|E_{ln}|(|T_i| + |T_j|)}.$$

Using the same idea, we recompute normal fluxes for all edges. At each step of this procedure the value of $\|\text{div}\, y + f\|_\Omega$ decreases and after several cycles of minimization we obtain a vector-valued field, which is equilibrated much better than the original one.

Minimization of the majorant with respect to the two variables v and y. Estimate (3.1.12) implies a new variational statement of the problem (3.0.1):

$$\overline{\mathfrak{M}}_\Delta(u,p) = \inf_{\substack{v\in V_0,\\ y\in H(\Omega,\text{div})}} \overline{\mathfrak{M}}_\Delta(v,y), \tag{3.6.4}$$

which is generated by the error majorant $\overline{\mathfrak{M}}_\Delta$.

Another variational statement follows from (3.2.4). Indeed, for any $\beta > 0$ we have

$$\overline{\mathfrak{M}}^{\,2}_{\beta,\Delta}(u,p) = \inf_{\substack{v\in V_0,\\ y\in H(\Omega,\text{div})}} \overline{\mathfrak{M}}^{\,2}_{\beta,\Delta}(v,y). \tag{3.6.5}$$

It is easy to note that $\overline{\mathfrak{M}}_\Delta(v,y)$ (and $\overline{\mathfrak{M}}_{\beta,\Delta}(v,y)$) equals zero if and only if the arguments coincide with the exact solution u and the exact flux, respectively. This means that we have a new variational statements of the problem (3.0.1).

It is worth noting that there is a significant difference between the primal variational problem $\mathcal{P}$ and problems (3.6.4) and (3.6.5). The value of $\inf \mathcal{P}$ is unknown an depends on f and Ω. The functional $J(v)$ may be positive, as well as negative, and the quantity $J(v)$ does not indicate the accuracy of v. In opposite, the exact lower bound in (3.6.4) (and (3.6.5)) is known: it is equal to zero regardless of the problem data (f and Ω). Moreover, the functional $\overline{\mathfrak{M}}_\Delta(v,y)$ gives a *guaranteed* upper bound of the error. It vanishes if and only if $v = u$ and $y = p$. Thus, the value of $\overline{\mathfrak{M}}_\Delta(v,y)$ supplies a measure of the quality for the approximations v and y.

In principle, one can use the above-discussed properties and solve the problem by directly minimizing $\overline{\mathfrak{M}}_\Delta(v,y)$ with respect to both variables v and y using two sequence of subspaces

$$\{V_{hk}\} \in V_0 \quad \text{and} \quad \{Y_{hk}\} \in H(\Omega,\text{div}).$$

For this purpose one can use the methods developed in the theory of *least square* mixed methods (e.g., see, J. H. Bramble, R. D. Lazarov, and J. E. Pasciak [72], G. F. Carey and A. I. Pehlivanov [85], and the references therein). Assume that the finite dimensional problem

$$\inf_{\substack{v\in V_{hk},\beta>0,\\ y\in Y_{hk},}} M_\oplus(v,y,\beta,C_{F\Omega},f) = M_\oplus(v_k,y_k,\beta_k,C_{F\Omega},f) := \epsilon_k$$

is solved. The quantity ϵ_k shows the accuracy achieved at the step k. If the subspaces are limit dense in the respective functional spaces, then it is easy to prove that approximate solutions (v_k, y_k) tend to (u,p) and the sequence of numbers ϵ_k tends to zero.

Sometimes, this method may be rather expensive and it may be more efficient to find first v_{hk} (using Problem $\mathcal{P}$) and after that y_{hk}. However, if v_{hk} and y_{hk} are defined with the help of a mixed method, then the respective ϵ_k is directly computable by the majorant.

3.6.4 Error indicators generated by error majorants

The theory considered in this chapter, was focused on getting guaranteed bounds of approximation errors. In practice, it is also important to have easily computable functions that furnish information on the overall error and adequately reproduce the error function

$$|e(x)| := |\nabla(u - v)|.$$

Such functions are called *error indicators.* In Chapter 2, we discussed some of them in the context of finite element approximations. Below, we introduce several error indicators, which are generated by error majorants.

1. Let y_τ be a vector-valued function found by minimization of $\overline{\mathfrak{M}}_\Delta(v, y)$ with respect to y on a certain finite-dimensional space Y_τ. Then a simple indicator of the squared error $|e(x)|^2$ is as follows:

$$\mathcal{E}_1(v, y_\tau) = |\eta(x)|^2, \qquad \text{where} \quad \eta(x) := y_\tau - \nabla v. \tag{3.6.6}$$

Since

$$\|e - \eta\| = \|\nabla(u - v) - y_\tau + \nabla v\| = \|p - y_\tau\|, \tag{3.6.7}$$

we see that the indicator $\mathcal{E}_1(v, y_\tau)$ is sharp (i.e., the computable function η is close to $|e|$), if y_τ is close to p.

Let $v = u_h$, where u_h is a finite element approximation computed on $\mathcal{T}_h$. Assume that $\{y_{\tau_k}\}$ is a sequence of fluxes computed by minimization of $\overline{\mathfrak{M}}_\Delta(v, y)$ on expanding spaces $\{Y_{\tau_k}\}$, which are limit dense in $H(\Omega, \mathrm{div})$. By Proposition 3.10 we know that

$$\overline{\mathfrak{M}}_\Delta(v, y_{\tau_k}) \to \|\nabla(u - v)\|. \tag{3.6.8}$$

Hence, the sequence $\{y_{\tau_k}\}$ is bounded in $H(\Omega, \mathrm{div})$ and a weak limit $\widetilde{y}$ of this sequence (or its subsequence) exists. Since $\overline{\mathfrak{M}}_\Delta(u_h, y)$ is convex and continuous with respect to y, we know that

$$\begin{aligned} \|\nabla(u - u_h)\| &= \lim_{k \to +\infty} \overline{\mathfrak{M}}_\Delta(u_h, y_{\tau_k}) \geq \overline{\mathfrak{M}}_\Delta(u_h, \widetilde{y}) \\ &= \|\nabla u_h - \widetilde{y}\| + C_{F\Omega}\|\mathrm{div}\, \widetilde{y} + f\| \geq \|\nabla(u - u_h)\|. \end{aligned} \tag{3.6.9}$$

Thus, we conclude that

$$\|\nabla u_h - \widetilde{y}\| + C_{F\Omega}\|\mathrm{div}\,\widetilde{y} + f\| = \|\nabla(u - u_h)\|$$

and, therefore, $\widetilde{y}$ minimizes the functional $\overline{\mathfrak{M}}_{\Delta}(u_h, y)$.

If $\nabla u_h \notin H(\Omega, \mathrm{div})$ (which is typical of FEM approximations), then one can prove that $\widetilde{y} = \nabla u$. Indeed, by Lemma 3.13, we know that $\widetilde{y} = \nabla \bar{u} \in H(\Omega, \mathrm{div})$, where $\bar{u} \in V_0$. Then,

$$\|\nabla(u_h - \bar{u})\| + C_{F\Omega}\|\Delta \bar{u} + f\| = \|e\|, \tag{3.6.10}$$

where $e = \nabla(u - u_h)$. On the other hand,

$$\|\nabla(u - \bar{u})\| \leq \|\nabla \bar{u} - y\| + C_{F\Omega}\|\mathrm{div}\, y + f\|$$

and, therefore,

$$C_{F\Omega}\|\Delta \bar{u} + f\| \geq \|\nabla(u - \bar{u})\|. \tag{3.6.11}$$

From (3.6.10) and (3.6.11) we conclude that

$$\|e\| \geq \|\nabla(u - \bar{u})\| + \|\nabla(u_h - \bar{u})\|. \tag{3.6.12}$$

By the triangle inequality,

$$\|e\| \leq \|\nabla(u - \bar{u})\| + \|\nabla(u_h - \bar{u})\|, \tag{3.6.13}$$

and, consequently, (3.6.12) and (3.6.13) result in the relation

$$\|e\| = \|\nabla(u - \bar{u})\| + \|\nabla(u_h - \bar{u})\|, \tag{3.6.14}$$

which implies

$$\int_{\Omega} \nabla(u - \bar{u}) \cdot \nabla(\bar{u} - u_h)\, dx = \|\nabla(u - \bar{u})\|\|\nabla(\bar{u} - u_h)\|. \tag{3.6.15}$$

Such a relation is true if (a) $\nabla(u - \bar{u}) = 0$, (b) $\nabla(u_h - \bar{u}) = 0$, or (c)

$$\nabla(\bar{u} - u_h) = \mu\nabla(u - \bar{u}) \qquad \text{for some } \mu \in \mathbb{R}\ (\mu \neq 0). \tag{3.6.16}$$

In view of the boundary conditions, the case (a) means that $u = \bar{u}$ holds. Since $\nabla u_h \notin H(\Omega, \mathrm{div})$, the case (b) is impossible. From (3.6.16), it follows that

$$\nabla u_h = (1 + \mu)\nabla \bar{u} - \mu\nabla u \in H(\Omega, \mathrm{div}),$$

so that if $\nabla u_h \notin H(\Omega, \mathrm{div})$, then this relation does not hold and (c) cannot be true. It remains to conclude that $\widetilde{y} = \nabla u$.

Then,

$$\begin{aligned}\|\nabla(u-u_h)\| &= \lim_{k\to+\infty} \overline{\mathfrak{M}}_{\Delta}(u_h, y_{\tau_k}) \geq \lim_{k\to+\infty} \|\nabla u_h - y_{\tau_k}\| \\ &\geq \|\nabla u_h - \widetilde{y}\| = \|\nabla(u-u_h)\|,\end{aligned}$$

so that

$$\|\nabla u_h - y_{\tau_k}\| \to \|\nabla(u_h - u)\| \quad \text{as } k \to +\infty.$$

From here, it follows that $\|y_{\tau_k}\| \to \|\nabla u\|$ and, consequently, y_{τ_k} tends to ∇u in $L^2(\Omega)$. Hence, $\|p - y_{\tau_k}\| \to 0$. By (3.6.7) we then conclude that the indicator $\eta_k := y_{\tau_k} - \nabla u_h$ tends to e as $k \to +\infty$.

The indicator $\mathcal{E}_1$ was verified in numerous tests not only for the Poisson's equation but also for diffusion, linear elasticity, Stokes, and Maxwell's problems (where analogs of this indicator were used). Experiments confirmed its efficiency and stability with respect to approximations of different types. Two other indicators discussed below are less studied numerically, but we believe that they will be also useful in practical computations.

2. Another error indicator follows from (3.5.10) if ϑ_y is replaced by a sufficiently accurate approximation. For example, we can define $\vartheta_{y\tau}$ as a function in a finite-dimensional space V_τ that satisfies the relation

$$\int_\Omega \nabla\vartheta_{y\tau}\cdot\nabla w_\tau\,dx = \int_\Omega (\operatorname{div} y + f)w_\tau\,dx, \qquad \forall w_\tau \in V_\tau. \tag{3.6.17}$$

Represent $\mathcal{F}_v(\vartheta_{y\tau})$ in the form

$$\begin{aligned}\mathcal{F}_v(\vartheta_y) &= \int_\Omega (f\vartheta_{y\tau} - \nabla v\cdot\nabla\vartheta_{y\tau})\,dx + \int_\Omega (\nabla v\cdot\nabla(\vartheta_{y\tau}-\vartheta_y) + f(\vartheta_y - \vartheta_{y\tau}))\,dx \\ &= \mathcal{F}_v(\vartheta_{y\tau}) + \int_\Omega \nabla(v-u)\cdot\nabla(\vartheta_{y\tau}-\vartheta_y)\,dx.\end{aligned}$$

Assume that v is the Galerkin solution u_h computed on V_h. Let G_h and G_τ be averaging operators on V_h and V_τ, respectively. Then the quantity

$$\begin{aligned}\mathcal{E}_2(u_h,\vartheta_{y\tau}) := \min_{\gamma>1/2}\Big\{&\tfrac{\gamma^2}{2\gamma-1}\|y-\nabla u_h\|^2 + \tfrac{2\gamma}{2\gamma-1}\Big(\mathcal{F}_{u_h}(\vartheta_{y\tau}) \\ &+ \int_\Omega (\nabla u_h - \mathrm{G}_h\nabla u_h)\cdot(\nabla\vartheta_{y\tau} - \mathrm{G}_\tau\nabla\vartheta_{y\tau})\Big)\,dx\Big\}\end{aligned} \tag{3.6.18}$$

is an indicator of the energy norm of the error. If $\mathcal{T}_\tau$ coincides with $\mathcal{T}_h$ and y_h is computed by a certain post-processing of ∇u_h on V_h (e.g., by quasi-equilibration), then the function ϑ_{yh} can be found with the help of the same solver that was used for finding u_h. In this case, the expenditures of the error indication are approximately the same as those required for getting u_h.

A somewhat different estimate follows from the relation

$$\int_\Omega \nabla(u-v)\cdot\nabla w\,dx = \int_\Omega (y+\nabla\vartheta_y-\nabla v)\cdot\nabla w\,dx.$$

Then,

$$\|\nabla(u-v)\| \le \|y+\nabla\vartheta_y-\nabla v\|. \tag{3.6.19}$$

Let $v = u_h$. From (3.6.19) we obtain a simple error indicator

$$\|\nabla(u-u_h)\| \cong \widetilde{\mathcal{E}}_2(u_h, y) = \|y-\nabla\vartheta_{y\tau}-\nabla u_h\|, \tag{3.6.20}$$

where $\vartheta_{y\tau}$ is an approximation of ϑ_y. Another version of the indicator arises if $\nabla\vartheta_{y\tau}$ is replaced by a post-processed (e.g., averaged) vector-valued function

$$\|\nabla(u-u_h)\| \cong \|y-\mathrm{G}_\tau\nabla\vartheta_{y\tau}-\nabla u_h\|. \tag{3.6.21}$$

Certainly the quality of indicators using approximations of ϑ_y depends on y. In practice, a suitable y can be found as follows. First, we post-process ∇u_h by a cheap procedure and obtain $y = G_h\nabla u_h \in H(\Omega,\mathrm{div})$. If the values of $\mathrm{div}\, y + f$ are large in some parts of the domain, then y should be modified to diminish them (exact equilibration is not required). After that, we solve (3.6.17) and find $\theta_{y\tau}$.

3. From (3.2.2) it follows that

$$\|\nabla(u-v)\|^2 \le \|y-\nabla v\|\|\nabla(u-v)\| + j_y(u-v), \tag{3.6.22}$$

where

$$j_y(u-v) := \int_\Omega (\mathrm{div}\, y+f)(u-v)\,dx.$$

By (3.6.22) we observe that

$$\|\nabla(u-v)\| \le \frac{1}{2}\|y-\nabla v\| + \sqrt{j_y(u-v)+\frac{1}{4}\|y-\nabla v\|^2}. \tag{3.6.23}$$

Since

$$\begin{aligned} j_y(u-v) &= \int_\Omega (\nabla u - y)\cdot\nabla(u-v)\,dx \\ &= \|\nabla(u-v)\|^2 - \int_\Omega (y-\nabla v)\cdot\nabla(u-v)\,dx \ge -\frac{1}{4}\|y-\nabla v\|^2, \end{aligned}$$

the determinant of (3.6.23) is nonnegative regardless of the sign of j_y. If $v = u_h$ and an advanced approximation $\mathcal{Q}_\tau u_h$ is computed by one of the post-processing procedures that we discussed in Chapter 2 (cf. (2.6.30)), then (3.6.23) shows that the quantity

$$\mathcal{E}_3(u_h, y) := \frac{1}{2}\|y-\nabla u_h\| + \sqrt{j_y(\mathcal{Q}_\tau u_h - u_h)+\frac{1}{4}\|y-\nabla u_h\|^2} \tag{3.6.24}$$

may also serve as an error indicator. Its quality depends on the choice of y and the efficiency of post-processing provided by $\mathcal{Q}_\tau$.

In particular, if $y = \mathrm{G}_h \nabla u_h$, then we arrive at the indicator

$$\mathcal{E}_3(u_h, \mathrm{G}_h \nabla u_h) := \|\mathrm{G}_h \nabla u_h - \nabla u_h\| + \mathcal{E}_{h,\tau}, \tag{3.6.25}$$

where

$$2\mathcal{E}_{h,\tau} = \Big(\int_\Omega \big(4(\operatorname{div} \mathrm{G}_h \nabla u_h + f)(\mathcal{Q}_\tau u_h - u_h)\, dx + |\mathrm{G}_h \nabla u_h - \nabla u_h|^2\big)\, dx \Big)^{1/2} - \|\mathrm{G}_h \nabla u_h - \nabla u_h\|.$$

The first term of $\mathcal{E}_3(u_h, \mathrm{G}_h \nabla u_h)$ is the standard gradient averaging indicator. The second term is obtained as a computable approximation of the term $j_y(u - u_h)$.

It is easy to see that the error arising if the exact upper bound (3.6.22) is replaced by the indicator (3.6.24) depends on the value of

$$\int_\Omega (\operatorname{div} y + f)(\mathcal{Q}_\tau u_h - u)\, dx.$$

If $\mathcal{Q}_\tau u_h$ provides a good approximation (in L^2-sense) of u and $\operatorname{div} y + f$ is small (in a weak integral sense), then (3.6.24) will give a correct representation of the error.

Finally, we note that (3.6.22) formally generates a simple indicator for the squared error norm $\|\nabla(u - u_h)\|^2$:

$$\|\mathrm{G}_h \nabla u_h - \nabla u_h\|^2 + 2 \int_\Omega (\operatorname{div} \mathrm{G}_h \nabla u_h + f)(\mathcal{Q}_\tau u_h - u_h)\, dx.$$

However, it is clear that the efficiency of such an indicator strongly depends on the efficiency of the averaging operators and deteriorates if $\mathrm{G}_h \nabla u_h$ does not properly reproduce ∇u and the residual $\operatorname{div} \mathrm{G}_h \nabla u_h + f$ is not small enough.

4 Linear elliptic problems

4.1 Two-sided estimates for stationary diffusion problem

4.1.1 Estimates for problems with mixed boundary conditions

Stationary diffusion problem. First, we consider the problem

$$\operatorname{div} A\nabla u + f = 0 \qquad \text{in } \Omega, \tag{4.1.1}$$

$$u = u_0 \qquad \text{on } \Gamma_1, \tag{4.1.2}$$

$$n \cdot A\nabla u = F \qquad \text{on } \Gamma_2, \tag{4.1.3}$$

where $\Omega \subset \mathbb{R}^d$ is a bounded connected domain with Lipschitz continuous boundary that consists of two measurable nonintersecting parts Γ_1 and Γ_2. We assume that $\operatorname{meas}_{d-1}\{\Gamma_1\} > 0$, $u_0 \in H^1(\Omega)$, and the matrix $A = \{a_{ij}\}$ is symmetric and satisfies the relation

$$c_1^2|\xi|^2 \leq A\xi \cdot \xi \leq c_2^2|\xi|^2, \qquad \forall \xi \in \mathbb{R}^d. \tag{4.1.4}$$

Also, we assume that $f \in L^2(\Omega)$ and $F \in L^2(\Gamma_2)$. Let

$$u_0 + V_0 := \{w = u_0 + w_0 \mid w_0 \in V_0(\Omega)\},$$

where

$$V_0 := \{w \in H^1(\Omega) \mid w = 0 \text{ on } \Gamma_1\}.$$

A generalized solution u of (4.1.1)–(4.1.3) is a function in $u_0 + V_0$ that meets the integral identity

$$\int_\Omega A\nabla u \cdot \nabla w \, dx = \int_\Omega f w \, dx + \int_{\Gamma_2} F w \, ds, \qquad \forall w \in V_0(\Omega). \tag{4.1.5}$$

It is well known that a generalized solution defined by (4.1.5) exists and unique.

Upper estimates of the error norm. Now, we use (4.1.5) in order to obtain an estimate of the difference between u and an approximation $v \in u_0 + V_0$ in the energy norm $|||\nabla(u - v)|||$, where

$$|||y|||^2 := \int_\Omega Ay \cdot y \, dx.$$

Also, we use another norm

$$|||y|||_*^2 := \int_\Omega A^{-1}y \cdot y \, dx.$$

In view of (4.1.4), these norms are equivalent to the natural norm of the space $Y := L^2(\Omega, \mathbb{R}^d)$.

Let y be a vector-valued function from the set

$$H_{\Gamma_2}(\Omega, \operatorname{div}) := \left\{ y \in H(\Omega, \operatorname{div}) \mid y \cdot n \in L^2(\Gamma_2) \right\}.$$

Then,

$$\int_\Omega ((\operatorname{div} y) w + \nabla w \cdot y)\, dx = \int_{\Gamma_2} (y \cdot n) w\, ds, \quad \forall w \in V_0. \tag{4.1.6}$$

By (4.1.5) and (4.1.6), we find that

$$\int_\Omega A\nabla(u - v) \cdot \nabla w\, dx = \int_\Omega (f + \operatorname{div} y) w\, dx + \int_\Omega (y - A\nabla v) \cdot \nabla w\, dx + \int_{\Gamma_2} (F - y \cdot n) w\, ds. \tag{4.1.7}$$

By recalling the Friedrichs type inequality

$$\|w\| \le C_{F\Gamma_1} \|\nabla w\|, \qquad \forall w \in V_0, \tag{4.1.8}$$

and the trace inequality

$$\|w\|_{\Gamma_2} \le C_{T\Gamma_2} \|\nabla w\|, \qquad \forall w \in V_0, \tag{4.1.9}$$

we conclude that there exists a positive constant $\lambda_1(\Omega, \Gamma_2)$ such that

$$\lambda_1^2(\Omega, \Gamma_2) = \inf_{w \in V_0} \frac{|||\nabla w|||^2}{\|w\|^2 + \|w\|^2_{\Gamma_2}}. \tag{4.1.10}$$

Since

$$\int_\Omega (A\nabla v - y) \cdot \nabla w\, dx \le |||A\nabla v - y|||_* |||\nabla w|||$$

and

$$\left| \int_\Omega (f + \operatorname{div} y) w\, dx + \int_{\Gamma_2} (F - y \cdot n) w\, ds \right| \le \left(\|f + \operatorname{div} y\|^2 + \|F - y \cdot n\|^2_{\Gamma_2} \right)^{1/2} \mathrm{C} |||\nabla w|||,$$

we arrive at the estimate

$$|||\nabla(u - v)||| \le |||A\nabla v - y|||_* + \mathrm{C}\sqrt{\|f + \operatorname{div} y\|^2 + \|F - y \cdot n\|^2_{\Gamma_2}}. \tag{4.1.11}$$

In (4.1.11), C is any constant greater than $\lambda_1^{-1}(\Omega, \Gamma_2)$. The right-hand side of (4.1.11) represents a computable error majorant $\overline{\mathfrak{M}}_{\rm DF}(v, y)$ for the diffusion problem (4.1.1)–(4.1.3).

A somewhat different form of the error bound follows from the estimate

$$\left| \int_\Omega (f + \operatorname{div} y) w \, dx + \int_{\Gamma_2} (F - y \cdot n) w \, ds \right| \\ \leq C_{F\Gamma_1} \| f + \operatorname{div} y \| \|\nabla w\| + C_{T\Gamma_2} \| F - y \cdot n \|_{\Gamma_2} \|\nabla w\|.$$

Applying it to (4.1.7), we obtain

$$|||\nabla(u - v)||| \leq |||A\nabla v - y|||_* + \frac{1}{c_1}\Big(C_{F\Gamma_1} \| f + \operatorname{div} y \| \\ + C_{T\Gamma_2} \| F - y \cdot n \|_{\Gamma_2} \Big). \tag{4.1.12}$$

Lower estimates of the error norm. A lower bound of $|||\nabla(u - v)|||$ can be derived as follows. Note that

$$\begin{aligned} \frac{1}{2}|||\nabla(u - v)|||^2 &= \sup_{y \in L^2(\Omega, \mathbb{R}^d)} \int_\Omega \left(A\nabla(u - v) \cdot y - \frac{1}{2} A y \cdot y \right) dx \\ &\geq \sup_{w \in V_0} \int_\Omega \left(A\nabla(u - v) \cdot \nabla w - \frac{1}{2} A\nabla w \cdot \nabla w \right) dx \\ &\geq \int_\Omega \left(A\nabla(u - v) \cdot \nabla(u - v) - \frac{1}{2} A\nabla(u - v) \cdot \nabla(u - v) \right) dx \\ &= \frac{1}{2}|||\nabla(u - v)|||^2 \end{aligned}$$

and we conclude that

$$\begin{aligned} \frac{1}{2}|||\nabla(u - v)|||^2 &= \sup_{w \in V_0} \int_\Omega \left(A\nabla(u - v) \cdot \nabla w - \frac{1}{2} A\nabla w \cdot \nabla w \right) dx \\ &= \sup_{w \in V_0} \left\{ \mathcal{F}_v(w) - \int_\Omega \frac{1}{2} A\nabla w \cdot \nabla w \, dx \right\}, \end{aligned}$$

where

$$\mathcal{F}_v(w) = \int_\Omega (f w - A\nabla v \cdot \nabla w) \, dx + \int_{\Gamma_2} F w \, ds.$$

It is easy to see that the lower bound given by the left-hand side of the above estimate is sharp (set $w = u - v$). Thus, the minorant is defined by the relation

$$\underline{\mathfrak{M}}^2_{\rm DF}(v, w) := 2\mathcal{F}_v(w) - \int_\Omega A\nabla w \cdot \nabla w \, dx. \tag{4.1.13}$$

4.1.2 Modifications of estimates

Quadratic form of the error majorant. Square both parts of (4.1.11) and apply Young's inequality. We obtain

$$\begin{aligned} |||\nabla(u-v)|||^2 &\leq \overline{\mathfrak{M}}^{\,2}_{\mathrm{DF}}(v,y,\beta) \\ &:= (1+\beta)|||A\nabla v - y|||_*^2 \\ &\quad + \frac{1+\beta}{\beta}\mathrm{C}^2\left(\|f+\operatorname{div} y\|^2 + \|F - y\cdot n\|^2_{\Gamma_2}\right), \end{aligned} \tag{4.1.14}$$

where β is an arbitrary positive number. For any $\beta > 0$, the right-hand side of (4.1.14) is a quadratic functional with respect to y the minimization of which on a finite-dimensional subspace is equivalent to solving a system of linear simultaneous equations. This estimate is exact in the sense that, by choosing proper β and y, one can make the right-hand side as arbitrarily close to the left-hand side.

Galerkin approximations. Let $v = u_h \in u_0 + V_{0h}$, where

$$\int_\Omega A\nabla u_h \cdot \nabla w_h \, dx = \int_\Omega f w_h \, dx + \int_{\Gamma_2} F w_h \, ds, \qquad \forall w_h \in V_{0h} \subset V_0.$$

In this case,

$$\int_\Omega A\nabla(u_h-u)\cdot\nabla w_h\,dx = \int_\Omega \nabla(u_h-u)\cdot A\nabla w_h\,dx = 0. \tag{4.1.15}$$

With the help of (4.1.15) we rewrite (4.1.7) in the form

$$\begin{aligned} \int_\Omega A\nabla(u-u_h)\cdot\nabla(u-u_h)\,dx &= \int_\Omega (f+\operatorname{div} y)(u-u_h)\,dx \\ &\quad + \int_\Omega (y - A\nabla u_h - A\nabla w_h)\cdot\nabla(u-u_h)\,dx \\ &\quad + \int_{\Gamma_2} (F - y\cdot n)(u-u_h)\,ds. \end{aligned} \tag{4.1.16}$$

From (4.1.16), we deduce the estimate

$$\begin{aligned} |||\nabla(u-u_h)||| &\leq |||y - A\nabla u_h - A\nabla w_h|||_* \\ &\quad + \mathrm{C}\left(\|f+\operatorname{div} y\|^2 + \|F - y\cdot n\|^2_{\Gamma_2}\right)^{1/2}. \end{aligned} \tag{4.1.17}$$

Here, w_h is an arbitrary function in V_{0h}, which can be used to reduce the value of the first term of the error majorant.

For $y = A\nabla u$, the estimate (4.1.17) reads

$$|||\nabla(u-u_h)||| \leq |||A\nabla(u-u_h-w_h)|||_* = |||\nabla(u-u_h-w_h)|||. \tag{4.1.18}$$

Since w_h is an arbitrary function in V_{0h}, we can take it as $w_h = \widetilde{u}_h + w_h$, where $\widetilde{w}_h \in V_{0h}$. Then (4.1.18) implies the projection error estimate

$$\|\nabla(u - u_h)\| \leq \frac{c_2}{c_1} \inf_{\widetilde{w}_h \in V_{0h}} \|\nabla(u - \widetilde{w}_h)\|. \tag{4.1.19}$$

An advanced form of the error majorant. In view of (4.1.7),

$$|||\nabla(u - v)|||^2 = \int_\Omega (y - A\nabla v) \cdot \nabla(u - v)\, dx + \int_\Omega \mathrm{r}_\Omega(y)(u - v)\, dx + \int_{\Gamma_2} \mathrm{r}_{\Gamma_2}(y)(u - v)\, ds, \tag{4.1.20}$$

where

$$\mathrm{r}_\Omega(y) := f + \operatorname{div} y \quad \text{and} \quad \mathrm{r}_{\Gamma_2}(y) = F - y \cdot n.$$

Let $\vartheta \in V_0(\Omega)$ be a function such that

$$A\nabla\vartheta \in H_{\Gamma_2}(\Omega, \operatorname{div}).$$

Since $u - v$ vanishes on Γ_1, we note that

$$\begin{aligned} \int_\Omega \operatorname{div} A\nabla\vartheta (u - v)\, dx &= -\int_\Omega A\nabla\vartheta \cdot \nabla(u - v)\, dx + \int_{\Gamma_2} (A\nabla\vartheta \cdot n)(u - v)\, ds \\ &= \int_{\Gamma_2} (A\nabla\vartheta \cdot n)(u - v)\, ds - \mathcal{F}_v(\vartheta), \end{aligned} \tag{4.1.21}$$

where the functional $\mathcal{F}_v$ is defined above.

By (4.1.21) we rewrite (4.1.20) in the form

$$\begin{aligned} |||\nabla(u - v)|||^2 = \int_\Omega (y - A\nabla v) \cdot \nabla(u - v)\, dx + \int_\Omega (\mathrm{r}_\Omega(y) + \operatorname{div} A\nabla\vartheta)(u-v)\, dx \\ + \int_{\Gamma_2} (\mathrm{r}_{\Gamma_2}(y) - A\nabla\vartheta \cdot n)(u-v)\, ds + \mathcal{F}_v(\vartheta). \end{aligned} \tag{4.1.22}$$

Hence,

$$|||\nabla(u - v)|||^2 \leq \overline{\mathfrak{M}}^2_{DF}(v, y, \vartheta, \gamma), \tag{4.1.23}$$

where γ is a positive constant and

$$\begin{aligned} \overline{\mathfrak{M}}^2_{DF}(v, y, \vartheta, \gamma) := \frac{2\gamma}{2\gamma - 1} \mathcal{F}_v(\vartheta) + \frac{\gamma^2}{2\gamma - 1} \Big(|||A\nabla v - y|||_* \\ + \mathrm{C}\big(\|\mathrm{r}_\Omega(y) + \operatorname{div} A\nabla\vartheta\|^2 + \|\mathrm{r}_{\Gamma_2}(y) - A\nabla\vartheta \cdot n\|^2_{\Gamma_2}\big)^{1/2} \Big)^2 \end{aligned}$$

represents an advanced form of the error majorant for the linear diffusion problem with mixed boundary conditions.

Decomposition of Ω. Assume that $\bar{\Omega} = \bigcup_{i=1}^{N} \bar{\Omega}_i$, where Ω_i are nonintersecting domains with Lipschitz continuous boundaries.

Assume that

$$\{\mathrm{r}_\Omega(y) + \operatorname{div} A\nabla\vartheta\}_{\Omega_i} = 0, \qquad i = 1, 2, \dots, N,$$
$$\mathrm{r}_{\Gamma_2}(y) - A\nabla\vartheta \cdot n = 0, \qquad \text{on } \Gamma_2.$$

In this case, the equation (4.1.22) yields an upper bound that contains constants $C_{P\Omega_i}$, $i = 1, 2, \dots, N$, instead of C (we recall that $C_{P\Omega_i}$ is the constant in the Poincaré inequality for Ω_i).

Indeed, we have

$$\begin{aligned}
|||\nabla(u-v)|||^2 &= \int_\Omega (y - A\nabla v)\cdot\nabla(u-v)\,dx \\
&\quad + \sum_{i=1}^{N} C_{P\Omega_i} \|\mathrm{r}_\Omega(y) + \operatorname{div} A\nabla\vartheta\|_{\Omega_i} \|\nabla(u-v)\|_{\Omega_i} + \mathcal{F}_v(\vartheta) \\
&\le |||A\nabla v - y|||_* |||\nabla(u-v)||| + \mathcal{F}_v(\vartheta) \\
&\quad + \sqrt{\sum_{i=1}^{N} C^2_{P\Omega_i} \|\mathrm{r}_\Omega(y) + \operatorname{div} A\nabla\vartheta\|^2_{\Omega_i}}\, \|\nabla(u-v)\|.
\end{aligned}$$

Let α and β be positive numbers such that $\alpha + \beta < 2$. Then, we arrive at the estimate

$$\begin{aligned}
&(2-\alpha-\beta)|||\nabla(u-v)|||^2 \\
&\le \frac{1}{\alpha}|||A\nabla v - y|||_*^2 + \frac{1}{c_1^2\beta}\sum_{i=1}^{N} C^2_{P\Omega_i}\|\mathrm{r}_\Omega(y) + \operatorname{div} A\nabla\vartheta\|^2_{\Omega_i} + \mathcal{F}_v(\vartheta).
\end{aligned} \tag{4.1.24}$$

4.1.3 Estimates for problems with Neumann boundary condition

Let $\Gamma_1 = \emptyset$ and $\Gamma_2 = \Gamma$. In this case, the energy space is

$$\widetilde{V} := \left\{ w \in H^1(\Omega) \;\middle|\; \int_\Omega w\,dx = 0 \right\}$$

and the equilibrium condition

$$\int_\Omega f\,dx + \int_\Gamma F\,ds = 0 \tag{4.1.25}$$

must be satisfied. The solution is defined by the integral identity

$$\int_\Omega A\nabla u\cdot\nabla w\,dx + \int_\Omega fw\,dx + \int_\Gamma Fw\,ds = 0, \quad \forall w \in \widetilde{V}. \tag{4.1.26}$$

Let $v \in \widetilde{V}$ be an approximate solution and $y \in H(\Omega, \mathrm{div})$. A transformation of (4.1.26) yields

$$\int_\Omega A\nabla(u-v)\cdot\nabla(u-v)\,dx = \int_\Omega (f+\mathrm{div}\,y)(u-v)\,dx \\ + \int_\Omega (y - A\nabla v)\cdot\nabla(u-v)\,dx + \int_\Gamma (F - y\cdot n)(u-v)\,ds.$$

Since $u - v \in \widetilde{V}$, we find that

$$\begin{aligned}\int_\Omega (f+\mathrm{div}\,y)(u-v)\,dx &= \int_\Omega (f+\mathrm{div}\,y - \{f+\mathrm{div}\,y\}_\Omega)(u-v)\,dx \\ &= \int_\Omega \widetilde{(f+\mathrm{div}\,y)}(u-v)\,dx \\ &\le \|\widetilde{f+\mathrm{div}\,y_\Omega}\| C_{P\Omega}\|\nabla(u-v)\|.\end{aligned}$$

Another term on the right-hand side is estimated by the trace inequality. We have

$$\begin{aligned}|||\nabla(u-v)|||^2 &\le |||y - A\nabla v|||_* |||\nabla(u-v)||| + (C_{P\Omega}\|\widetilde{f+\mathrm{div}\,y}\| \\ &\quad + C_{T\Gamma}\|F - y\cdot n\|_\Gamma)\|\nabla(u-v)\| \\ &\le c_1^{-1}(C_{P\Omega}\|\widetilde{f+\mathrm{div}\,y}\| + C_{T\Gamma}\|F-y\cdot n\|_\Gamma)|||\nabla(u-v)||| \end{aligned} \tag{4.1.27}$$

and arrive at the estimate

$$|||\nabla(u-v)||| \\ \le |||A\nabla v - y|||_* + c_1^{-1}(C_{P\Omega}\|\widetilde{f+\mathrm{div}\,y}\| + C_{T\Gamma}\|F - y\cdot n\|_\Gamma). \tag{4.1.28}$$

4.2 The stationary reaction-diffusion problem

Diffusion problem with mixed Dirichlet–Neumann boundary conditions. The reaction-diffusion problem is represented by the system

$$-\mathrm{div}\,p + \varrho^2 u = f \quad \text{in } \Omega, \tag{4.2.1}$$

$$p = A\nabla u \quad \text{in } \Omega, \tag{4.2.2}$$

$$u = u_0 \quad \text{on } \Gamma. \tag{4.2.3}$$

We assume that ϱ is a nonnegative function of x and the matrix A satisfies (4.1.4).

Now $V_0 = \overset{\circ}{H}{}^1(\Omega)$ and the generalized solution $u \in u_0 + V_0$ of (4.2.1)–(4.2.3) is defined by the integral identity

$$\int_\Omega (A\nabla u\cdot\nabla w + \varrho^2 u\, w)\,dx = \int_\Omega f w\,dx, \qquad w \in V_0. \tag{4.2.4}$$

It minimizes the functional

$$I(w) = \int_\Omega \left(\frac{1}{2} A\nabla w \cdot \nabla w + \frac{\varrho^2}{2} |w|^2 - f w \right) dx \tag{4.2.5}$$

on the set $u_0 + V_0$.

Let $v \in u_0 + V_0$. Then (4.2.4) implies the relation

$$\int_\Omega (\nabla(u-v) \cdot \nabla w + \varrho^2 (u-v)\, w)\, dx = \int_\Omega (f w - \varrho^2 v w - \nabla v \cdot \nabla w)\, dx, \tag{4.2.6}$$

which holds for any $w \in V_0$. Since w vanishes on the boundary, we rewrite (4.2.6) in the form

$$\int_\Omega (A\nabla(u-v) \cdot \nabla w + \varrho^2 (u-v)\, w)\, dx$$
$$= \int_\Omega ((f - \varrho^2 v + \operatorname{div} y) w + (y - A\nabla v) \cdot \nabla w)\, dx, \tag{4.2.7}$$

where y is a vector-valued function in the space $H(\Omega, \operatorname{div})$. The second term on the right-hand side is estimated as in Section 4.1, but the first one has two different upper bounds:

$$\int_\Omega (f - \varrho^2 v + \operatorname{div} y) \cdot w\, dx \le \|\tfrac{1}{\varrho}(f - \varrho^2 v + \operatorname{div} y)\| \|\varrho w\|, \tag{4.2.8}$$

$$\int_\Omega (f - \varrho^2 v + \operatorname{div} y) \cdot w\, dx \le \mathrm{C} \|f - \varrho^2 v + \operatorname{div} y\| |||\nabla w|||, \tag{4.2.9}$$

where C is a constant in the inequality

$$\|w\| \le \mathrm{C} |||\nabla w|||, \qquad \forall w \in V_0. \tag{4.2.10}$$

Note that $\mathrm{C} \le c_1^{-1} C_{F\Omega}$. By (4.2.7) and (4.2.8), we deduce the estimate

$$|[u-v]| \le \left(|||A\nabla v - y|||_*^2 + \|\frac{1}{\varrho} \mathrm{r}_\Omega(v, y)\|^2 \right)^{1/2}, \tag{4.2.11}$$

where

$$|[w]|^2 := |||\nabla w|||^2 + \|\varrho\, w\|^2$$

is the energy norm related to the problem and

$$\mathrm{r}_\Omega(v, y) := f - \varrho^2 v + \operatorname{div} y.$$

From (4.2.7) and (4.2.9), we obtain another estimate

$$|||\nabla(u-v)||| \le |||A\nabla v - y|||_* + \mathrm{C} \|\mathrm{r}_\Omega(v, y)\|. \tag{4.2.12}$$

Let us denote the majorants in (4.2.11) and (4.2.12) by $\overline{\mathfrak{M}}_{\mathrm{RD1}}(v, y)$ and $\overline{\mathfrak{M}}_{\mathrm{RD0}}(v, y)$, respectively (the motivation of this will be given below). Note that the majorant $\overline{\mathfrak{M}}_{\mathrm{RD1}}(v, y)$ was earlier derived by the variational method (see [282]). It is easy to show that

$$\inf_{y \in H(\Omega,\mathrm{div})} \overline{\mathfrak{M}}_{\mathrm{RD1}}(v, y) = |[u - v]|.$$

This fact follows from the relation

$$\begin{aligned} \overline{\mathfrak{M}}^2_{\mathrm{RD1}}(v, p) &= |||A\nabla(v - u)|||^2_* + \|\tfrac{1}{\varrho}\mathrm{r}_\Omega(v, p)\|^2 \\ &= |||\nabla(v - u)|||^2 + \int_\Omega \varrho(u - v)^2\, dx = |[u - v]|^2. \end{aligned}$$

However, $\overline{\mathfrak{M}}_{\mathrm{RD1}}(v, y)$ has an essential drawback: if ϱ is small, then the second term has a large multiplier that makes the whole estimate sensitive to the residual $\mathrm{r}_\Omega(v, y)$. In the problems where ϱ is small (or zero) in one part of Ω and large in the other one, the majorant $\overline{\mathfrak{M}}_{\mathrm{RD1}}(v, y)$ may lead to a considerable overestimation of the error. On the contrary, $\overline{\mathfrak{M}}_{\mathrm{RD0}}(v, y)$ is robust with respect to small ϱ but it may have an inherent gap between the left-hand and right-hand sides of (4.2.12).

An advanced form of the error majorant. In order to overcome the above difficulties and to obtain an estimate that possesses positive features of the above estimates we apply another modus operandi for the deviation of an upper bound of $u - v$ suggested in [305].

Represent the first term on the right-hand side of (4.2.7) in the form

$$\int_\Omega \mathrm{r}_\Omega(v, y) w\, dx = \int_\Omega \alpha \mathrm{r}_\Omega(v, y)\varrho w\, dx + \int_\Omega (1 - \alpha)\mathrm{r}_\Omega(v, y) w\, dx,$$

where

$$\alpha \in L^\infty_{[0,1]}(\Omega) := \{\alpha \in L^\infty(\Omega) \mid 0 \le \alpha(x) \le 1\}.$$

Then, we have

$$\left|\int_\Omega \mathrm{r}_\Omega(v, y) w\, dx\right| \le \left\|\frac{\alpha}{\varrho}\mathrm{r}_\Omega(v, y)\right\| \|\varrho w\| + \mathrm{C}\,\|(1 - \alpha)\mathrm{r}_\Omega(v, y)\|\; |||\nabla w|||.$$

Setting $w = u - v$, we arrive at the estimate

$$\begin{aligned} |[u - v]|^2 &\le (\mathrm{C}\|(1 - \alpha)\mathrm{r}_\Omega(v, y)\| + |||A\nabla v - y|||_*)^2 + \left\|\frac{\alpha}{\varrho}\mathrm{r}_\Omega(v, y)\right\|^2 \\ &=: \overline{\mathfrak{M}}^2_{\mathrm{DF}\alpha}(v, y). \end{aligned} \tag{4.2.13}$$

It is easy to see that (4.2.11) and (4.2.12) are special cases of (4.2.13).

Also, we can represent (4.2.13) in the form

$$[\![u - v]\!]^2 \le C^2(1+\beta)\|(1-\alpha)\mathrm{r}_\Omega(v,y)\|^2 + \frac{1+\beta}{\beta}|\!|\!|A\nabla v - y|\!|\!|_*^2 + \|\frac{\alpha}{\varrho}\mathrm{r}_\Omega(v,y)\|^2, \quad (4.2.14)$$

where β is an arbitrary positive number.

The minimization of the right-hand side of (4.2.14) with respect to α is reduced to the following auxiliary variational problem: Find $\widehat{\alpha} \in L^\infty_{[0,1]}(\Omega)$ such that

$$g(\widehat{\alpha}) = \inf_{\alpha \in L^\infty_{[0,1]}(\Omega)} g(\alpha), \qquad g(\alpha) := \int_\Omega \left(\alpha^2 S(x) + (1-\alpha)^2 T(x)\right) dx, \quad (4.2.15)$$

where S and T are nonnegative integrable functions that do not vanish simultaneously. It is easy to find that for almost all x,

$$\widehat{\alpha}(x) = \frac{T}{S+T} \in [0,1], \qquad g(\widehat{\alpha}) = \frac{ST}{S+T}.$$

In our case, $S = \rho^{-2}\mathrm{r}^2_\Omega(v,y)$ and $T = C^2(1+\beta)\mathrm{r}^2_\Omega(v,y)$. Therefore, we obtain

$$\begin{aligned}[\![u - v]\!]^2 &\le \int_\Omega \frac{C^2(1+\beta)}{C^2\varrho^2(1+\beta)+1}\mathrm{r}^2_\Omega(v,y)\,dx + \frac{1+\beta}{\beta}|\!|\!|A\nabla v - y|\!|\!|_*^2 \\ &=: \overline{\mathfrak{M}}^2_{\mathrm{RD}}(v,y,\beta). \end{aligned} \quad (4.2.16)$$

Since

$$\overline{\mathfrak{M}}^2_{\mathrm{RD}}(v,p,\beta) = \int_\Omega \left(\frac{C^2(1+\beta)}{C^2\varrho^2(1+\beta)+1}\varrho^4(v-u)^2\right) dx + \frac{1+\beta}{\beta}|\!|\!|\nabla(v-u)|\!|\!|^2,$$

we find that

$$\inf_{y \in H(\Omega,\mathrm{div}),\, \beta>0} \overline{\mathfrak{M}}^2_{\mathrm{RD}}(v,y,\beta) \le \inf_{\beta>0} \overline{\mathfrak{M}}^2_{\mathrm{RD}}(v,p,\beta) = [\![u-v]\!]^2.$$

Therefore, (4.2.16) has no "gap". At the same time the structure of the first term of (4.2.16) is such that it is not sensitive to small values of ϱ. Moreover, if $\varrho = 0$, then (4.2.16) implies the estimate

$$|\!|\!|\nabla(u-v)|\!|\!|^2 \le C^2(1+\beta)\|f + \operatorname{div} y\|^2 + \frac{1+\beta}{\beta}|\!|\!|A\nabla v - y|\!|\!|_*^2 \quad (4.2.17)$$

for the diffusion problem without convection.

If $\varrho = \mathrm{const}$ then the value of β that minimizes the right-hand side of (4.2.16) can be found analytically. At the end of Section 6.4, this question is discussed with the paradigm of the generalized Stokes problem where a similar functional arises.

Diffusion problem with mixed Dirichlet–Robin boundary conditions. Diffusion problems are often considered with the Robin boundary condition

$$n \cdot A\nabla u + \kappa(x)u = 0 \quad \text{on } \Gamma_2, \tag{4.2.18}$$

where $\kappa(x) \geq 0$ and $\kappa(x) \not\equiv 0$. For this case, error estimates can be derived from the integral identity

$$\int_\Omega (A\nabla u \cdot \nabla w + \varrho^2 u\, w)\, dx + \int_{\Gamma_2} kuw\, ds = \int_\Omega f w\, dx, \qquad \forall w \in V_0, \tag{4.2.19}$$

by the method discussed in the previous section.

Let $v \in u_0 + V_0$. Then (4.1.6) and (4.2.19) imply the relation

$$\begin{aligned}
&\int_\Omega (A\nabla(u-v) \cdot \nabla w + \varrho^2(u-v)\, w)\, dx + \int_{\Gamma_2} k(u-v)w\, ds \\
&\quad = \int_\Omega (fw - \varrho^2 vw - A\nabla v \cdot \nabla w)\, dx - \int_{\Gamma_2} kv\, ds \\
&\quad = \int_\Omega \mathrm{r}_\Omega(v,y)w\, dx + \int_\Omega (y - A\nabla v) \cdot \nabla w\, dx - \int_{\Gamma_2} (y \cdot n + kv)w\, ds.
\end{aligned} \tag{4.2.20}$$

Set $w = u - v$; by (4.2.20) we have the estimate

$$|[u-v]| \leq \left(|||A\nabla v - y|||_*^2 + \left\| \frac{\mathrm{r}_\Omega(v,y)}{\varrho} \right\|^2 + \left\| \frac{y \cdot n + kv}{\kappa^{1/2}} \right\|_{\Gamma_2} \right)^{1/2}, \tag{4.2.21}$$

where

$$|[w]|^2 := |||\nabla w|||^2 + \|\varrho w\|^2 + \|\kappa^{1/2} w\|_{\Gamma_2}^2.$$

Estimate (4.2.21) is the simplest error majorant for the problem with Robin boundary conditions. In Chapter 7, we deduce it by a different method, using general results of the variational approach.

In view of the estimates

$$\int_\Omega \mathrm{r}_\Omega(v,y)w\, dx \leq \mathrm{C}\|\mathrm{r}_\Omega(v,y)\|\, |||\nabla w|||,$$

$$\int_{\Gamma_2} (y \cdot n + kv)w\, ds \leq c_1^{-1} C_{T\Gamma_2} \|y \cdot n + kv\|\, |||\nabla w|||,$$

we obtain another upper bound:

$$|[u-v]| \leq |||A\nabla v - y|||_* + \mathrm{C}\, \|\mathrm{r}_\Omega(v,y)\| + \frac{C_{T\Gamma_2}}{c_1} \|y \cdot n + kv\|_{\Gamma_2}, \tag{4.2.22}$$

which is not sensitive to small values of ϱ and κ.

By combining the methods used for the derivation of (4.2.21) and (4.2.22), one can deduce a more general estimate (an analog of (4.2.13)) for the reaction-diffusion problem with the Robin boundary condition.

Remark 4.1. For the problem $\Delta u + \lambda u = 0$ in $\Omega \subset \mathbb{R}^2$ with boundary condition $\nabla u \cdot n + ku = 0$ in which k does not depend on x, it is known that among all domains with given area, the circle yields the lowest principle eigenvalue (e.g., see [65]) This fact can be used for getting sharp estimates of the constant in the error majorant if $\Gamma = \Gamma_2$.

A problem generated by the Sturm–Liouville operator. Consider the boundary value problem

$$-(a(x)u')' + b(x)u = f(x) \qquad x \in \Omega := (\xi_1, \xi_2), \tag{4.2.23}$$

$$u(\xi_1) = u_1, \qquad u(\xi_2) = u_2, \quad \xi_2 > \xi_1, \tag{4.2.24}$$

generated by the Sturm–Liouville operator with bounded coefficients a and b. Also, we assume that

$$a(x) \geq a_0 > 0, \qquad b(x) \geq 0, \qquad \text{and } f \in L^2(\Omega).$$

It can be viewed as the 1D form of (4.2.1)–(4.2.3). In this case,

$$|||u - v|||^2 = \int_{\xi_1}^{\xi_2} a(x)(u - v)^2\, dx \geq a_0 \|u - v\|^2,$$

$$|||u - v|||_*^2 = \int_{\xi_1}^{\xi_2} a^{-1}(x)(u - v)^2\, dx.$$

If we set $\mathrm{C} = \frac{\xi_2 - \xi_1}{a_0 \pi}$, $\varrho = \sqrt{b}$, and $\mathrm{r}_\Omega(v, y) = y' - by + f$, then (4.2.14) provides an upper bound of the error.

If the boundary conditions (4.2.24) are represented in a more general form, namely

$$u' + c_i u = u_i \quad \text{at} \quad \xi_i, \qquad i = 1, 2, \tag{4.2.25}$$

then the corresponding estimate is obtained by the same arguments we used for the diffusion problem with Robin type boundary condition. We recommend the reader to derive it as an exercise.

Decomposition of Ω. Assume that $\bar{\Omega} = \bigcup_{i=1}^{N} \bar{\Omega}_i$, where Ω_i are nonintersecting domains with Lipschitz continuous boundaries and $y \in H(\Omega, \operatorname{div})$ is balanced in the subdomains, i.e.,

$$\{\mathrm{r}_\Omega(v, y)\}_{\Omega_i} = 0, \qquad i = 1, 2, \ldots, N, \tag{4.2.26}$$

$$y \cdot n + kv = 0, \qquad \text{on } \Gamma_2. \tag{4.2.27}$$

Then instead of (4.2.22), we can use the estimate (which follows from (4.2.20))

$$|[u-v]| \leq |||A\nabla v - y|||_* + \frac{1}{c_1}\sqrt{\sum_{i=1}^{N} C^2_{P\Omega_i}\,\|\mathrm{r}_\Omega(v,y)\|^2_{\Omega_i}}, \qquad (4.2.28)$$

where $C_{P\Omega_i}$ are constants in the Poincaré inequalities associated with Ω_i.

If (4.2.26) does not hold, then we can apply the same method as in Section 3.5.3. We have

$$\begin{aligned}\int_\Omega \mathrm{r}_\Omega(v,y)w\,dx &= \sum_{i=1}^{N}\int_{\Omega_i} w(\widetilde{\mathrm{r}_\Omega(v,y)})_{\Omega_i}\,dx + \sum_{i=1}^{N}\{\mathrm{r}_\Omega(v,y)\}_{\Omega_i}\int_{\Omega_i} w\,dx\\ &\leq \sum_{i=1}^{N}\int_{\Omega_i} w(\widetilde{\mathrm{r}_\Omega(v,y)})_{\Omega_i}\,dx + \sum_{i=1}^{N}\frac{1}{\varrho_i}\{\mathrm{r}_\Omega(v,y)\}_{\Omega_i}|\Omega_i|^{1/2}\|\varrho w\|_{\Omega_i},\end{aligned}$$

where $\varrho_i = \min_{x\in\Omega_i}\varrho(x)$. Note that

$$\sum_{i=1}^{N}\int_{\Omega_i} w(\widetilde{\mathrm{r}_\Omega(v,y)})_{\Omega_i}\,dx \leq R_1|||\nabla w|||,$$

$$\sum_{i=1}^{N}\frac{1}{\varrho_i}\{\mathrm{r}_\Omega(v,y)\}_{\Omega_i}|\Omega_i|^{1/2}\|\varrho w\|_{\Omega_i} \leq R_2\|\varrho w\|,$$

where

$$R_1 = \frac{1}{c_1}\sqrt{\sum_{i=1}^{N} C^2_{P\Omega_i}\,\|(\widetilde{\mathrm{r}_\Omega(v,y)})_{\Omega_i}\|^2_{\Omega_i}} \quad\text{and}\quad R_2 = \sqrt{\sum_{i=1}^{N}\frac{|\Omega_i|}{\varrho_i^2}\{\mathrm{r}_\Omega(v,y)\}^2_{\Omega_i}}.$$

We set $w = u - v$ and use (4.2.20), which implies the estimate

$$|[u-v]|^2 \leq (|||A\nabla v - y|||_* + R_1)|||\nabla(u-v)||| + R_2\|\varrho(u-v)\|.$$

Hence, we conclude that

$$|[u-v]|^2 \leq (|||A\nabla v - y|||_* + R_1)^2 + R_2^2. \qquad (4.2.29)$$

We outline that this estimate contains only Poincaré constants, so that if Ω can be decomposed into a set of "simple" subdomains (for which $C_{P\Omega_i}$ are known) then the respective upper bound of the error is easily computable.

4.3 Diffusion problems with convective term

In this section, we analyze diffusion problems with convective term. First, we consider the simplest convection-diffusion problem with homogeneous Dirichlet boundary conditions. This problem is used to discuss transparently modifications of the method, which are due to the presence of the convective term. Subsequently, a more general class of problems is analyzed.

4.3.1 The stationary convection-diffusion problem

Consider the simplest model involving the convective term.

$$-\operatorname{div} A\nabla u + \mathrm{a}\cdot\nabla u = f \qquad \text{in } \Omega, \tag{4.3.1}$$

$$u = 0 \qquad \text{on } \Gamma. \tag{4.3.2}$$

Here a is a given vector-valued function satisfying the conditions

$$\mathrm{a} \in L^\infty(\Omega, \mathbb{R}^d), \quad \operatorname{div}\mathrm{a} \in L^\infty(\Omega), \quad \operatorname{div}\mathrm{a} \le 0. \tag{4.3.3}$$

The generalized solution $u \in V_0$ meets the integral identity

$$\int_\Omega (A\nabla u\cdot\nabla w + (\mathrm{a}\cdot\nabla u)w)\,dx = \int_\Omega f w\,dx, \quad \forall w \in V_0. \tag{4.3.4}$$

Again, an upper bound of the error is derived by transformations of the integral identity that defines the solution. As before, we take $v \in V_0$ and insert it into (4.3.4), which yields the relation

$$\int_\Omega (A\nabla(u-v)\cdot\nabla w + (\mathrm{a}\cdot\nabla(u-v))w)\,dx$$
$$= \int_\Omega (f w - A\nabla v\cdot\nabla w - (\mathrm{a}\cdot\nabla v)w)\,dx. \tag{4.3.5}$$

Since (cf. (1.4.10))

$$\operatorname{div}((u-v)\,\mathrm{a}) = (u-v)\operatorname{div}\mathrm{a} + \mathrm{a}\cdot\nabla(u-v) \tag{4.3.6}$$

and w vanishes at the boundary, we have

$$\int_\Omega w\mathrm{a}\cdot\nabla(u-v)\,dx = \int_\Omega \Big(\operatorname{div}((u-v)\,\mathrm{a}) - (\operatorname{div}\mathrm{a})(u-v)\Big)w\,dx$$
$$= -\int_\Omega \Big((u-v)\mathrm{a}\cdot\nabla w + (\operatorname{div}\mathrm{a})(u-v)\,w\Big)\,dx.$$

From here, we obtain

$$\frac{1}{2}\int_\Omega (\operatorname{div}\mathrm{a})(u-v)^2\,dx = -\int_\Omega (u-v)\mathrm{a}\cdot\nabla(u-v)\,dx. \tag{4.3.7}$$

Set $w = u - v$ and rearrange (4.3.5), using (4.3.7). We arrive at the relation

$$|||\nabla(u-v)|||^2 - \frac{1}{2}\int_\Omega (\operatorname{div}\mathrm{a})(u-v)^2\,dx$$
$$= \int_\Omega \Big(\mathrm{r}_\Omega(v,y)\,(u-v) + (y - A\nabla v)\cdot\nabla(u-v)\Big)\,dx, \tag{4.3.8}$$

where $y(x)$ is an arbitrary function in $H(\Omega, \mathrm{div})$ and

$$\mathrm{r}_\Omega(v, y) = f - \mathrm{a} \cdot \nabla v + \operatorname{div} y.$$

Introduce the norm

$$|[u - v]|^2 := |||\nabla(u - v)|||^2 + \|\delta(u - v)\|^2,$$

where

$$\delta^2 = -\tfrac{1}{2}\operatorname{div} \mathrm{a} \geq 0.$$

If $\delta(x) > 0$ for almost all $x \in \Omega$, then, by the estimate

$$\begin{aligned}\left\|\tfrac{1}{\delta}\mathrm{r}_\Omega(v, y)\right\| \|\delta(u - v)\| + |||y - A\nabla v|||_* |||\nabla(u - v)||| \\ \leq \left(|||\tfrac{1}{\delta}\mathrm{r}_\Omega(v, y)|||^2 + |||y - A\nabla v|||_*^2\right)^{1/2} |[u - v]|,\end{aligned}$$

we deduce the first error majorant:

$$|[u - v]|^2 \leq \left\|\tfrac{1}{\delta}\mathrm{r}_\Omega(v, y)\right\|^2 + |||y - A\nabla v|||_*^2 =: \overline{\mathfrak{M}}{}^2_{\mathrm{CD1}}(v, y). \tag{4.3.9}$$

For small values of δ and large values of a this estimate may be coarse.

Another estimate that clearly follows from (4.3.8) is

$$|[u - v]| \leq |||y - A\nabla v|||_* + \mathrm{C}\|\mathrm{r}_\Omega(v, y)\| =: \overline{\mathfrak{M}}_{\mathrm{CD2}}(v, y). \tag{4.3.10}$$

If Ω is decomposed into a collection of subdomains Ω_i and $\{\mathrm{r}_\Omega(v, y)\}_{\Omega_i} = 0$, then (4.3.8) implies the estimate similar to (4.2.26), namely

$$|[u - v]| \leq |||y - A\nabla v|||_* + \sqrt{\sum_i^N \frac{C^2_{P\Omega_i}}{c_1^2} \|\mathrm{r}_\Omega(v, y)\|^2_{\Omega_i}}. \tag{4.3.11}$$

If $\{\mathrm{r}_\Omega(v, y)\}_{\Omega_i} \neq 0$, then repeating the arguments of the previous section we derive an upper bound analogous to (4.2.29).

Consider a special but important case of the problem where the convection term is dominant:

$$-\epsilon\Delta u + \mathrm{a} \cdot \nabla u = f \qquad \text{in } \Omega, \tag{4.3.12}$$

$$u = u_0 \qquad \text{on } \Gamma. \tag{4.3.13}$$

Here ϵ is a small positive number and (for the sake of simplicity) it is assumed that $\Gamma = \Gamma_D$. Also, we assume that δ is a positive constant and $\|\mathrm{a}\|$ is of the order 1. In this case,

$$\mathrm{C} = \frac{C_{F\Omega}}{\sqrt{\epsilon}}, \qquad A = \epsilon I, \qquad A^{-1} = \frac{1}{\epsilon} I,$$

$$|||y|||^2 = \int_\Omega \epsilon |y|^2\, dx = \epsilon \|y\|^2, \qquad |||y|||_*^2 = \int_\Omega \frac{1}{\epsilon}|y|^2\, dx = \frac{1}{\epsilon}\|y\|^2,$$

$$|||y - A\nabla v|||_*^2 = \int_\Omega \left(\epsilon|\nabla v|^2 + \epsilon^{-1}|y|^2 - 2y\cdot\nabla v\right) dx = \left\|\sqrt{\epsilon}\nabla v - \frac{1}{\sqrt{\epsilon}} y\right\|^2.$$

We use the estimate (4.3.9) that does not include negative powers of ϵ and obtain the following upper bound of the error:

$$\begin{aligned} |[u - v]|^2 &:= \epsilon\|\nabla(u - v)\|^2 + \delta^2\|u - v\|^2 \le \overline{\mathfrak{M}}^2_{\mathrm{CD1}}(v, y) \\ &= \left\|\frac{1}{\delta}(f - \mathrm{a}\cdot\nabla v + \operatorname{div} y)\right\|^2 + \left\|\frac{1}{\sqrt{\epsilon}}(y - \epsilon\nabla v)\right\|^2. \end{aligned} \tag{4.3.14}$$

Set $y = \epsilon\nabla u$. Then,

$$f + \operatorname{div} y = f + \epsilon\Delta u = \mathrm{a}\cdot\nabla u,$$

and we find that

$$\mathcal{E}_1^2(v) \le \frac{1}{\delta^2}\|\mathrm{a}\cdot\nabla(u - v)\|^2 + \epsilon\|\nabla(u - v)\|^2, \tag{4.3.15}$$

where

$$\mathcal{E}_1(v) := \inf_{q\in H(\Omega,\mathrm{div})} \overline{\mathfrak{M}}_{\mathrm{CD1}}(v, q).$$

By (4.3.14) and (4.3.15), we conclude that

$$0 \le \mathcal{E}_1^2(v) - |[u - v]|^2 \le \frac{1}{\delta^2}\|\mathrm{a}\cdot\nabla(u - v)\|^2 - \delta^2\|u - v\|^2.$$

Hence,

$$|[u - v]|^2 \le \mathcal{E}_1^2(v) \le |[u - v]|^2 + \frac{\|\mathrm{a}\|^2}{\delta^2}\|\nabla(u - v)\|^2.$$

Since $\|\nabla(u - v)\|^2 \le \epsilon^{-1}|[u - v]|^2$, we deduce the estimate

$$|[u - v]| \le \mathcal{E}_1(v) \le \sqrt{1 + \frac{\|\mathrm{a}\|^2}{\delta^2\epsilon}}\,|[u - v]|. \tag{4.3.16}$$

If $\delta = 0$, then we use (4.3.10), which shows that

$$\mathcal{E}_2(v) := \inf_{q \in H(\Omega,\mathrm{div})} \overline{\mathfrak{M}}_{\mathrm{CD2}}(v, q) \leq \mathrm{C}\|\mathrm{a} \cdot \nabla(u - v)\| + |[u - v]|,$$

where $|[u - v]| = \sqrt{\epsilon}\|\nabla(u - v)\|$. Therefore,

$$0 \leq \mathcal{E}_2(v) - |[u - v]| \leq \mathrm{C}\|\mathrm{a} \cdot \nabla(u - v)\| \leq \mathrm{C}\frac{\|\mathrm{a}\|}{\sqrt{\epsilon}}|[u - v]|,$$

and we find that

$$|[u - v]| \leq \mathcal{E}_2(v) \leq \Big(1 + \frac{\|\mathrm{a}\|\mathrm{C}}{\sqrt{\epsilon}}\Big)|[u - v]|. \tag{4.3.17}$$

It is not surprising that the quality of the upper bounds in (4.3.16) and (4.3.17) is negatively affected by ϵ. Moreover, properties of the exact solution u, as well as the quality of approximations, usually deteriorates if ϵ goes to zero. Such difficulties are predictable and typical for all problems with small parameter.

Assume that $v = u_h$ where u_h is computed on the finite-dimensional space V_h and the error majorant is computed on a finite-dimensional subspace Q_τ that satisfies the condition

$$\exists p_\tau \in Q_\tau, \quad \|p - p_\tau\| + \|\mathrm{div}\,(p - p_\tau)\| \leq \mu_\tau, \tag{4.3.18}$$

where μ_τ is a small positive number. Then, we have

$$\begin{aligned} |[u - u_h]|^2 &\leq \mathcal{E}^2_{1\tau}(u_h) := \inf_{q_h \in Q_\tau} \overline{\mathfrak{M}}^{\,2}_{\mathrm{CD1}}(u_h, q_h) \leq \overline{\mathfrak{M}}^{\,2}_{\mathrm{CD1}}(u_h, p_\tau) \\ &\leq 2\bigg(\Big\|\frac{1}{\sqrt{\delta}}(f - a \cdot \nabla u_h + \mathrm{div} p)\Big\|^2 + \Big\|\frac{1}{\sqrt{\delta}}\mathrm{div}(p - p_\tau)\Big\|^2 \\ &\quad + \Big\|\frac{1}{\sqrt{\epsilon}}(p - \epsilon\nabla u_h)\Big\|^2 + \frac{1}{\epsilon}\|p - p_\tau\|^2\bigg) \\ &\leq 2\Big(1 + \frac{\|\mathrm{a}\|^2}{\delta\epsilon}\Big)\,|[u - u_h]|^2 + 2\mu_\tau^2\left(\frac{1}{\epsilon} + \frac{1}{\delta}\right). \end{aligned} \tag{4.3.19}$$

If the subspaces V_h and Q_τ are constructed such that the p_τ approximates p with the same accuracy as u_h approximates u in the energy norm (i.e., if (4.3.18) holds with μ_τ, which is of the same order as $|[u - u_h]|$), then we conclude that the computable quantity $\mathcal{E}_{1\tau}(u_h)$ is equivalent to the error and has the same rate with respect to ϵ as the quantity $\mathcal{E}_1(u_h)$ in (4.3.16).

A lower bound of the approximation error in terms of a different quantity can be obtained by the method we used for the derivation of (4.1.13). Assume that $\mathrm{div}\,\mathrm{a} = 0$.

Then,

$$\sup_{w\in V_0}\left\{\int_\Omega (\epsilon\nabla(u-v)\cdot\nabla w + \mathrm{a}\cdot\nabla(u-v)w)\,dx - \frac{\epsilon}{2}\|\nabla w\|^2\right\}$$
$$= \sup_{w\in V_0}\left\{\int_\Omega (\epsilon\nabla(u-v) - (u-v)\mathrm{a})\cdot\nabla w\,dx - \frac{\epsilon}{2}\|\nabla w\|^2\right\}$$
$$\leq \sup_{\tau\in L^2}\int_\Omega (\epsilon\nabla(u-v) - (u-v)\mathrm{a})\cdot\tau - \frac{\epsilon}{2}|\tau|^2)\,dx$$
$$= \frac{1}{2\epsilon}\|\epsilon\nabla(u-v) - (u-v)\mathrm{a}\|^2$$
$$= \frac{\epsilon}{2}\|\nabla(u-v)\|^2 + \frac{1}{2\epsilon}\|\mathrm{a}(u-v)\|^2 - \int_\Omega \nabla(u-v)\cdot\mathrm{a}(u-v)\,dx.$$

The last integral on the right-hand side is equal to zero. Since

$$\int_\Omega (\epsilon\nabla u\cdot\nabla w + w\mathrm{a}\cdot\nabla u)\,dx = \int_\Omega f w\,dx,$$

we conclude that

$$\frac{\epsilon}{2}\|\nabla(u-v)\|^2 + \frac{1}{2\epsilon}\|\mathrm{a}(u-v)\|^2$$
$$\geq \sup_{w\in V_0}\left\{-\frac{\epsilon}{2}\|\nabla w\|^2 - \int_\Omega (\epsilon\nabla v\cdot\nabla w + w\mathrm{a}\cdot\nabla v - fw)\,dx\right\}$$
$$= \sup_{w\in V_0} \underline{\mathfrak{M}}^2_{\mathrm{CD}}(v,w). \tag{4.3.20}$$

Note that

$$\sup_{w\in V_0} \underline{\mathfrak{M}}^2_{\mathrm{CD}}(v,w) \geq \underline{\mathfrak{M}}^2_{\mathrm{CD}}(v,u-v) = \frac{\epsilon}{2}\|\nabla(u-v)\|^2,$$

so that for small ϵ the quality of the lower bound deteriorates no faster than $\|\mathrm{a}\|\epsilon^{-1}$.

4.3.2 The reaction-convection-diffusion problem

Now, we consider the most general statement of a linear elliptic problem related to diffusion type models. It is the *reaction-convection-diffusion* problem

$$-\mathrm{div}\,A\nabla u + \mathrm{a}\cdot\nabla u + \varrho^2 u = f \quad \text{in } \Omega, \tag{4.3.21}$$
$$u = u_0 \quad \text{on } \Gamma_1, \tag{4.3.22}$$
$$A\nabla u\cdot n = F \quad \text{on } \Gamma_2, \tag{4.3.23}$$

where the data satisfy the same assumptions as in the problem (4.1.1)–(4.1.3) and

$$\varrho \in L^\infty(\Omega), \qquad \varrho \le \varrho_\oplus, \tag{4.3.24}$$

$$-\frac{1}{2}\operatorname{div} \mathrm{a} + \varrho^2 := \delta^2 \ge \delta_0^2. \tag{4.3.25}$$

One more assumption is that the function $\kappa(x) := \frac{1}{2}(\mathrm{a} \cdot n)(x)$ is defined at almost all points of Γ and that the inflow part of the boundary is a subset of Γ_1, i.e.,

$$\Gamma^- := \{x \in \Gamma \mid \kappa(x) < 0\} \subset \Gamma_1. \tag{4.3.26}$$

Now

$$V_0 := \{w \in H^1(\Omega) \mid w = 0 \text{ on } \Gamma_1\}$$

and u is defined as a function in $u_0 + V_0$ that meets the integral identity

$$\int_\Omega (A\nabla u \cdot \nabla w + (a \cdot \nabla u)w + \varrho^2 w)\, dx$$
$$= \int_\Omega f w\, dx + \int_{\Gamma_2} F w\, ds, \qquad \forall w \in V_0(\Omega). \tag{4.3.27}$$

The existence of a generalized solution u follows from the known results in the theory of partial differential equations (e.g., see [151, 217]).

The assertion below suggests a general form of a computable upper bound of the error measured in terms of a natural energy type norm.

Theorem 4.2. *Let the above made assumptions on the problem data be fulfilled. Then, for any $\alpha, \beta \in L^\infty_{[0,1]}$, $v \in u_0 + V_0$, and $y \in H_{\Gamma_2}(\Omega, \operatorname{div})$ the following estimate holds:*

$$|[u - v]| \le \overline{\mathfrak{M}}_{\mathrm{RCD}}(\alpha, \beta, v, y) :=$$
$$\le \Big(\big\| \tfrac{\alpha}{\delta} \mathrm{r}_\Omega(v, y) \big\|^2_{\Omega_+} + |||y - A\nabla v|||_*^2 + \Big\| \tfrac{\beta}{\sqrt{\kappa}}(F - y \cdot n) \Big\|_{\Gamma_2^+} \Big)^{1/2}$$
$$+ \frac{1}{c_1}\Big(C_{F\Omega} \|(1-\alpha)\mathrm{r}_\Omega(v, y)\| + C_{T\Gamma_2}\|(1-\beta)(F - y \cdot n)\| \Big), \tag{4.3.28}$$

where

$$\mathrm{r}_\Omega(v, y) := f - \mathrm{a} \cdot \nabla v - \varrho^2 v + \operatorname{div} y,$$
$$|[u - v]|^2 := |||\nabla(u - v)|||^2 + \int_\Omega \delta^2 (u - v)^2\, dx + \int_{\Gamma_2} \kappa (u - v)^2\, ds,$$
$$\Omega^+ := \{x \in \Omega \mid \delta(x) \ne 0\}, \qquad \Gamma_2^+ := \{x \in \Gamma_2 \mid \kappa(x) \ne 0\},$$

$\alpha = 0$ on $\Omega \setminus \Omega^+$, $\beta = 0$ on $\Gamma_2 \setminus \Gamma_2^+$, and $C_{F\Gamma_1}$ and $C_{T\Gamma_2}$ are constants in (4.1.8) and (4.1.9), respectively.

Proof. Rewrite (4.3.27) in the form

$$\int_\Omega \left(A\nabla(u-v)\cdot\nabla(u-v) + (\mathrm{a}\cdot\nabla(u-v))(u-v) + \varrho^2(u-v)^2\right) dx$$
$$= \int_\Omega (f - (\mathrm{a}\cdot\nabla v) - \varrho^2 v)(u-v)\,dx - \int_\Omega A\nabla v\cdot\nabla(u-v)\,dx$$
$$+ \int_{\Gamma_2} F(u-v)\,ds. \qquad (4.3.29)$$

Note that

$$\frac{1}{2}\int_\Omega (\operatorname{div}\mathrm{a})(u-v)^2\,dx = -\frac{1}{2}\int_\Omega \mathrm{a}\cdot\nabla(u-v)^2\,dx + \frac{1}{2}\int_{\Gamma_2}\mathrm{a}\cdot n(u-v)^2\,ds$$
$$= -\int_\Omega (u-v)\mathrm{a}\cdot\nabla(u-v)\,dx + \int_{\Gamma_2}\kappa(u-v)^2\,ds. \quad (4.3.30)$$

Thus, the left-hand side of (4.3.29) is converted into the norm $|[u-v]|^2$, which is a natural measure of the difference between u and v.

Now, we have

$$|[u-v]|^2 = \Re(v,y) := \int_\Omega \mathrm{r}_\Omega(v,y)(u-v)\,dx + \int_{\Gamma_2}(F - y\cdot n)(u-v)\,ds$$
$$+ \int_\Omega (y - A\nabla v)\cdot\nabla(u-v)\,dx, \qquad (4.3.31)$$

where $y(x)$ is an arbitrary function in $H_{\Gamma_2}(\Omega,\operatorname{div})$.

Set $\alpha(x) = 0$ on $\Omega\setminus\Omega^+$ and represent the first term on the right-hand side of (4.3.31) in the form

$$\int_\Omega \mathrm{r}_\Omega(v,y)(u-v)\,dx = \int_\Omega \alpha\mathrm{r}_\Omega(v,y)(u-v)\,dx + \int_\Omega (1-\alpha)\mathrm{r}_\Omega(v,y)(u-v)\,dx$$
$$\leq \left\|\frac{\alpha}{\delta}\mathrm{r}_\Omega(v,y)\right\|_{\Omega+}\|\delta(u-v)\|_{\Omega+}$$
$$+ C_{F\Omega}c_1^{-1}\|(1-\alpha)\mathrm{r}_\Omega(v,y)\|\,|||\nabla(u-v)|||.$$

Similarly,

$$\left|\int_{\Gamma_2}(F - y\cdot n)(u-v)\,ds\right| \leq \|\frac{\beta}{\sqrt{\kappa}}(F - y\cdot n)\|_{\Gamma_2^+}\|\sqrt{\kappa}(u-v)\|_{\Gamma_2^+}$$
$$+ C_{T\Gamma_2}c_1^{-1}\|(1-\beta)(F - y\cdot n)\|_{\Gamma_2}|||\nabla(u-v)|||,$$

where $\beta \in L^{\infty}_{[0,1]}$ and $\beta(x) = 0$ on $\Gamma_2 \setminus \Gamma_2^+$. From the above relations, we obtain

$$\begin{aligned}
\Re(v,y) &\le \left\| \frac{\alpha}{\delta} \mathrm{r}_\Omega(v,y) \right\|_{\Omega+} \|\delta(u-v)\| + |||y - A\nabla v|||_* \, \|\nabla(u-v)\| \\
&\quad + \left\| \frac{\beta}{\sqrt{\kappa}}(F - y\cdot n) \right\|_{\Gamma_2^+} \|\sqrt{\kappa}(u-v)\|_{\Gamma_2} + \frac{1}{c_1}\Big(C_{F\Omega} \|(1-\alpha)\mathrm{r}_\Omega(v,y)\| \\
&\quad + C_{T\Gamma_2}\|(1-\beta)(F - y\cdot n)\|_{\Gamma_2}\Big) |||\nabla(u-v)||| \\
&\le \left(\left\| \frac{\alpha}{\delta} \mathrm{r}_\Omega(v,y) \right\|^2_{\Omega+} + |||y - A\nabla v|||^2_* \right. \\
&\quad \left. + \left\| \frac{\beta}{\sqrt{\kappa}}(F - y\cdot n) \right\|^2_{\Gamma_2^!} \right)^{1/2} |[u-v]| + \frac{1}{c_1}\Big(\mathrm{C} \|(1-\alpha)\mathrm{r}_\Omega(v,y)\| \\
&\quad + C_{T\Gamma_2}\|(1-\beta)(F - y\cdot n)\|_{\Gamma_2}\Big) |||\nabla(u-v)|||. \qquad (4.3.32)
\end{aligned}$$

Estimate (4.3.28) follows from (4.3.31) and (4.3.32). □

Remark 4.3. Estimate (4.3.28) shows that the upper bound is given by the norms of residuals of three relations

$$y = A\nabla v \qquad \text{in } \Omega, \qquad (4.3.33)$$

$$\operatorname{div} y - \mathrm{a}\cdot\nabla v - \varrho^2 v + f = 0 \qquad \text{in } \Omega, \qquad (4.3.34)$$

$$y \cdot n = F \qquad \text{on } \Gamma_2, \qquad (4.3.35)$$

which jointly define the reaction-convection-diffusion problem. Therefore, the majorant $\overline{\mathfrak{M}}_{\mathrm{RCD}}(\alpha,\beta,v,y)$ vanishes if and only if $u = v$ and $y = p$.

4.3.3 Special cases and modifications

If $\alpha = \beta = 0$, then we arrive at the estimate

$$|[u-v]| \le |||y - A\nabla v|||_* + \frac{C_{F\Omega}}{c_1}\|\mathrm{r}_\Omega(v,y)\| + \frac{C_{T\Gamma_2}}{c_1}\|F - y\cdot n\|_{\Gamma_2}. \qquad (4.3.36)$$

It includes two global constants $C_{F\Omega}$ and $C_{T\Gamma_2}$ and does not contain δ and κ in negative powers. Therefore, it is stable with respect to small values of parameters.

For $\alpha = 1$ and $\beta = 0$ we obtain another estimate

$$|[u-v]| \le \frac{C_{T\Gamma_2}}{c_1}\|F - y\cdot n\|_{\Gamma_2} + \left(\left\| \frac{1}{\delta}\mathrm{r}_\Omega(v,y) \right\|^2 + |||y - A\nabla v|||^2_* \right)^{1/2}, \qquad (4.3.37)$$

which involves only the trace constant.

Set $\alpha = \beta = 1$. Then we obtain such a form of the upper bound that contains neither $C_{F\Omega}$ nor $C_{T\Gamma_2}$:

$$|[u-v]|^2 \le \left\| \frac{\mathrm{r}_\Omega(v,y)}{\delta} \right\|^2 + \left\| \frac{F - y\cdot n}{\sqrt{\kappa}} \right\|^2_{\Gamma_2} + |||y - A\nabla v|||^2_*. \qquad (4.3.38)$$

Remark 4.4. If F is a simple (e.g., affine) function, then it is not difficult to satisfy the condition Γ_2 exactly. For

$$y \in H^0_{\Gamma_2} := \{y \in H_{\Gamma_2}(\Omega, \mathrm{div}) \mid y \cdot n - F = 0 \text{ on } \Gamma_2\},$$

the estimate (4.3.28) has a simplified form

$$|[u - v]| \le \left(\left\|\tfrac{\alpha}{\delta}\mathrm{r}_\Omega(v, y)\right\|^2_{\Omega+} + |||y - A\nabla v|||^2_*\right)^{1/2} + \frac{C_{F\Omega}}{c_1}\|(1-\alpha)\mathrm{r}_\Omega(v, y)\|. \tag{4.3.39}$$

Remark 4.5. In (4.3.28), we can set

$$\alpha = \min\{\delta, 1\}, \qquad \beta = \min\{\kappa, 1\}. \tag{4.3.40}$$

This choice may be efficient if δ and κ are small in some parts of Ω and Γ_2, respectively. However, the best upper bound is attained if (4.3.28) is minimized with respect to α, β, and y.

Set in (4.3.28) $y = \mathrm{a}v + \tau$, where $\tau \in H(\Omega, \mathrm{div})$. Then, it has the form

$$\begin{aligned}|[u - v]| &\le \overline{\mathfrak{M}}_{\mathrm{RCD}}(\alpha, \beta, v, \tau) \\ &:= \Big(\left\|\tfrac{\alpha}{\delta}\widehat{\mathrm{r}_\Omega}(v, \tau)\right\|^2_{\Omega+} + |||\tau + \mathrm{a}v - A\nabla v|||^2_* \\ &\quad + \left\|\tfrac{\beta}{\sqrt{\kappa}}(F - \tau\cdot n - (\mathrm{a}\cdot n)v)\right\|_{\Gamma_2^+}\Big)^{1/2} + \frac{1}{c_1}\Big(C_{F\Omega}\|(1-\alpha)\widehat{\mathrm{r}_\Omega}(v, \tau)\| \\ &\quad + C_{T\Gamma_2}\|(1-\beta)(F - \tau\cdot n - (\mathrm{a}\cdot n)v)\|\Big),\end{aligned} \tag{4.3.41}$$

where $\widehat{\mathrm{r}_\Omega}(v, \tau) := f - \hat{\varrho}^2 v + \mathrm{div}\,\tau$, $\hat{\varrho}^2 = \varrho^2 - \mathrm{div}\,\mathrm{a}$. This majorant is related to the representation

$$\hat{p} = A\nabla u - \mathrm{a}u \qquad \text{in } \Omega, \tag{4.3.42}$$

$$\mathrm{div}\,\hat{p} - \hat{\varrho}^2 v + f = 0 \qquad \text{in } \Omega, \tag{4.3.43}$$

$$\hat{p}\cdot n = F - (\mathrm{a}\cdot n)\,u \qquad \text{on } \Gamma_2. \tag{4.3.44}$$

In particular, for $\alpha = \beta = 0$ we obtain

$$|[u - v]| \le |||\mathrm{a}v + \tau - A\nabla v|||_* + C_{F\Omega}c_1^{-1}\|f - \hat{\varrho}^2 v + \mathrm{div}\,\tau\| + C_{T\Gamma_2}c_1^{-1}\|F - (\tau + \mathrm{a}v)\cdot n\|_{\Gamma_2}. \tag{4.3.45}$$

It is easy to see that the majorants defined by (4.3.41) and (4.3.45) possess the same properties as (4.3.28) and (4.3.36); namely, they vanish at the exact solution and tend to zero if v and y tend to u and p (or $\hat{p}$).

We can obtain a better estimate, using the method based on introducing a function $\vartheta \in V_0$ (cf. (4.1.21)–(4.1.23)) such that $A\nabla\vartheta \in H_{\Gamma_2}(\Omega, \mathrm{div})$. Since $u - v$ vanishes on Γ_1, we note that

$$\begin{aligned}
&\int_\Omega \mathrm{div}\left(A\nabla\vartheta + \vartheta \mathrm{a}\right)(u-v)\,dx \\
&\quad = -\int_\Omega \left(A\nabla\vartheta + \vartheta \mathrm{a}\right)\cdot\nabla(u-v)\,dx + \int_{\Gamma_2}\left(A\nabla\vartheta + \vartheta \mathrm{a}\right)\cdot n(u-v)\,ds \\
&\quad = \int_\Omega \left((A\nabla\vartheta + \vartheta \mathrm{a})\cdot\nabla v - \varrho^2\vartheta - f\vartheta\right)dx - \int_{\Gamma_2} F\vartheta\,ds \\
&\qquad + \int_{\Gamma_2}\left(A\nabla\vartheta + \vartheta \mathrm{a}\right)\cdot n(u-v)\,ds
\end{aligned} \tag{4.3.46}$$

where we have used the relation

$$\int_\Omega (A\nabla u\cdot\nabla\vartheta + (a\cdot\nabla u)\vartheta + \varrho^2\vartheta)\,dx = \int_\Omega f\vartheta\,dx + \int_{\Gamma_2} F\vartheta\,ds.$$

By (4.3.46) we rearrange (4.3.31) as follows:

$$\begin{aligned}
\|[u-v]\|^2 = {} & \int_\Omega \mathrm{r}_\Omega(v,y,\vartheta)(u-v)\,dx + \int_{\Gamma_2}\mathrm{r}_{\Gamma_2}(y,\vartheta)(u-v)\,ds \\
& + \int_\Omega (y - A\nabla v)\cdot\nabla(u-v)\,dx + \mathcal{F}_v(\vartheta),
\end{aligned} \tag{4.3.47}$$

where

$$\begin{aligned}
\mathrm{r}_\Omega(v,y,\vartheta) &:= \mathrm{r}_\Omega(v,y) - \mathrm{div}\,(A\nabla\vartheta + \vartheta \mathrm{a}), \\
\mathrm{r}_{\Gamma_2}(y,\vartheta) &:= F - (y - A\nabla\vartheta - \vartheta \mathrm{a})\cdot n,
\end{aligned}$$

and $\mathcal{F}_v(\vartheta) = \int_\Omega \left(f\vartheta - (A\nabla\vartheta + \vartheta \mathrm{a})\cdot\nabla v - \varrho^2\vartheta\right)dx + \int_{\Gamma_2} F\vartheta\,ds$.

As before, let α and β be functions with values in $[0,1]$ such that $\alpha(x) = 0$ on $\Omega\setminus\Omega^+$ and $\beta(x) = 0$ on $\Gamma_2\setminus\Gamma_2^+$. Then,

$$\begin{aligned}
\int_\Omega \mathrm{r}_\Omega(v,y,\vartheta)(u-v)\,dx \le {} & \left\|\frac{\alpha}{\delta}\mathrm{r}_\Omega(v,y,\vartheta)\right\|_{\Omega+}\|\delta(u-v)\|_{\Omega+} \\
& + C_{F\Omega}c_1^{-1}\|(1-\alpha)\mathrm{r}_\Omega(v,y,\vartheta)\|\,|\!|\!|\nabla(u-v)|\!|\!|
\end{aligned}$$

and

$$\begin{aligned}
\left|\int_{\Gamma_2}\mathrm{r}_{\Gamma_2}(y,\vartheta)(u-v)\,ds\right| \le {} & \left\|\frac{\beta}{\sqrt{\kappa}}\mathrm{r}_{\Gamma_2}(y,\vartheta)\right\|_{\Gamma_2^+}\|\sqrt{\kappa}(u-v)\|_{\Gamma_2^+} \\
& + C_{T\Gamma_2}c_1^{-1}\|(1-\beta)\mathrm{r}_{\Gamma_2}(y,\vartheta)\|_{\Gamma_2}|\!|\!|\nabla(u-v)|\!|\!|.
\end{aligned}$$

From the above relations, we obtain

$$
\begin{aligned}
|[u-v]|^2 \leq\ & \frac{1}{4\mu_1}\left\|\frac{\alpha}{\delta}\mathrm{r}_\Omega(v,y,\vartheta)\right\|^2_{\Omega+} + \mu_1\|\delta(u-v)\|^2_{\Omega+} \\
& + \frac{1}{4\mu_2}\frac{C^2_{F\Omega}}{c_1^2}\|(1-\alpha)\mathrm{r}_\Omega(v,y,\vartheta)\|^2 + \mu_2|||\nabla(u-v)|||^2 \\
& + \frac{1}{4\mu_3}\left\|\frac{\beta}{\sqrt{\kappa}}\mathrm{r}_{\Gamma_2}(y,\vartheta)\right\|^2_{\Gamma_2^+} + \mu_3\|\sqrt{\kappa}(u-v)\|^2_{\Gamma_2^+} \\
& + \frac{1}{4\mu_4}\frac{C^2_{T\Gamma_2}}{c_1^2}\|(1-\beta)\mathrm{r}_{\Gamma_2}(y,\vartheta)\|^2_{\Gamma_2} + \mu_4|||\nabla(u-v)|||^2 \\
& + \frac{1}{4\mu_5}|||y-A\nabla v|||^2_* + \mu_5|||\nabla(u-v)|||^2 + \mathcal{F}_v(\vartheta).
\end{aligned}
\tag{4.3.48}
$$

Here, $\mu_1, \mu_3 \in [0,1)$ and μ_2, μ_4, and μ_5 are positive numbers such that

$$\mu_2 + \mu_4 + \mu_5 \leq 1.$$

Now, we obtain an upper bound of the error evaluated in terms of a weighted norm:

$$
\begin{aligned}
(1-\mu_2-\mu_4-\mu_5)|||\nabla(u-v)|||^2 + (1-\mu_1)\|\delta(u-v)\|^2_{\Omega} + (1-\mu_3)\|\sqrt{\kappa}(u-v)\|^2_{\Gamma_2} \\
\leq \tfrac{1}{4\mu_1}\left\|\frac{\alpha}{\delta}\mathrm{r}_\Omega(v,y,\vartheta)\right\|^2_{\Omega+} + \tfrac{C^2_{F\Omega}}{4\mu_2 c_1^2}\|(1-\alpha)\mathrm{r}_\Omega(v,y,\vartheta)\|^2 \\
+ \tfrac{1}{4\mu_3}\|\frac{\beta}{\sqrt{\kappa}}\mathrm{r}_{\Gamma_2}(y,\vartheta)\|^2_{\Gamma_2^+} + \tfrac{C^2_{T\Gamma_2}}{4\mu_4 c_1^2}\|(1-\beta)\mathrm{r}_{\Gamma_2}(y,\vartheta)\|^2_{\Gamma_2} \\
+ \tfrac{1}{4\mu_5}|||y-A\nabla v|||^2_* + \mathcal{F}_v(\vartheta).
\end{aligned}
\tag{4.3.49}
$$

If δ and κ are positive, then we can set $\alpha = \beta = 1$, $\mu_2 = \mu_4 = 0$, and $\mu_1 = \mu_3 = \frac{1}{2}$. In this case, the estimate (4.3.49) has a simplified form:

$$
\begin{aligned}
|[u-v]|^2 \leq \overline{\mathfrak{M}}^2_{\mathrm{RCD}}(v,y,\vartheta) := \left\|\frac{1}{\delta}\mathrm{r}_\Omega(v,y,\vartheta)\right\|^2_{\Omega+} + \left\|\frac{1}{\sqrt{\kappa}}\mathrm{r}_{\Gamma_2}(y,\vartheta)\right\|^2_{\Gamma_2^+} \\
+ |||y-A\nabla v|||^2_* + \mathcal{F}_v(\vartheta).
\end{aligned}
\tag{4.3.50}
$$

Set $y = A\nabla u$ and define ϑ as a solution of the problem

$$\operatorname{div}\big(A\nabla\vartheta + \vartheta \mathrm{a}\big) = \mathrm{a}\cdot\nabla(u-v) + \varrho^2(u-v), \tag{4.3.51}$$

$$\vartheta = 0 \qquad \text{on } \Gamma_1, \tag{4.3.52}$$

$$(A\nabla\vartheta + \vartheta a)\cdot n = 0 \qquad \text{on } \Gamma_2. \tag{4.3.53}$$

Then,

$$\begin{aligned}
\mathrm{r}_\Omega(v,y) - \operatorname{div}\big(A\nabla\vartheta - \vartheta \mathrm{a}\big) &= 0,\\
|||y - A\nabla v|||_*^2 &= |||\nabla(u-v)|||^2,\\
\mathrm{r}_{\Gamma_2}(y,\vartheta) &= F - (A\nabla u - A\nabla\vartheta - \vartheta\mathrm{a})\cdot n = 0.
\end{aligned}$$

Since

$$\int_\Omega \big(A\nabla\vartheta + \vartheta\mathrm{a}\big)\cdot\nabla(u-v)\,dx = \int_\Omega \big(\mathrm{a}\cdot\nabla(u-v)(u-v) + \varrho^2(u-v)^2\big)\,dx,$$

we have

$$\begin{aligned}
\mathcal{F}_v(\vartheta) &= \int_\Omega (A\nabla\vartheta + \vartheta\mathrm{a})\cdot\nabla(u-v)\,dx\\
&= \int_\Omega (u-v)\mathrm{a}\cdot\nabla(u-v)\,dx + \int_\Omega \varrho^2(u-v)^2\,dx\\
&= \int_\Omega \delta^2(u-v)^2\,dx + \int_{\Gamma_2} \kappa(u-v)^2\,ds.
\end{aligned}$$

Hence, the right-hand side of (4.3.50) coincides with the left-hand side.

Remark 4.6. Other modifications of the estimates (which are similar to those considered in Sections 3.5 and 4.2) can be also derived using the same arguments. In particular, if Ω is decomposed into Ω_i, then the estimates analogous to (4.2.28) and (4.2.29) take place under the same assumptions on $\mathrm{r}_\Omega(v,y)$. We leave this task for the reader as a good exercise.

Remark 4.7. In Section 8.5 we discuss another (rather general) way to obtain two-sided estimates that have no gap between upper (lower) bounds and error measures. For reaction-convection-diffusion problems it yields other modifications of error estimates.

4.3.4 Estimates for fluxes

Theorem 4.2 allows us to derive computable bounds of errors measured in terms of the dual variable (flux). Let $y \in H_{\Gamma_2}(\Omega)$ be an approximation of the exact flux p. For any $v \in u_0 + V_0$, we have

$$\begin{aligned}
|||p - y|||_* &\le |||A\nabla(u-v)|||_* + |||A\nabla v - y|||_* \\
&= |||\nabla(u-v)||| + |||A\nabla v - y|||_*\\
&\le \overline{\mathfrak{M}}_{\mathrm{RCD}}(\alpha,\beta,v,y) + |||A\nabla v - y|||_*.
\end{aligned} \tag{4.3.54}$$

By (4.3.28) and (4.3.54), we obtain an upper bound for a combined primal-dual norm

$$|[u-v]| + |||p-y|||_* \le 2\overline{\mathfrak{M}}_{\mathrm{RCD}}(\alpha,\beta,v,y) + |||A\nabla v - y|||_*. \tag{4.3.55}$$

We can use (4.3.50) instead of (4.3.28). Then, in (4.3.55) $\overline{\mathfrak{M}}_{\mathrm{RCD}}(\alpha,\beta,v,y)$ must be replaced by $\overline{\mathfrak{M}}_{\mathrm{RCD}}(v,y,\vartheta)$.

The norm

$$|[y]|_* := |||y|||_* + \|\mathrm{div}\, y\| + \|y\cdot n\|_{\Gamma_2}$$

is another quantity, which is natural to use for fluxes. For this norm, we have

$$\begin{aligned} |[p-y]|_* &\le |||A\nabla(u-v)|||_* + |||A\nabla v - y|||_* + \|(p-y)\cdot n\|_{\Gamma_2} + \|\mathrm{div}\,(p-y)\| \\ &= |||\nabla(u-v)||| + |||A\nabla v - y|||_* + \|y\cdot n - F\|_{\Gamma_2} \\ &\quad + \|\mathrm{div}\,(p-y)\|. \end{aligned} \tag{4.3.56}$$

By (4.3.28) and (4.3.56), we find that

$$\begin{aligned} |[p-y]|_* \le \overline{\mathfrak{M}}_{\mathrm{RCD}}(\alpha,\beta,v,y) + |||A\nabla v - y|||_* + \|y\cdot n - F\|_{\Gamma_2} \\ + \|\mathrm{div}\, y - \mathrm{a}\cdot\nabla u - \varrho^2 u + f\|. \end{aligned}$$

Assume that $\delta_0 > 0$. Then,

$$\begin{aligned} &\|\mathrm{div}\, y - \mathrm{a}\cdot\nabla u - \varrho^2 u + f\| \\ &\quad\le \|\mathrm{r}_\Omega(v,y)\| + \|\mathrm{a}\|\,\|\nabla(u-v)\| + \left\|\frac{\varrho^2}{\delta}\right\|\|\delta(u-v)\| \\ &\quad\le \|\mathrm{r}_\Omega(v,y)\| + \sqrt{\left(\frac{\|\mathrm{a}\|^2}{c_1^2} + \left\|\frac{\varrho^2}{\delta}\right\|\right)\left(|||\nabla(u-v)|||^2 + \|\delta(u-v)\|^2\right)} \end{aligned}$$

and we obtain

$$\begin{aligned} |[p-y]|_* \le \overline{\mathfrak{M}}_{\mathrm{RCD}}(\alpha,\beta,v,y) + |||A\nabla v - y|||_* + \|\mathrm{r}_\Omega(v,y)\| \\ + \|y\cdot n - F\|_{\Gamma_2} + \sqrt{\frac{\|\mathrm{a}\|^2}{c_1^2} + \left\|\frac{\varrho^2}{\delta}\right\|}\,|[u-v]|, \end{aligned}$$

which yields the estimate

$$\begin{aligned} |[p-y]|_* \le \zeta\overline{\mathfrak{M}}_{\mathrm{RCD}}(\alpha,\beta,v,y) + \|\mathrm{r}_\Omega(v,y)\| + |||A\nabla v - y|||_* \\ + \|y\cdot n - F\|_{\Gamma_2}, \end{aligned} \tag{4.3.57}$$

where $\zeta = 1 + \sqrt{\frac{\|\mathrm{a}\|^2}{c_1^2} + \left\|\frac{\varrho^2}{\delta}\right\|}$.

If $y \in H^0_{\Gamma_2}(\Omega)$, then we have a simpler estimate

$$|[p-y]|_* \leq \zeta \overline{\mathfrak{M}}_{\mathrm{RCD}}(\alpha,\beta,v,y) + \|\mathrm{r}_\Omega(v,y)\| + |\!|\!| A\nabla v - y |\!|\!|_*. \tag{4.3.58}$$

Hence, for the pair $(v,y) \in (u_0+V_0) \times H^0_{\Gamma_2}$ an upper bound in the full primal-dual norm has the form

$$\begin{aligned}|[u-v]| + |[p-y]|_* \leq (1+\zeta)\overline{\mathfrak{M}}_{\mathrm{RCD}}(\alpha,\beta,v,y) + \|\mathrm{r}_\Omega(v,y)\| \\ + |\!|\!| A\nabla v - y |\!|\!|_*.\end{aligned} \tag{4.3.59}$$

In particular, for $\alpha = \beta = 0$ we have

$$\begin{aligned}&|[u-v]| + |[p-y]|_* \\ &\quad \leq \left(1 + (1+\zeta)\frac{C_{F\Omega}}{c_1}\right)\|\mathrm{r}_\Omega(v,y)\| + (2+\zeta)|\!|\!| A\nabla v - y |\!|\!|_*.\end{aligned} \tag{4.3.60}$$

Equivalence to the error. By Theorem 4.2, we observe that the quantity

$$\mathcal{E}(v) := \inf_{y \in H_{\Gamma_2}(\Omega)} \overline{\mathfrak{M}}_{\mathrm{RCD}}(\alpha,\beta,v,y)$$

yields an upper bound of the error. The proposition below shows that $\mu\mathcal{E}(v)$ also furnishes a lower bound, where μ does not depend on v and is explicitly estimated by the problem data.

Proposition 4.8. *For any $\alpha,\beta \in L^\infty_{[0,1]}$ and any $v \in u_0 + V_0$,*

$$\mathcal{E}(v) \leq c_\oplus(\gamma)|[u-v]|, \tag{4.3.61}$$

where γ is an arbitrary positive number, and $c_\oplus$ is defined by (4.3.62).

Proof. We have

$$\inf_{y \in H_{\Gamma_2}(\Omega)} \overline{\mathfrak{M}}_{\mathrm{RCD}}(\alpha,\beta,v,y) \leq \overline{\mathfrak{M}}_{\mathrm{RCD}}(\alpha,\beta,v,p).$$

Note that

$$\begin{aligned}\|\mathrm{r}_\Omega(\alpha,\beta,v,p)\| &= \|f - \mathrm{a}\cdot\nabla v - \varrho^2 v + \operatorname{div} p\| \\ &\leq \|\mathrm{a}\cdot\nabla(u-v)\| + \|\varrho^2(u-v)\| \\ &\leq \sqrt{2\frac{\|a\|^2}{c_1^2}|\!|\!|\nabla(u-v)|\!|\!|^2 + 2\frac{\varrho^2{}_\oplus}{\delta_0^2}\int_\Omega \delta^2(u-v)^2\,dx}.\end{aligned}$$

Since $|||p - A\nabla v|||_* = |||\nabla(u-v)|||$ and $p \cdot n = F$ we find that

$$\begin{aligned}
&\frac{1}{(1+\gamma)}\overline{\mathfrak{M}}^2_{\mathrm{RCD}}(\alpha,\beta,v,p)\\
&\quad\le \left(\frac{1}{\delta_0^2}+\frac{C^2_{F\Omega}}{\gamma c_1^2}\right)\|\mathrm{r}_\Omega(v,y)\|^2 + |||\nabla(u-v)|||^2\\
&\quad\le 2\left(\frac{1}{\delta_0^2}+\frac{C^2_{F\Omega}}{\gamma c_1^2}\right)\left(\frac{\|a\|^2}{c_1^2}|||\nabla(u-v)|||^2 + \frac{\varrho^2{}_\oplus}{\delta_0^2}\int_\Omega \delta^2(u-v)^2\,dx\right)\\
&\qquad + |||\nabla(u-v)|||^2\\
&\quad= \left(1+\left(\frac{1}{\delta_0^2}+\frac{C^2_{F\Omega}}{\gamma c_1^2}\right)\frac{2\|a\|^2}{c_1^2}\right)|||\nabla(u-v)|||^2\\
&\qquad + \left(\frac{1}{\delta_0^2}+\frac{C^2_{F\Omega}}{\gamma c_1^2}\right)\frac{2\varrho^2{}_\oplus}{\delta_0^2}\int_\Omega \delta^2(u-v)^2\,dx.
\end{aligned}$$

Define

$$c^2_\oplus(\gamma) := (1+\gamma)\max\left\{\left(1+\left(\frac{1}{\delta_0^2}+\frac{2C^2_{F\Omega}}{\gamma c_1^2}\right)\frac{2\|a\|^2}{c_1^2}\right), \left(\frac{1}{\delta_0^2}+\frac{C^2_{F\Omega}}{\gamma c_1^2}\right)\frac{2\varrho^2{}_\oplus}{\delta_0^2}\right\}. \tag{4.3.62}$$

Then, $\overline{\mathfrak{M}}_{\mathrm{RCD}}(\alpha,\beta,v,p) \le c_\oplus(\gamma)|[u-v]|$, and we obtain (4.3.61). □

It is worth noting that, in the above consideration, we have overestimated the value of $\overline{\mathfrak{M}}_{\mathrm{RCD}}(\alpha,\beta,v,p)$ at several places, so that the constant $c_\oplus(\gamma)$ defined by the relation (4.3.62) is rather coarse. Anyway it gives an idea how $c_\oplus$ depends on $\|a\|$.

Also, we can establish the equivalence of $\overline{\mathfrak{M}}$ and the error computed in the full primal-dual norm.

Proposition 4.9. *For any $v \in u_0 + V_0$ and $y \in H^0_{\Gamma_2}(\Omega)$, the following two-sided estimate holds:*

$$C_\ominus\overline{\mathfrak{M}}_{\mathrm{RCD}}(\widehat{\alpha},\beta,v,y) \le |[u-v]| + |[p-y]|_* \le C_\oplus\overline{\mathfrak{M}}_{\mathrm{RCD}}(\widehat{\alpha},\beta,v,y), \tag{4.3.63}$$

where $\widehat{\alpha} = \min\{1,\delta\}$, $C_\ominus = \frac{1}{\mu}$,

$$\mu = \max\left\{1, \bar{C}_{F\Omega}, \sqrt{2}\left(1+\bar{C}_{F\Omega}\frac{\|\mathrm{a}\|}{c_1}\right), \sqrt{2}\bar{C}_{F\Omega}\frac{\varrho^2{}_\oplus}{\delta_0}\right\},\quad \bar{C}_{F\Omega} := 1+\frac{C_{F\Omega}}{c_1},$$

and $C_\oplus = 3 + \sqrt{\frac{\|\mathrm{a}\|^2}{c_1^2}+\frac{\varrho^2{}_\oplus}{\delta_0}}$.

Proof. The right-hand side inequality follows from the structure of $|[u-v]|$ and (4.3.59).

To prove the left-hand side, we note that $\widehat{\alpha}/\delta \leq 1$. Therefore,

$$\overline{\mathfrak{M}}_{\mathrm{RCD}}(\widehat{\alpha},\beta,v,y) \leq \left(\|\mathrm{r}_\Omega(v,y)\|^2 + |||y - A\nabla v|||_*^2\right)^{1/2} + \frac{C_{F\Omega}}{c_1}\|\mathrm{r}_\Omega(v,y)\|$$
$$\leq \bar{C}_{F\Omega}\|\mathrm{r}_\Omega(v,y)\| + |||y - A\nabla v|||_*.$$

We have

$$\|\mathrm{r}_\Omega(v,y)\| \leq \|\mathrm{div}\,(p-y)\| + \|\mathrm{a}\cdot\nabla(u-v)\| + \|\varrho^2(u-v)\|$$
$$\leq \|\mathrm{div}\,(p-y)\| + \tfrac{\|\mathrm{a}\|}{c_1}|||\nabla(u-v)||| + \tfrac{\varrho^2{}_\oplus}{\delta_0}\|\delta(u-v)\|,$$
$$|||y - A\nabla v|||_* \leq |||\nabla(u-v)||| + |||p-y|||_*.$$

Thus,

$$\overline{\mathfrak{M}}_{\mathrm{RCD}}(\widehat{\alpha},\beta,v,y) \leq \left(1 + \bar{C}_{F\Omega}\frac{\|\mathrm{a}\|}{c_1}\right)|||\nabla(u-v)||| + \bar{C}_{F\Omega}\frac{\varrho^2{}_\oplus}{\delta_0}\|\delta(u-v)\|$$
$$+ |||p-y|||_* + \bar{C}_{F\Omega}\|\mathrm{div}\,(p-y)\|$$
$$\leq \sqrt{2}\max\left\{1 + \bar{C}_{F\Omega}\frac{\|\mathrm{a}\|}{c_1}, \bar{C}_{F\Omega}\frac{\varrho^2{}_\oplus}{\delta_0}\right\}|[u-v]|$$
$$+ \max\{1,\bar{C}_{F\Omega}\}|[p-y]|_*. \tag{4.3.64}$$

Now (4.3.64) implies the left-hand side of (4.3.63). □

4.4 Notes for the chapter

For linear and nonlinear diffusion problems, the first a posteriori estimates of the functional type were obtained in the context of the general variational method, basic principles of which are briefly discussed at the beginning of this chapter (see [276]–[282] and [244] for a consequent exposition). The nonvariational approach to the derivation of a posteriori estimates for stationary diffusion problems (and other linear elliptic problems having the divergent form) was suggested in the author's paper [286]. Section 4.1 is based on these results. Later, a posteriori estimates for the convection-diffusion and reaction-convection-diffusion problems were derived (see [293, 300]). A posteriori estimates for the reaction-diffusion problem discussed in Section 4.2 were obtained in the paper by S. Repin and S. Sauter [305]. Functional type a posteriori estimates for the stationary reaction-convection-diffusion problem were analyzed in S. Nicaise and S. Repin [248]. In [287], the nonvariational method was applied to the evolutionary diffusion equation. The latter results are discussed in Chapter 9.

5 Elasticity

5.1 The linear elasticity problem

Consider an elastic body that occupies a domain $\Omega \in \mathbb{R}^d$ the boundary Γ of which consists of two disjoint parts Γ_1 and Γ_2 ($|\Gamma_1| > 0$). The classical statement of the linear elasticity problem is as follows: Find a tensor-valued function σ (stress) and a vector-valued function u (displacement) that satisfy the system of equations

$$\sigma = \mathbb{L}\varepsilon(u) \qquad \text{in } \Omega, \tag{5.1.1}$$

$$\operatorname{Div}\sigma + f = 0 \qquad \text{in } \Omega, \tag{5.1.2}$$

$$u = u_0 \qquad \text{on } \Gamma_1, \tag{5.1.3}$$

$$\sigma n = F \qquad \text{on } \Gamma_2. \tag{5.1.4}$$

Here f and F are given forces and $\mathbb{L} = \{L_{ijkm}\}$ is the tensor of elasticity constants, which is subject to the conditions

$$c_1^2|\varepsilon|^2 \leq \mathbb{L}\,\varepsilon : \varepsilon \leq c_2^2|\varepsilon|^2, \qquad \forall \varepsilon \in \mathbb{M}_s^{d\times d}. \tag{5.1.5}$$

Henceforth, we assume that the coefficients of the elasticity tensor are bounded and possess natural symmetry, i.e.,

$$\mathbb{L}_{ijkm} = \mathbb{L}_{jikm} = \mathbb{L}_{kmij}, \quad i,j,k,m = 1,\dots,d, \tag{5.1.6}$$

$$\mathbb{L}_{ijkm} \in L^\infty(\Omega). \tag{5.1.7}$$

Moreover, we assume that

$$f \in L^2(\Omega,\mathbb{R}^d), \quad F \in L^2(\Gamma_2,\mathbb{R}^d), \quad u_0 \in H^1(\Omega,\mathbb{R}^d). \tag{5.1.8}$$

A function $u \in u_0 + V_0$ is a generalized solution of (5.1.1)–(5.1.5) if it satisfies the integral relation

$$\int_\Omega \mathbb{L}\,\varepsilon(u) : \varepsilon(w)\,dx = \langle \ell, w\rangle \tag{5.1.9}$$

for all $w \in V_0$. In (5.1.9), the right-hand side is determined by the volume and surface loads

$$\langle \ell, w\rangle = \int_\Omega f\cdot w\,dx + \int_{\Gamma_2} F\cdot w\,ds,$$

and

$$V_0 := \{w \in H^1(\Omega,\mathbb{R}^d) \mid w = 0 \quad \text{on } \Gamma_1\}.$$

The linear elasticity problem has a variational statement: Find $u \in u_0 + V_0$ such that

$$J(u) = \inf_{w \in u_0 + V_0} J(v), \quad J(v) := \frac{1}{2} \int_\Omega \mathbb{L}\varepsilon(w) : \varepsilon(w)\, dx - \langle \ell, w \rangle. \tag{5.1.10}$$

The existence and uniqueness of u are easy to prove if the coercivity of

$$|||\varepsilon(w)||| := \left(\int_\Omega \mathbb{L}\varepsilon(w) : \varepsilon(w)\, dx \right)^{1/2}$$

on the space V_0 is established. This fact follows from Korn's inequality (1.4.32), which shows that $\|w\|^2_{1,2,\Omega}$ is equivalent to

$$[|w|]^2_{1,2,\Omega} := \int_\Omega (|\varepsilon(w)|^2 + |w|^2)\, dx.$$

Proposition 5.1. *Let Ω be a Lipschitz domain in $\mathbb{R}^d$, and let Γ_1 satisfy the following property:*

$$\exists\, x_0 \in \Gamma_1 \text{ and } \epsilon > 0 \text{ such that } \Gamma \cap B(x_0, \varepsilon) \subset \Gamma_1. \tag{5.1.11}$$

Then

$$\|w\|_{1,2,\Omega} \leq \mu_{\Omega\Gamma_1} \|\varepsilon(w)\|, \qquad \forall w \in V_0, \tag{5.1.12}$$

where $\mu_{\Omega\Gamma_1}$ is a positive constant independent of w.

Proof. Assume that the assertion is wrong. Then, for any $m \in \mathbb{N}$ there exists $v^{(m)} \in V_0$ such that

$$\|v^{(m)}\|_{1,2,\Omega} > m \left(\int_\Omega |\varepsilon(v^{(m)})|^2\, dx \right)^{1/2}. \tag{5.1.13}$$

Consider the sequence of normalized functions $\bar{v}^m := \frac{v^{(m)}}{\|v^{(m)}\|_{1,2,\Omega}}$. Obviously, $\|\bar{v}^m\|_{1,2,\Omega} = 1$ and, by (5.1.13),

$$\frac{1}{m} \geq \left(\int_\Omega |\varepsilon(\bar{v}^m)|^2\, dx \right)^{1/2}. \tag{5.1.14}$$

Hence, we can extract a subsequence of $\{\bar{v}^m\}$ that converges to a vector-valued function $\bar{v} \in V_0$ weakly in $H^1(\Omega, \mathbb{R}^d)$ and strongly in $L^2(\Omega, \mathbb{R}^d)$. Therefore, $\varepsilon(\bar{v}^m)$ weakly converges (in L^2) to $\varepsilon(\bar{v})$. By (5.1.13), we know that

$$\|\varepsilon(\bar{v}^m)\| \to 0. \tag{5.1.15}$$

Thus,

$$0 = \liminf_{m \to +\infty} \|\varepsilon(\bar{v}^m)\| \geq \|\varepsilon(\bar{v})\|,$$

and, therefore, $\varepsilon(\bar{v}) = 0$. Then $\bar{v} \in \mathbf{R}(\Omega) \cap V_0$ (cf. 1.4.3). In view of (5.1.11), the intersection of these sets contains only the zero function, so that $\bar{v} = 0$. However, in such a case we arrive at a contradiction. Indeed, on the one hand, by Korn's inequality we conclude that

$$\|\bar{v}^{(m)}\|_{1,2,\Omega} \leq C_{K\Omega} \left(\int_\Omega (\|\varepsilon(\bar{v}^m)\|^2 + \|\bar{v}^m\|^2)\, dx \right)^{1/2} \underset{m\to\infty}{\longrightarrow} 0.$$

On the other hand, $\|\bar{v}^m\|_{1,2,\Omega} = 1$ for any m and such a sequence cannot tend to zero. It remains to conclude that (5.1.13) is wrong and the constant $\mu_{\Omega\Gamma_1}$ exists. □

Corollary 5.2. *There exists a constant* C *such that*

$$\int_\Omega |w|^2\, dx + \int_{\Gamma_2} |w|^2\, ds \leq \mathrm{C}^2 |||\varepsilon(w)|||^2, \qquad \forall w \in V_0. \tag{5.1.16}$$

Proof. The proof follows from Proposition 5.1 and trace inequalities. Indeed, by a Friedrichs type inequality for the functions in V_0 and the trace inequality, we conclude that

$$\int_\Omega |w|^2\, dx + \int_{\Gamma_2} |w|^2\, ds \leq C_\Omega^2 \|w\|_{1,2,\Omega}^2, \qquad \forall w \in V_0. \tag{5.1.17}$$

Now (5.1.16) follows from (5.1.5), (5.1.12), and (5.1.17). □

Note that on V_0 the norms $\|\nabla w\|$ and $\|w\|_{1,2,\Omega}$ are equivalent. Therefore, instead (5.1.17) we can use the inequality

$$\int_\Omega |w|^2\, dx + \int_{\Gamma_2} |w|^2\, ds \leq \widehat{C}_\Omega^2 \|\nabla w\|^2, \qquad \forall w \in V_0, \tag{5.1.18}$$

where $\widehat{C}_\Omega$ is a somewhat different constant (which also depends only on Ω and Γ_2). Also, (5.1.12) means that

$$\|\nabla w\| \leq c_{\Omega\Gamma_1} \|\varepsilon(w)\|, \qquad \forall w \in V_0, \tag{5.1.19}$$

with a constant $c_{\Omega\Gamma_1} \leq \mu_{\Omega\Gamma_1}$.

Remark 5.3. In practice, the value of C can be estimated by minimizing the quotient

$$\frac{|||\varepsilon(w)|||^2}{\int_\Omega |w|^2\, dx + \int_{\Gamma_2} |w|^2\, ds}$$

over a sufficiently representative finite-dimensional space $V_{0h} \subset V_0$.

The estimation of constants arising in various versions of Korn's inequality is a more complicated task. This question was investigated by a number of authors (e.g., see C. Horgan [179]).

5.2 Estimates for displacements

Let $v \in u_0 + V_0$. Represent (5.1.9) in the form

$$\int_\Omega \mathbb{L}\,\varepsilon(u-v) : \varepsilon(w)\,dx = \int_\Omega (f \cdot w - \mathbb{L}\,\varepsilon(v) : \varepsilon(w))\,dx + \int_{\Gamma_2} F \cdot w\,ds, \quad (5.2.1)$$

where $w \in V_0$. Let $\Sigma(\Omega) := L^2(\Omega, \mathbb{M}^{d\times d})$ and

$$\tau \in H_{\Gamma_2}(\Omega, \mathrm{Div}) := \{\tau \in \Sigma(\Omega) \mid \mathrm{Div}\,\tau \in L^2(\Omega, \mathbb{R}^d),\ \tau n \in L^2(\Gamma_2, \mathbb{R}^d)\}.$$

Since

$$\mathrm{Div}\,(\tau w) = w \cdot \mathrm{Div}\,\tau + \tau^{\mathrm{T}} : \nabla w, \quad (5.2.2)$$

we have (for $\tau \in \Sigma_s(\Omega)$)

$$\int_{\Gamma_2} \tau n \cdot w\,ds = \int_\Omega (w \cdot \mathrm{Div}\,\tau + \tau : \varepsilon(w))\,dx. \quad (5.2.3)$$

By (5.2.1) and (5.2.3), we obtain

$$\int_\Omega \mathbb{L}\,\varepsilon(u-v) : \varepsilon(w)\,dx = \int_\Omega (f + \mathrm{Div}\,\tau) \cdot w\,dx + \int_\Omega (\tau - \mathbb{L}\,\varepsilon(v)) : \varepsilon(w)\,dx$$
$$+ \int_{\Gamma_2} (F - \tau n) \cdot w\,ds. \quad (5.2.4)$$

Set $w = u - v$ and use (5.1.16). We arrive at the estimate

$$|||\varepsilon(u-v)||| \le \overline{\mathfrak{M}}_{\mathrm{EL}}(v,\tau)$$
$$:= |||\tau - \mathbb{L}\varepsilon(v)|||_* + \mathrm{C}\left(\|\mathrm{Div}\,\tau + f\|^2 + \|F - \tau n\|^2_{\Gamma_2}\right)^{1/2}, \quad (5.2.5)$$

which is quite analogous to the estimate (4.1.11). Squaring both parts of (5.2.5), we obtain an analog of (4.1.15):

$$|||\varepsilon(u-v)|||^2 \le \overline{\mathfrak{M}}^2_{\mathrm{EL}}(v,\tau,\beta) := (1+\beta)\int_\Omega \left(\varepsilon(v) - \mathbb{L}^{-1}\tau\right) : \left(\mathbb{L}\,\varepsilon(v) - \tau\right)dx$$
$$+ \frac{1+\beta}{\beta}\mathrm{C}^2\left\{\|\mathrm{Div}\,\tau + f\|^2 + \|F - \tau n\|^2_{\Gamma_2}\right\}. \quad (5.2.6)$$

In (5.2.6), τ is a symmetric tensor-valued function and β is an arbitrary positive constant. It is easy to note that the first term is nonnegative and vanishes if and only if

$$\tau = \mathbb{L}\varepsilon(v).$$

It penalizes violations of the *Hooke's law*. The meaning of the second and third terms is obvious: they represent other two relations and penalize errors in the *equilibrium equation* (5.1.2) and *boundary condition* (5.1.4). Thus, the majorant not only provides a bound of the overall value of the error, but also shows its physically sensible parts. Latter information suggests a correct way for finding a better approximation.

In (5.2.5) and (5.2.6), the error is measured in terms of the norm generated by the tensor ε. For this reason, the Korn's constant does not explicitly occur in the estimates. However, a tensor-valued function τ in $\overline{\mathfrak{M}}_{\mathrm{EL}}(v,\tau)$ must be symmetric. This requirement is easy to fulfill if τ is approximated with the help of finite element approximations having degrees of freedom at nodes (if the nodal tensors are symmetric then an extension of them is also symmetric). For other approximations (e.g., for mixed approximations degrees of freedom of which are associated with faces) the symmetry condition may be rather burdensome. Below we discuss a way of avoiding it.

Take an arbitrary tensor-valued function $\widehat{\tau} \in H_{\Gamma_2}(\Omega, \mathrm{Div})$ and rearrange (5.2.1) with the help of (5.2.2) as follows:

$$\int_\Omega \mathbb{L}\,\varepsilon(u-v) : \varepsilon(w)\,dx = \int_\Omega \big((\mathrm{Div}\,\widehat{\tau} + f)\cdot w + (\widehat{\tau}^{\mathrm{T}} - \mathbb{L}\varepsilon(v)) : \nabla w\big)\,dx$$

$$+ \int_{\Gamma_2} (F - \widehat{\tau}^{\mathrm{T}} n)\cdot w\,ds, \qquad \forall w \in V_0, \quad (5.2.7)$$

Decompose $\widehat{\tau}$ into symmetric and skew-symmetric parts $\widehat{\tau}_{\mathrm{sm}}$ and $\widehat{\tau}_{\mathrm{sk}}$. Then,

$$\widehat{\tau}^{\mathrm{T}} : \nabla w = \widehat{\tau}_{\mathrm{sm}} : \varepsilon(w) - \widehat{\tau}_{\mathrm{sk}} : \nabla w.$$

Note that $\mathbb{L}\varepsilon(v)$ is a symmetric tensor. Therefore, (5.2.7) implies the estimate

$$\int_\Omega \mathbb{L}\,\varepsilon(u-v) : \varepsilon(w)\,dx \le \|\mathrm{Div}\,\widehat{\tau} - f\|\,\|w\| + \|\widehat{\tau}^{\mathrm{T}} n + F\|_{\Gamma_2}\|w\|_{\Gamma_2}$$

$$+ |||\widehat{\tau}_{\mathrm{sm}} - \mathbb{L}\varepsilon(v)|||_*\,|||\varepsilon(w)||| + \|\widehat{\tau}_{\mathrm{sk}}\|\,\|\nabla w\|. \quad (5.2.8)$$

By (5.1.16), we find that

$$\|\mathrm{Div}\,\widehat{\tau} - f\|\,\|w\| + \|\widehat{\tau}^{\mathrm{T}} n + F\|_{\Gamma_2}\|w\|_{\Gamma_2}$$

$$\le \mathrm{C}\Big\{\|\mathrm{Div}\,\widehat{\tau} - f\|^2 + \|\widehat{\tau}^{\mathrm{T}} n + F\|^2_{\Gamma_2}\Big\}^{1/2} |||\varepsilon(w)|||.$$

Set $w = u - v$ and use (5.1.19). We arrive at the estimate

$$|||\varepsilon(u-v)||| \le \mathrm{C}\Big\{\|\mathrm{Div}\,\widehat{\tau} - f\|^2 + \|\widehat{\tau}^{\mathrm{T}} n + F\|^2_{\Gamma_2}\Big\}^{1/2}$$

$$+ |||\widehat{\tau}_{\mathrm{sm}} - \mathbb{L}\varepsilon(v)|||_* + \frac{c_{\Omega\Gamma_1}}{c_1}\|\widehat{\tau}_{\mathrm{sk}}\|. \quad (5.2.9)$$

This estimate includes an additional term which is positive if the tensor-valued function $\widehat{\tau}$ is nonsymmetric. If $\widehat{\tau}$ is a symmetric tensor (i.e., $\widehat{\tau}_{\mathrm{sk}} = 0$), then (5.2.9) is equivalent to (5.2.5).

Remark 5.4. We note that the right-hand sides of (5.2.5) and (5.2.9) are equal to $|||\varepsilon(u-v)|||$ if τ and $\widehat{\tau}$ coincide with σ.

By the same arguments as for the diffusion equation, one can prove that

$$\int_\Omega \mathbb{L}\varepsilon(v-u) : \varepsilon(v-u)dx \geq \underline{\mathfrak{M}}^2_{\mathrm{EL}}(v,w) := 2\mathcal{F}_v(w) - |||\varepsilon(w)|||^2, \tag{5.2.10}$$

where $w \in V_0$ and

$$\mathcal{F}_v(w) := \int_\Omega f\cdot w\,dx + \int_{\Gamma_2} F\cdot w\,ds - \int_\Omega \mathbb{L}\varepsilon(v) : \varepsilon(w)\,dx.$$

Also, we can show that

$$|||\varepsilon(v-u)|||^2 = \sup_{w\in V_0} \underline{\mathfrak{M}}^2_{\mathrm{EL}}(v,w).$$

By the maximization of the functional $\underline{\mathfrak{M}}_{\mathrm{EL}}(v,w)$ on a sequence of finite-dimensional spaces $V_{0k} \subset V_0$, we obtain a sequence of computable lower bounds

$$M^k_\ominus = \sup_{w\in V_{0k}} \underline{\mathfrak{M}}_{\mathrm{EL}}(v,w_k).$$

If the spaces V_{0k} satisfy the limit density condition stated, then the sequence $\{M^k_\ominus\}$ tends to $|||\varepsilon(v-u)|||^2$.

5.3 Estimates for stresses

Assume that $\tau \in H_{\Gamma_2}(\Omega,\mathrm{Div})$ is a symmetric tensor-valued function that approximates σ. By (5.1.1), for any $v \in u_0 + V_0$ we have

$$\begin{aligned}|||\sigma-\tau|||_* &= |||\mathbb{L}\varepsilon(u)-\tau|||_* \leq |||\mathbb{L}\varepsilon(u-v)|||_* + |||\mathbb{L}\varepsilon(v)-\tau|||_* \\ &= |||\varepsilon(u-v)||| + |||\mathbb{L}\varepsilon(v)-\tau|||_* \leq \overline{\mathfrak{M}}_{\mathrm{EL}}(v,\tau) + |||\mathbb{L}\varepsilon(v)-\tau|||_*.\end{aligned} \tag{5.3.1}$$

By (5.2.5) and (5.3.1), we obtain an upper bound for a combined primal-dual norm

$$|||\varepsilon(u-v)||| + |||\sigma-\tau|||_* \leq 2\overline{\mathfrak{M}}_{\mathrm{EL}}(v,\tau) + |||\mathbb{L}\varepsilon(v)-\tau|||_*. \tag{5.3.2}$$

For another norm

$$|[\tau]|_* := |||\tau|||_* + \|\mathrm{Div}\,\tau\| + \|\tau\cdot n\|_{\Gamma_2},$$

we have

$$\begin{aligned}|[\sigma-\tau]|_* &\leq \overline{\mathfrak{M}}_{\mathrm{EL}}(v,\tau) + |||\mathbb{L}\varepsilon(v)-\tau|||_* + \|\tau\cdot n - F\|_{\Gamma_2} + \|\mathrm{Div}\,(\sigma-\tau)\| \\ &= \overline{\mathfrak{M}}_{\mathrm{EL}}(v,\tau) + |||\mathbb{L}\varepsilon(v)-\tau|||_* + \|\tau\cdot n - F\|_{\Gamma_2} + \|\mathrm{Div}\,\tau + f\|.\end{aligned} \tag{5.3.3}$$

Hence,

$$
\begin{aligned}
&|||\varepsilon(u-v)||| + |[\sigma-\tau]|_* \\
&\qquad \le 2\overline{\mathfrak{M}}_{\mathrm{EL}}(v,\tau) + |||\mathbb{L}\varepsilon(v)-\tau|||_* + \|\tau\cdot n - F\|_{\Gamma_2} + \|\mathrm{Div}\,\tau + f\| \\
&\qquad \le \bar{c}\,\overline{\mathfrak{M}}_{\mathrm{EL}}(v,\tau),
\end{aligned}
\tag{5.3.4}
$$

where $\bar{c} = 3 + \sqrt{2}/\mathrm{C}$. On the other hand,

$$
\begin{aligned}
\overline{\mathfrak{M}}_{\mathrm{EL}}(v,\tau) &\le |||\mathbb{L}\varepsilon(v-u)|||_* + |||\sigma-\tau|||_* + \mathrm{C}\left(\|\mathrm{Div}\,(\tau-\sigma)\|^2 + \|(\sigma-\tau)n\|^2_{\Gamma_2}\right)^{1/2} \\
&\le |||\varepsilon(u-v)||| + |||\sigma-\tau|||_* + \mathrm{C}\left(\|\mathrm{Div}\,(\tau-\sigma)\| + \|(\sigma-\tau)n\|_{\Gamma_2}\right) \\
&\le \underline{c}(|||\varepsilon(u-v)||| + |[\sigma-\tau]|_*),
\end{aligned}
\tag{5.3.5}
$$

where $\underline{c} = \max\{1, \mathrm{C}\}$. Thus, we note that $\overline{\mathfrak{M}}_{\mathrm{EL}}(v,\tau)$ is equivalent to the error evaluated in the combined stress-strain norm.

5.4 Isotropic linear elasticity

In the important case of isotropic elastic media, the components of $\mathbb{L}$ and $\mathbb{L}^{-1}$ depend on two material constants only. Below we present respective forms of a posteriori estimates derived in the previous section.

5.4.1 3D problems

For $d = 3$, we can represent the elasticity tensor in the form

$$
\mathbb{L}\,\varepsilon = K_0\,\mathrm{tr}\,\varepsilon\,\mathbb{I} + 2\mu\varepsilon^{\mathrm{D}}, \tag{5.4.1}
$$

$$
\mathbb{L}^{-1}\,\tau = \frac{1}{9K_0}\,\mathrm{tr}\,\tau\mathbb{I} + \frac{1}{2\mu}\tau^{\mathrm{D}}. \tag{5.4.2}
$$

In (5.4.1)–(5.4.2), K_0 and μ are positive (elasticity) constants that depend on properties of media to resist compression and shear forces, respectively.

In this case, the norms are defined by the relations

$$
|||\varepsilon(w)|||^2 := \int_\Omega \left(K_0\,\mathrm{tr}\,\varepsilon(w)^2 + 2\mu\left|\varepsilon^{\mathrm{D}}(w)\right|^2\right) dx, \tag{5.4.3}
$$

$$
|||\tau|||^2_* := \int_\Omega \left(\frac{1}{9K_0}\,\mathrm{tr}(\tau)^2 + \frac{1}{2\mu}\left|(\tau)^{\mathrm{D}}\right|^2\right) dx. \tag{5.4.4}
$$

Since

$$
\varepsilon(v) - \mathbb{L}^{-1}\tau = \left(\frac{1}{3}\,\mathrm{tr}(\varepsilon(v)) - \frac{1}{9K_0}\,\mathrm{tr}(\tau)\right)\mathbb{I} + \left(\varepsilon^{\mathrm{D}}(v) - \frac{1}{2\mu}\,\mathrm{tr}\,\tau\right), \tag{5.4.5}
$$

$$
\mathbb{L}\,\varepsilon(v) - \tau = \left(K_0\,\mathrm{tr}(\varepsilon(v)) - \frac{1}{3}\,\mathrm{tr}(\tau)\right)\mathbb{I} + \left(2\mu(\varepsilon)^{\mathrm{D}}(v) - \tau^{\mathrm{D}}\right), \tag{5.4.6}
$$

we find that

$$\left(\varepsilon(v) - \mathbb{L}^{-1}\tau\right) : (\mathbb{L}\,\varepsilon(v) - \tau) \\ = K_0 \left(\operatorname{div} v - \frac{1}{3K_0} \operatorname{tr}\tau\right)^2 + 2\mu \left|\varepsilon^{\mathrm{D}}(v) - \frac{1}{2\mu}\operatorname{tr}\tau\right|^2. \quad (5.4.7)$$

Hence, for isotropic media the estimate (5.1.6) reads as follows:

$$\int_\Omega \left(K_0|\operatorname{div}(u-v)|^2 + 2\mu\left|\varepsilon^{\mathrm{D}}(u-v)\right|^2\right) dx \\ \leq (1+\beta)\int_\Omega \left(K_0\left(\operatorname{div} v - \frac{1}{3K_0}\operatorname{tr}\tau\right)^2 + 2\mu\left|\varepsilon^{\mathrm{D}}(v) - \frac{1}{2\mu}\operatorname{tr}\tau\right|^2\right) dx \\ + \frac{1+\beta}{\beta}\,\mathrm{C}^2\left\{\|\operatorname{Div}\tau + f\|^2 + \|F - \tau n\|^2_{\Gamma_2}\right\}. \quad (5.4.8)$$

Instead of K_0 and μ, one cane use another pair of constants and represent (5.4.1) and (5.4.2) in the form

$$\mathbb{L}\,\varepsilon = \lambda\, \operatorname{tr}(\varepsilon)\,\mathbb{I} + 2\mu\varepsilon, \quad (5.4.9)$$

$$\mathbb{L}^{-1}\,\tau = \frac{1}{2\mu}\left(\tau - \frac{\lambda}{3\lambda+2\mu}\operatorname{tr}(\tau)\,\mathbb{I}\right). \quad (5.4.10)$$

Here $\lambda = K_0 - \frac{2\mu}{3}$. In this case,

$$|||\varepsilon(v)|||^2 := \int_\Omega \left(\lambda|\operatorname{div} v|^2 + 2\mu\,|\varepsilon(v)|^2\right) dx, \quad (5.4.11)$$

$$|||\tau|||^2_* := \int_\Omega \frac{1}{2\mu}\left(|\tau|^2 - \frac{\lambda}{3\lambda+2\mu}\operatorname{tr}(\tau)^2\right) dx, \quad (5.4.12)$$

and

$$\left(\varepsilon(v) - \mathbb{L}^{-1}\tau\right) : (\mathbb{L}\,\varepsilon(v) - \tau) \\ = \lambda|\operatorname{div} v|^2 + 2\mu\,|\varepsilon(v)|^2 + \frac{1}{2\mu}\left(|\tau|^2 - \frac{\lambda}{3\lambda+2\mu}\operatorname{tr}(\tau)^2\right) - 2\varepsilon(v) : \tau.$$

5.4.2 The plane stress problem

The plane stress problem arises if a thin elastic plate is subject to an action of forces lying in its plane. Let

$$\Omega := \widehat{\Omega} \times \left(-\frac{h}{2}, \frac{h}{2}\right),$$

where $\widehat{\Omega} \in \mathbb{R}^2$ is the so-called "middle surface" Γ_0. It is assumed that this surface contains x_1 and x_2 axes of a Cartesian coordinate system and x_3 is perpendicular to Γ_0. Throughout this chapter the symbol $\widehat{\ }$ is used to mark "plane" components of vectors and tensors. In particular, we write $\widehat{x} = (x_1, x_2)$ for the plane coordinate vector, $\widehat{v} = (v_1, v_2)$ is a plane vector, and $\widehat{\tau} := \{\tau_{st}\}$, $s, t = 1, 2$, is a plane tensor.

Assume that h is small with respect to the character size of $\widehat{\Omega}$. In general, it may be difficult to exactly define this notion, but for sufficiently regular domains we can compare h with the diameter $\operatorname{diam} \widehat{\Omega}$. Let $f = \{\widehat{f}_1(\widehat{x}), \widehat{f}_2(\widehat{x}), 0\}$ and assume that the surfaces

$$\Gamma_{\pm} := \left\{ x = (\widehat{w}, x_3) \mid \widehat{x} \in \widehat{\Omega},\ x_3 = \pm \frac{h}{2} \right\}$$

are free from loads. Also, we assume that the boundary conditions on

$$\Gamma_i := \gamma_i \times \left(-\frac{h}{2}, \frac{h}{2} \right)$$

are defined by the functions $u_0 = \{\widehat{u}_0(\widehat{x}), 0\}$ and $F = \{\widehat{F}(\widehat{x}), 0\}$. Here the γ_i denote two parts of γ, which is the boundary of $\widehat{\Omega}$.

In the plane stress model, it is suggested to consider the stress tensor as a plane tensor. In other words, we set

$$\sigma_{i3} = 0, \quad i = 1, 2, 3. \tag{5.4.13}$$

In addition, in the plane stress model it is assumed that σ insignificantly depends on x_3, so that

$$\widehat{\sigma} = \widehat{\sigma}(\widehat{x}). \tag{5.4.14}$$

Strictly speaking these assumptions are not true and violate the 3D relations of linear elasticity problem. However, they lead to a 2D problem which is much simpler. The error arising owing to this dimension reduction procedure is the *modeling error*. In the next section, we derive an upper bound for this error. Now, our goal is to present the error majorant for the plane stress problem.

In view of (5.4.13), $\varepsilon_{13}(u) = \varepsilon_{23}(u) = 0$. For the component $\varepsilon_{33}(u)$ we have the relation (which comes from the Hooke's law)

$$0 = \sigma_{33} = K_0(\varepsilon_{11}(u) + \varepsilon_{22}(u) + \varepsilon_{33}(u)) + 2\mu \frac{2\varepsilon_{33}(u) - \varepsilon_{11}(u) - \varepsilon_{22}(u)}{3},$$

which implies

$$\operatorname{tr} \varepsilon = \frac{6\mu}{3K_0 + 4\mu} \widehat{\operatorname{tr}\,\varepsilon}.$$

Therefore, 2D form of the Hooke's law is

$$\widehat{\sigma} = \widehat{\mathbb{L}}\widehat{\varepsilon} = \widehat{K_0}\,\widehat{\operatorname{tr}}\,\widehat{\varepsilon}\,\widehat{\mathbb{I}} + 2\mu\,(\widehat{\varepsilon})^{\mathrm{D}}, \tag{5.4.15}$$

$$\widehat{\varepsilon} = \widehat{\mathbb{L}}^{-1}\,\widehat{\sigma} = \frac{1}{4\widehat{K_0}}\,\widehat{\operatorname{tr}}\,\widehat{\sigma}\,\widehat{\mathbb{I}} + \frac{1}{2\mu}\,\widehat{\sigma^D}, \tag{5.4.16}$$

where $\widehat{K_0} = \frac{9K_0\mu}{3K_0+4\mu}$ and $\widehat{\operatorname{tr}}\,\widehat{\varepsilon} := \widehat{\varepsilon}_{11} + \widehat{\varepsilon}_{22}$. The other relations that define the solution are as follows:

$$\widehat{\operatorname{Div}\sigma} + \widehat{f} = 0 \qquad \text{in } \widehat{\Omega}, \tag{5.4.17}$$

$$\widehat{u} = \widehat{u}_0 \qquad \text{on } \gamma_1, \tag{5.4.18}$$

$$\widehat{\sigma}\,\widehat{n} = \widehat{F} \qquad \text{on } \gamma_2. \tag{5.4.19}$$

We repeat the arguments adduced for deriving the error majorant in the 3D elasticity model and arrive at an estimate similar to (5.4.8):

$$\begin{aligned}
&\int_\Omega \left(\widehat{K}_0|\widehat{\operatorname{tr}}\,\widehat{\varepsilon}(u-v)|^2 + 2\mu\left|\widehat{\varepsilon}^{\mathrm{D}}(u-v)\right|^2\right) dx \\
&\quad \le (1+\beta)\int_{\widehat{\Omega}} \left(\widehat{K}_0\left(\operatorname{div} v - \frac{1}{2\widehat{K}_0}\widehat{\operatorname{tr}}\,\widehat{\tau}\right)^2 + 2\mu\left|\widehat{\varepsilon}^{\mathrm{D}}(v) - \frac{1}{2\mu}\widehat{\operatorname{tr}}\,\widehat{\tau}\right|^2\right) d\widehat{x} \\
&\quad\quad + \frac{1+\beta}{\beta}\,\widehat{\mathrm{C}}^2\left\{\|\widehat{\operatorname{Div}\tau} + \widehat{f}\|^2_{\widehat{\Omega}} + \|\widehat{F} - \widehat{\tau}\,\widehat{n}\|^2_{\gamma_2}\right\}.
\end{aligned} \tag{5.4.20}$$

5.4.3 The plane strain problem

The deformation is called plane if at any point the displacement vector is collinear to a certain plane. Assume that this plane is Ox_1x_2 and

$$u_1 = u_1(\widehat{x}), \quad u_2 = \widehat{u}_2(\widehat{x}), \quad u_3 = 0.$$

In this case, only the plane part of the deformation tensor $\widehat{\varepsilon}(u)$ has nonzero components. By the Hooke's law, we conclude that $\sigma_{13} = \sigma_{23} = 0$ and σ_{33} is excluded by the relation

$$\sigma_{33} = \frac{\lambda}{2(\lambda+\mu)}\widehat{\operatorname{tr}}(\widehat{\sigma}),$$

which also follows from the Hooke's law. In view of the above relations, the plane parts of stresses and strains are connected by the relations

$$\widehat{\sigma} = \lambda\,\widehat{\operatorname{tr}}(\widehat{\varepsilon}(\widehat{u}))\,\widehat{\mathbb{I}} + 2\mu\,\widehat{\varepsilon}(\widehat{u}), \tag{5.4.21}$$

$$\widehat{\varepsilon}(\widehat{u}) = \frac{1}{2\mu}\left(\widehat{\sigma} - \frac{\lambda}{2(\lambda+\mu)}\widehat{\operatorname{tr}}(\widehat{\sigma})\,\widehat{\mathbb{I}}\right). \tag{5.4.22}$$

In view of (5.4.22), the term of the majorant related to the constitutive relations of the plane strain problem has the form

$$\int_{\widehat{\Omega}} \Big(\lambda |\widehat{\operatorname{div} v}|^2 + 2\mu\, |\widehat{\varepsilon}(\widehat{v})|^2 + \frac{1}{2\mu}\, |\widehat{\tau}|^2 - \frac{\lambda}{4\mu(\lambda+\mu)} \widehat{\operatorname{tr}}(\widehat{\tau})^2 - 2\widehat{\varepsilon}(\widehat{v}) \,:\, \widehat{\tau}\Big)\, d\widehat{x}.$$

5.4.4 Error of the plane stress model

Estimate (5.4.8) allows one to measure the error that arises if the 3D elasticity problem is replaced by a simplified 2D one. Let $\widehat{u}$ and $\widehat{\sigma}$ be the displacement vector and the stress tensor satisfying (5.4.17)–(5.4.19) and the relations

$$\widehat{\sigma}_{11} = \widehat{K_0}(\widehat{u}_{1,1} + \widehat{u}_{2,2}) + \mu(\widehat{u}_{1,1} - \widehat{u}_{2,2}), \tag{5.4.23}$$

$$\widehat{\sigma}_{22} = \widehat{K_0}(\widehat{u}_{1,1} + \widehat{u}_{2,2}) + \mu(\widehat{u}_{2,2} - \widehat{u}_{1,1}), \tag{5.4.24}$$

$$\widehat{\sigma}_{12} = \mu(\widehat{u}_{1,2} + \widehat{u}_{2,1}). \tag{5.4.25}$$

With the help of $\widehat{u}$ and $\widehat{\sigma}$ we construct $(\widetilde{u}, \widetilde{\sigma})$, which yields an approximate solution of the original (3D) problem. For this purpose, we set

$$\widetilde{u} = (\widehat{u}_1, \widehat{u}_2, \phi),$$
$$\widetilde{\sigma}_{\alpha\beta} = \widehat{\sigma}_{\alpha\beta}, \quad \widetilde{\sigma}_{3\alpha} = 0,$$

where $\phi \in \Phi(\Omega) := \{\phi \in H^1(\Omega) \mid \phi = 0 \text{ on } \Gamma_1\}$.

Since $\widehat{u} = \widehat{u}_0(\widehat{x})$ on γ_1, the function $\widetilde{u}$ belongs to $u_0 + V_0$. Therefore, we can use (5.4.8) with $v = \widetilde{u}$ and $\tau = \widetilde{\sigma}$. Note that

$$\widetilde{\sigma} n = 0 \text{ on } \Gamma_\pm \quad \text{and} \quad \widetilde{\sigma} n = \{\widehat{\sigma}\,\widehat{n}; 0\} = F \text{ on } \Gamma_2, \tag{5.4.26}$$

$$\operatorname{Div} \widetilde{\sigma} = \{\widehat{\operatorname{Div}} \,\widehat{\sigma};\, 0\} = \{-\widehat{f};\, 0\} = -f. \tag{5.4.27}$$

Therefore, the last term of (5.4.8) vanishes and the whole error is contained in the first one. To evaluate it, we use the relations

$$\operatorname{tr}(\varepsilon(\widetilde{u})) = \widehat{u}_{1,1} + \widehat{u}_{2,2} + \phi_{,3}, \tag{5.4.28}$$

$$\operatorname{tr}(\widetilde{\sigma}) = \widehat{\operatorname{tr}}(\widehat{\sigma}) = \widehat{\sigma}_{11} + \widehat{\sigma}_{22} = 2\widehat{K_0}(\widehat{u}_{1,1} + \widehat{u}_{2,2}) \tag{5.4.29}$$

and find that

$$\operatorname{tr}(\varepsilon(\widetilde{u})) - \tfrac{1}{3K_0} \operatorname{tr}(\widetilde{\sigma}) = \rho\, (\widehat{u}_{1,1} + \widehat{u}_{2,2}) + \phi_{,3},$$

where

$$\rho = 1 - \frac{2\widehat{K_0}}{3K_0} = \frac{3K_0 - 2\mu}{3K_0 + 4\mu} = \frac{1}{2}\left(\frac{\widehat{K_0}}{\mu} - 1\right).$$

Let us estimate the quantity $\left|\frac{1}{2\mu}\widetilde{\sigma} - (\varepsilon)^{\mathrm{D}}(\widetilde{u})\right|^2$.

In view of (5.4.23)–(5.4.25), we have

$$\frac{2\widehat{\sigma}_{11}-\widehat{\sigma}_{22}}{3}=\frac{\widehat{K_0}}{3}(\widehat{u}_{1,1}+\widehat{u}_{2,2})+\mu(\widehat{u}_{1,1}-\widehat{u}_{2,2}),$$
$$\frac{2\widehat{\sigma}_{22}-\widehat{\sigma}_{11}}{3}=\frac{\widehat{K_0}}{3}(\widehat{u}_{1,1}+\widehat{u}_{2,2})+\mu(\widehat{u}_{2,2}-\widehat{u}_{1,1}),$$
$$-\frac{\widehat{\sigma}_{11}+\widehat{\sigma}_{22}}{3}=\frac{2\widehat{K_0}}{3}(\widehat{u}_{1,1}+\widehat{u}_{2,2}).$$

Therefore,

$$\frac{1}{2\mu}\widetilde{\sigma}^{\mathrm{D}}=\begin{bmatrix}\frac{\widehat{K_0}}{6\mu}(\widehat{u}_{1,1}+\widehat{u}_{2,2})+\frac{\widehat{u}_{1,1}-\widehat{u}_{2,2}}{2} & \frac{\widehat{u}_{1,2}+\widehat{u}_{2,1}}{2} & 0\\ \frac{\widehat{u}_{1,2}+\widehat{u}_{2,1}}{2} & \frac{\widehat{K_0}}{6\mu}(\widehat{u}_{1,1}+\widehat{u}_{2,2})+\frac{\widehat{u}_{2,2}-\widehat{u}_{1,1}}{2} & 0\\ 0 & 0 & -\frac{\widehat{K_0}}{3\mu}(\widehat{u}_{1,1}+\widehat{u}_{2,2})\end{bmatrix}.$$

Next,

$$\varepsilon^{\mathrm{D}}(\widetilde{u})=\begin{bmatrix}\frac{\widehat{u}_{1,1}+\widehat{u}_{2,2}}{6}+\frac{\widehat{u}_{1,1}-\widehat{u}_{2,2}}{2}-\frac{\phi_{,3}}{3} & \frac{\widehat{u}_{1,2}+\widehat{u}_{2,1}}{2} & \frac{\phi_{,1}}{2}\\ \frac{\widehat{u}_{1,2}+\widehat{u}_{2,1}}{2} & \frac{\widehat{u}_{1,1}+\widehat{u}_{2,2}}{6}+\frac{\widehat{u}_{2,2}-\widehat{u}_{1,1}}{2}-\frac{\phi_{,3}}{3} & \frac{\phi_{,2}}{2}\\ \frac{\phi_{,1}}{2} & \frac{\phi_{,2}}{2} & \frac{2\phi_{,3}}{3}-\frac{\widehat{u}_{1,1}+\widehat{u}_{2,2}}{3}\end{bmatrix}$$

and we obtain

$$\frac{1}{2\mu}\widetilde{\sigma}-\varepsilon^{\mathrm{D}}(\widetilde{u})=\begin{bmatrix}\frac{\rho(\widehat{u}_{1,1}+\widehat{u}_{2,2})+\phi_{,3}}{3} & 0 & -\frac{\phi_{,1}}{2}\\ 0 & \frac{\rho(\widehat{u}_{1,1}+\widehat{u}_{2,2})+\phi_{,3}}{3} & -\frac{\phi_{,2}}{2}\\ -\frac{\phi_{,1}}{2} & -\frac{\phi_{,2}}{2} & -\frac{2(\rho(\widehat{u}_{1,1}+\widehat{u}_{2,2})+\phi_{,3})}{3}\end{bmatrix}.$$

Hence,

$$\left|\frac{1}{2\mu}\widetilde{\sigma}-(\varepsilon)^{\mathrm{D}}(\widetilde{u})\right|^2=\frac{2}{3}(\rho(\widehat{u}_{1,1}+\widehat{u}_{2,2})+\phi_{,3})^2+\frac{1}{2}(\phi_{,1}^2+\phi_{,2}^2),$$

and we find that

$$\int_\Omega \left(K_0|\operatorname{div}(u-\widetilde{u})|^2 + 2\mu\,|\varepsilon^{\mathrm{D}}(u-\widetilde{u})|^2\right) dx$$

$$\leq \int_\Omega \left(K_0\left(\operatorname{div}\widetilde{u} - \frac{1}{3K_0}\operatorname{tr}\widetilde{\sigma}\right)^2 + 2\mu\left|\varepsilon^{\mathrm{D}}(\widetilde{u}) - \frac{1}{2\mu}\operatorname{tr}\widetilde{\sigma}\right|^2\right) dx$$

$$\leq \left(K_0 + \frac{4\mu}{3}\right)\int_\Omega \left(\rho\,(\widehat{u}_{1,1} + \widehat{u}_{2,2}) + \phi_{,3}\right)^2 dx + \mu\int_\Omega (\phi_{,1}^2 + \phi_{,2}^2)\, dx. \quad (5.4.30)$$

It is easy to see that the right-hand side of the above estimate is positive. Indeed, if the second integral is equal to zero then $\phi = \phi(x_3)$. Then the first integral is positive (because $\widehat{u}_1$ and $\widehat{u}_2$ depend only on $\widehat{x}$). The only one exception is the case $\rho = 0$ or $(\widehat{\operatorname{div}}\,\widehat{u} = 0)$ Since $\rho = \frac{\nu}{1-\nu}$ we observe that ρ can be equal to zero only if the Poisson coefficient ν is equal to zero. In all other cases, the modelling error related to the plane stress model is positive.

5.5 Notes for the chapter

The reader can find a more detailed discussion of a posteriori estimates for linear elasticity problems author's papers [280, 282, 284] and in the book P. Neittaanmäki and S. Repin [244]. Numerical tests and extensions of the above theory to thermoelastic problems are presented in A. Muzalevskii and S. Repin [240, 241].

In the last section of this chapter, we have touched an important problem: estimation of modeling errors. This problem deserves a special consideration, which is beyond the framework of the book. In the context of functional error majorants, a posteriori error estimates for modeling errors arising in dimension reduction models of diffusion type problems were derived in the papers by S. Repin, S. Sauter, and A. Smolianski [310, 310, 312] and for elasticity problems in [284, 285].

6 Incompressible viscous fluids

6.1 The Stokes problem

Statement of the problem. The Stokes model is one of the simplest models in the theory of viscous incompressible fluids. It is represented by the relations

$$u_t - \nu\Delta u = f - \nabla \mathfrak{p} \qquad \text{in } \Omega, \tag{6.1.1}$$

$$\operatorname{div} u = 0, \tag{6.1.2}$$

$$u(x,0) = \widehat{u}(x), \tag{6.1.3}$$

$$u = u_0 \qquad \text{on } \Gamma, \tag{6.1.4}$$

where $u(x,t)$ is the velocity field, $\mathfrak{p}(x,t)$ is the pressure function, $\nu > 0$ is the viscosity parameter (or a positive function), and $\widehat{u}(x)$ and u_0 are solenoidal functions that define the initial and boundary conditions, respectively.

In the stationary case, u does not depend on t. Then, the problem is to find $u(x)$ and $\mathfrak{p}(x)$ such that

$$-\nu\Delta u = f - \nabla \mathfrak{p} \qquad \text{in } \Omega, \tag{6.1.5}$$

$$\operatorname{div} u = 0, \tag{6.1.6}$$

$$u = u_0 \qquad \text{on } \Gamma. \tag{6.1.7}$$

In the case of mixed boundary conditions, (6.1.7) holds only on a part $\Gamma_1 \subset \Gamma$, and on another part Γ_2, the Neumann boundary condition

$$\sigma n = F \quad \text{on } \Gamma_2, \tag{6.1.8}$$

(where $\sigma = \nu\nabla u - \mathfrak{p}$) is considered. We assume that

$$f \in L_2(\Omega, \mathbb{R}^n) \quad \text{and} \quad u_0 \in \overset{\circ}{\mathrm{S}}{}^1(\Omega), \tag{6.1.9}$$

where $\overset{\circ}{\mathrm{S}}{}^1(\Omega)$ denotes the closure of smooth solenoidal functions with compact supports in Ω with respect to the norm of $H^1(\Omega, \mathbb{R}^d)$. Henceforth, we denote $H^1(\Omega, \mathbb{R}^d)$ by V and define V_0 as the subspace of V containing the functions with zero traces on Γ (for problems with mixed boundary conditions V_0 contains functions vanishing on the Dirichlet part of the boundary). We recall that the (Friedrichs) inequality

$$\|w\| \le C_{F\Omega}\|\nabla w\| \tag{6.1.10}$$

holds for $w \in V_0$. The set $u_0 + \overset{\circ}{\mathrm{S}}{}^1(\Omega)$ consists of functions $u_0 + w$, where $w \in \overset{\circ}{\mathrm{S}}{}^1(\Omega)$. The space of square summable functions with zero mean is denoted by $\widetilde{L}^2(\Omega)$.

A generalized solution of the stationary Stokes problem (6.1.5)–(6.1.7) is defined by the integral identity

$$\int_\Omega \nu \nabla u : \nabla w \, dx = \int_\Omega f \cdot w \, dx, \qquad \forall w \in \overset{\circ}{\mathrm{S}}{}^1(\Omega). \tag{6.1.11}$$

Remark 6.1. Stokes problem can be stated using the tensor of small strains

$$\varepsilon(u) = \{\varepsilon_{ij}(u)\}, \quad \varepsilon_{ij}(u) := \frac{1}{2}\left(u_{i,j} + u_{j,i}\right),$$

instead of ∇u. In view of the relation $\mathrm{Div}\, \varepsilon(v) = \frac{1}{2}(\Delta v + \nabla \mathrm{div}\, v)$, for solenoidal fields we can write (6.1.11) in terms of the operator ε (instead of ∇). Such a statement is equivalent to (6.1.11) if ν is multiplied by 2.

A generalized solution can also be defined as a function $u \in u_0 + \overset{\circ}{\mathrm{S}}{}^1(\Omega)$ satisfying the relation

$$\int_\Omega \nu \nabla u : \nabla w \, dx = \int_\Omega (f \cdot w + \mathfrak{p}\, \mathrm{div}\, w)\, dx, \qquad w \in V_0, \tag{6.1.12}$$

on a wider set V_0. In this case, the respective $\mathfrak{p}$ is to be defined with the help of (6.1.6).

Inf-Sup condition. First, we recall an important result in the theory of functions related to the operator div.

Lemma 6.2. *Let Ω be a bounded domain with Lipschitz continuous boundary. Then, for any function $f \in \widetilde{L}^2(\Omega)$ one can find a function $w_f \in V_0$ such that* $\mathrm{div}\, w_f = f$ *and*

$$\|\nabla w_f\| \le \mathrm{c}_\Omega \|f\|, \tag{6.1.13}$$

where c_Ω is a positive constant dependent only on Ω.

The reader can find a proof in I. Babuška and A. K. Aziz [22] (for $d = 2$) and O. Ladyzhenskaya and V. Solonnikov [213]. Also, Lemma 6.2 can be viewed as a special case of the *closed range lemma* (e.g., see F. Brezzi and M. Fortin [79] and K. Yosida [374]). Lemma 6.2 implies several important results.

First, it leads to the key condition in the mathematical theory of incompressible materials known in the literature as the Inf-Sup (or *Ladyzhenskaya–I. Babuška–Brezzi* (LBB)) condition. The latter reads: there exists a positive constant $\mathbb{C}_{\mathrm{LBB}}$ such that

$$\inf_{\substack{\mathfrak{q} \in \widetilde{L}^2(\Omega) \\ \mathfrak{q} \neq 0}} \sup_{\substack{w \in V_0 \\ w \neq 0}} \frac{\int_\Omega \mathfrak{q}\, \mathrm{div}\, w \, dx}{\|\mathfrak{q}\| \, \|\nabla w\|} \ge \mathbb{C}_{\mathrm{LBB}}. \tag{6.1.14}$$

In fact, by Lemma 6.2 we know that for any $\mathfrak{q} \in \widetilde{L}^2(\Omega)$ one can find a function $v_{\mathfrak{q}} \in V_0$ satisfying the conditions

$$\operatorname{div} v_{\mathfrak{q}} = \mathfrak{q}, \quad \|\nabla v_{\mathfrak{q}}\| \leq c_\Omega \|\mathfrak{q}\|. \tag{6.1.15}$$

In this case,

$$\sup_{\substack{v \in V_0(\Omega) \\ w \neq 0}} \frac{\int_\Omega \mathfrak{q} \operatorname{div} v \, dx}{\|\nabla v\| \, \|\mathfrak{q}\|} \geq \frac{\int_\Omega \mathfrak{q} \operatorname{div} v_{\mathfrak{q}} \, dx}{\|\nabla v_{\mathfrak{q}}\| \, \|\mathfrak{q}\|} = \frac{\|\mathfrak{q}\|}{\|\nabla v_{\mathfrak{q}}\|} \geq \frac{1}{c_\Omega}$$

and, consequently, (6.1.14) holds with $\mathbb{C}_{\text{LBB}} = (c_\Omega)^{-1}$.

The condition (6.1.14) and its discrete analogs are used for proving the stability and convergence of numerical methods in various problems related to the theory of viscous incompressible fluids. In I. Babuška [21] and F. Brezzi [77], this condition was proved and used to justify the convergence of *mixed* methods, in which a boundary value problem is reduced to a saddle-point problem for a certain Lagrangian. It is worth noting, that conditions similar to (6.1.14) arise in many other problems if they are stated as saddle point problems.

Also, (6.1.14) follows from the Nečas inequality

$$\|\mathfrak{q}\| \leq c_\Omega \|\nabla \mathfrak{q}\|_{-1,\Omega} := \sup_{\substack{w \in V_0 \\ w \neq 0}} \frac{\int_\Omega \mathfrak{q} \operatorname{div} w \, dx}{\|\nabla w\|}, \qquad \forall \, \mathfrak{q} \in \widetilde{L}^2(\Omega). \tag{6.1.16}$$

A simple proof of the Inf-Sup condition for domains with Lipschitz boundaries can be found in J. Bramble [71]. Estimates of the value of $\mathbb{C}_{\text{LBB}}$ for various domains are discussed in, e.g., E. Chizhonkov and M. Olshanskii [105], M. Dobrowolski [116], and L. Halpern [165].

Saddle-point statement, the existence of a solution, and energy estimates. With the help of Lemma 6.2, it is not difficult to prove the existence of u, $\mathfrak{p}$, and σ that deliver a solution to the problem (6.1.5)–(6.1.7). For this purpose, we use general theorems in convex analysis concerning saddle-points of Lagrangians. Consider the Lagrangian $L : (u_0 + V_0) \times \widetilde{L}^2(\Omega) \to \mathbb{R}$ of the form

$$L(v, \mathfrak{q}) := \int_\Omega \left(\frac{\nu}{2} |\nabla v|^2 - \mathfrak{q} \operatorname{div} v \right) dx - \int_\Omega f \cdot v \, dx$$

and the saddle point problem

$$L(u, \mathfrak{q}) \leq L(u, \mathfrak{p}) \leq L(v, \mathfrak{p}), \qquad \forall v \in u_0 + V_0, \; \mathfrak{q} \in \widetilde{L}^2(\Omega). \tag{6.1.17}$$

It is not difficult to verify that a saddle point is formed by the velocity field u and the pressure function $\mathfrak{p}$ satisfying (6.1.5)–(6.1.7).

Indeed, the left-hand side of (6.1.16) reads

$$\int_\Omega (\mathfrak{q} - \mathfrak{p})\operatorname{div} u\, dx = 0, \qquad \forall\, \mathfrak{q} \in \widetilde{L}^2(\Omega). \tag{6.1.18}$$

Hence, $\operatorname{div} u$ is orthogonal to any function with zero mean. However,

$$\int_\Omega \operatorname{div} u\, dx = -\int_\Gamma u \cdot n\, ds = -\int_\Gamma u_0 \cdot n\, ds = \int_\Omega \operatorname{div} u_0\, dx = 0,$$

so that $\{\operatorname{div} u\}_\Omega = 0$. By (6.1.18), we now conclude that $\operatorname{div} u = 0$.

Set $v = u \pm \alpha w$, where $\alpha > 0$ and $w \in V_0$. Then, the right-hand side of (6.1.17) means that

$$\pm \int_\Omega (\nu \nabla u : \nabla w - \mathfrak{p} \operatorname{div} w - f \cdot w)\, dx \le \frac{\alpha\nu}{2} \int_\Omega |\nabla w|^2\, dx.$$

Since α can be taken arbitrarily small the above inequalities imply

$$\int_\Omega (\nu \nabla u : \nabla w - \mathfrak{p} \operatorname{div} w)\, dx = \int_\Omega f \cdot w\, dx, \qquad \forall w \in V_0. \tag{6.1.19}$$

If $w \in \mathring{S}^1(\Omega)$, then (6.1.11) follows from (6.1.19).

The saddle point problem (6.1.17) is equivalent to two variational problems

$$(\mathcal{P}_u) \quad \inf_{v \in u_0 + V_0} \sup_{\mathfrak{q} \in \widetilde{L}^2(\Omega)} L(v, \mathfrak{q}) \qquad \text{and} \qquad (\mathcal{P}_\mathfrak{p}) \quad \sup_{\mathfrak{q} \in \widetilde{L}^2(\Omega)} \inf_{v \in u_0 + V_0} L(v, \mathfrak{q}).$$

Since $\{\operatorname{div} v\}_\Omega = 0$ for any $v \in u_0 + V_0$ and

$$\sup_{\mathfrak{q} \in \widetilde{L}^2(\Omega)} \int_\Omega \mathfrak{q} \operatorname{div} v\, dx = \sup_{\mathfrak{q} \in L^2(\Omega)} \int_\Omega \mathfrak{q} \operatorname{div} v\, dx = \begin{cases} 0 & \text{if } \operatorname{div} v = 0, \\ +\infty & \text{if } \operatorname{div} v \neq 0, \end{cases}$$

we note that

$$\inf_{v \in u_0 + V_0} \sup_{\mathfrak{q} \in \widetilde{L}^2(\Omega)} L(v, \mathfrak{q}) = \inf_{v \in u_0 + \mathring{S}^1(\Omega)} I(v) = I(u),$$

where

$$I(w) = \int_\Omega \left(\frac{1}{2} |\nabla w|^2 - f \cdot w \right) dx. \tag{6.1.20}$$

The Problem $\mathcal{P}_\mathfrak{p}$ determines the pressure field, however the functional of this problem cannot be represented in explicit form.

The existence of u and $\mathfrak{p}$ follow from Lemma 6.2 and known theorems in the theory of saddle-points. Obviously, L is convex and continuous with respect to the first variable. Moreover, it is linear and continuous with respect to the second one. Therefore (e.g., see I. Ekeland and R. Temam [121], Chapter 4, §2), it suffices to show that

$$\exists \mathfrak{q}_* \in \widetilde{L}^2(\Omega) \quad \text{such that} \quad \lim_{\|v\| \to +\infty} L(v, \mathfrak{q}_*) = +\infty \tag{6.1.21}$$

and

$$\lim_{\|\mathrm{q}\|\to+\infty} \inf_{v\in u_0+V_0} L(v,\mathrm{q}) = -\infty. \tag{6.1.22}$$

Set $\mathrm{q}_* = 0$, then (6.1.21) is satisfied.

To prove (6.1.22), we use Lemma 6.2 and select $w_\mathrm{q} \in V_0$ such that

$$\operatorname{div} w_\mathrm{q} = \mathrm{q} \quad \text{and} \quad \|\nabla w_\mathrm{q}\| \le \mathrm{c}_\Omega \|\mathrm{q}\|. \tag{6.1.23}$$

Then

$$\begin{aligned}\inf_{v\in u_0+V_0} L(v,\mathrm{q}) &\le L(u_0+\lambda w_\mathrm{q},\mathrm{q})\\ &= \int_\Omega \left(\frac{\nu}{2}\lambda^2|\nabla w_\mathrm{q}|^2 - \lambda|\mathrm{q}|^2\right) dx + \int_\Omega \lambda\nu\nabla w_q : \nabla u_0\, dx\\ &\quad - \lambda\int_\Omega f\cdot w_\mathrm{q}\, dx + \mathcal{E}(u_0),\end{aligned}$$

where

$$\mathcal{E}(u_0) = \int_\Omega \left(\frac{\nu}{2}|\nabla(u_0)|^2 - f\cdot u_0\right) dx.$$

By (6.1.23), we conclude that

$$\left|\int_\Omega f\cdot w_\mathrm{q}\, dx\right| \le C_{F\Omega}\,\mathrm{c}_\Omega\|\mathrm{q}\|\,\|f\|$$

and

$$\int_\Omega \frac{\nu}{2}\lambda^2|\nabla w_\mathrm{q}|^2\, dx \le \int_\Omega \frac{\nu \mathrm{c}_\Omega^2}{2}\lambda^2|\mathrm{q}|^2\, dx.$$

Hence, we find that

$$\begin{aligned}\inf_{v\in u_0+V_0} L(v,\mathrm{q}) &\le \int_\Omega \left(\frac{\nu}{2}\lambda^2\mathrm{c}_\Omega^2 - \lambda\right)|\mathrm{q}|^2\, dx\\ &\quad + \lambda\mathrm{c}_\Omega\|q\|(\nu\|\nabla u_0\| + C_{F\Omega}\|f\|) + \mathcal{E}(u_0).\end{aligned}$$

Set $\lambda = \dfrac{1}{\nu c_\Omega^2}$. Then, we note that

$$\inf_{w\in V_0} L(w,\mathrm{q}) \le -\frac{\|\mathrm{q}\|^2}{2\nu\mathrm{c}_\Omega^2} + \mathcal{E}(u_0) \to -\infty \text{ as } \|\mathrm{q}\| \to +\infty.$$

Thus, (6.1.22) holds and the saddle point problem (6.1.16) has a solution.

Now we establish energy estimates for the velocity and pressure functions. Set $w = u - u_0$ in (6.1.11). We have the relation

$$\nu\|\nabla u\|^2 = \int_\Omega (\nu\nabla u : \nabla u_0 + f \cdot (u - u_0))\, dx$$
$$\leq (\nu\|\nabla u_0\| + C_{F\Omega}\|f\|)\|\nabla u\| + C_{F\Omega}\|\nabla u_0\|\|f\|.$$

From here, it follows that the velocity norm is bounded by $C(\|\nabla u_0\| + \|f\|)$, where C depends only on the given data. If $u_0 = 0$, then the energy estimate has the simplest form

$$\nu\|\nabla u\| \leq C_{F\Omega}\|f\|. \tag{6.1.24}$$

It is not difficult to show that a similar estimate holds for the pressure field $\mathfrak{p}$. Let $v_\mathfrak{p} \in V_0$ be the function defined as a counterpart of $\mathfrak{p}$ in Lemma 6.2, i.e.,

$$\operatorname{div} v_\mathfrak{p} = \mathfrak{p} \quad \text{and} \quad \|\nabla v_\mathfrak{p}\| \leq c_\Omega\|\mathfrak{p}\|.$$

Then, by (6.1.19), we have

$$\int_\Omega (\nu\nabla u : \nabla v_\mathfrak{p} - f \cdot v_\mathfrak{p})\, dx = \int_\Omega \mathfrak{p} \operatorname{div} v_\mathfrak{p}\, dx = \|\mathfrak{p}\|^2. \tag{6.1.25}$$

Hence,

$$\|\mathfrak{p}\|^2 \leq (\nu\|\nabla u\| + C_{F\Omega}\|f\|)\|\nabla v_\mathfrak{p}\|,$$

and we obtain

$$\|\mathfrak{p}\| \leq c_\Omega(C_{F\Omega}\|f\| + \nu\|\nabla u\|). \tag{6.1.26}$$

Since $\|\nabla u\|$ is bounded (by the norms of given data), we note that the same is true for $\|\mathfrak{p}\|$, so that the saddle point problem is stable with respect to both components.

Estimates of the distance to the set $\overset{\circ}{\mathrm{S}}{}^1(\Omega)$. Approximations computed by a numerical procedure may not belong to the space $\overset{\circ}{\mathrm{S}}{}^1(\Omega)$. Lemma 6.2 allows us to estimate the distance between such an approximation and the set of solenoidal fields. Subsequently, we will use such estimates in the derivation of a posteriori estimates valid for nonsolenoidal approximations.

First, we note that an estimate of the distance in L^2-norm follows from Lemma 3.2. However, in the case of flow problems we need an estimate in a stronger norm given by the lemma below.

Lemma 6.3. *For any function $\widehat{v} \in V_0$, there exists a function $v_0 \in \overset{\circ}{\mathrm{S}}{}^1(\Omega)$ such that*

$$\|\nabla(\widehat{v} - v_0)\| \leq c_\Omega\|\operatorname{div} \widehat{v}\|. \tag{6.1.27}$$

Proof. Let $f = \operatorname{div}\widehat{v}$. By Lemma 6.2, we find a function $w_f \in V_0$ such that

$$\operatorname{div} w_f = f$$

and

$$\|\nabla w_f\| \leq c_\Omega \|\operatorname{div}\widehat{v}\|.$$

Since $\operatorname{div}(\widehat{v} - w_f) = 0$, the function $v_0 := \widehat{v} - w_f$ belongs to $\overset{\circ}{\mathrm{S}}{}^1(\Omega)$. Moreover,

$$\|\nabla(\widehat{v} - v_0)\| = \|\nabla w_f\| \leq c_\Omega \|\operatorname{div}\widehat{v}\|$$

and the estimate (6.1.27) follows. □

In other words, the distance between $\widehat{v} \in V_0$ and the set of solenoidal fields $\overset{\circ}{\mathrm{S}}{}^1(\Omega)$ is estimated from above by the quantity $\|\operatorname{div}\widehat{v}\|$ with the multiplier c_Ω that comes from Lemma 6.2.

Corollary 6.4. *Let $u_\phi \in V$ be a vector-valued function such that*

$$\operatorname{div} u_\phi = \phi \in L^2(\Omega).$$

From Lemma 6.3 it follows that for any function $\widehat{v} \in u_\phi + V_0$ there exists a function $v_\phi \in u_\phi + \overset{\circ}{\mathrm{S}}{}^1(\Omega)$ satisfying the estimate

$$\|\nabla(\widehat{v} - v_\phi)\| \leq c_\Omega \|\operatorname{div}\widehat{v} - \phi\|. \tag{6.1.28}$$

Proof. Since $\widehat{v} - u_\phi \in V_0$, we can find a function $v_0 \in \overset{\circ}{\mathrm{S}}{}^1(\Omega)$ such that

$$\|\nabla(\widehat{v} - u_\phi - v_0)\| \leq c_\Omega \|\operatorname{div}(\widehat{v} - u_\phi)\| = c_\Omega \|\operatorname{div}\widehat{v} - \phi\|.$$

Hence, the function $v_\phi = u_\phi + v_0$ belongs to $u_\phi + \overset{\circ}{\mathrm{S}}{}^1(\Omega)$ and satisfies (6.1.28). □

6.2 A posteriori estimates for the stationary Stokes problem

In this section, we derive functional a posteriori estimates for the stationary Stokes problem. For the sake of simplicity, we assume that ν is constant. Estimates for problems with variable viscosity are considered in Section 6.3.

6.2.1 Estimates for the velocity field

Let $v \in u_0 + \overset{\circ}{\mathrm{S}}{}^1(\Omega)$. Then (6.1.8) implies the relation

$$\int_\Omega \nu\nabla(u - v) : \nabla w \, dx = \int_\Omega (f \cdot w - \nu\nabla v : \nabla w) \, dx, \qquad \forall w \in \overset{\circ}{\mathrm{S}}{}^1(\Omega). \tag{6.2.1}$$

For any tensor-valued function $\tau \in \Sigma(\Omega) = L^2(\Omega, \mathbb{M}^{d\times d})$, the functional

$$\mathcal{F}_{\tau,f}(w) := \int_\Omega (f \cdot w - \tau : \nabla w)\, dx$$

is linear and continuous on V_0, and its norm can be characterized by the quantity

$$| \mathcal{F}_{\tau,f} | := \sup_{w \in V_0,\, w \neq 0} \frac{\left|\int_\Omega (f \cdot w - \tau : \nabla w)\, dx\right|}{\|\nabla w\|}.$$

The set

$$Q_{\tau,f}(\Omega) := \left\{ \tau \in \Sigma(\Omega) \,\middle|\, \int_\Omega \tau : \nabla w \, dx = \int_\Omega f \cdot w \, dx, \;\; \forall w \in V_0 \right\}$$

defines the kernel of $\mathcal{F}_{\tau,f}$. It contains the tensor-valued functions that satisfy (in a generalized sense) the equilibrium equation $\operatorname{Div} \tau + f = 0$.

Represent the integral identity (6.2.1) in the form

$$\int_\Omega \nu \nabla(u - v) : \nabla w \, dx = \mathcal{F}_{\tau,f}(w) + \int_\Omega (\tau - \nu \nabla v) : \nabla w \, dx.$$

Since $|\mathcal{F}_{\tau,f}(w)| \leq | \mathcal{F}_{\tau,f} | \|\nabla w\|$ and

$$\int_\Omega (\tau - \nu \nabla v) : \nabla w \, dx = \int_\Omega (\tau + \mathrm{q}\,\mathbb{I} - \nu \nabla v) : \nabla w \, dx \leq \|\tau + \mathrm{q}\,\mathbb{I} - \nu \nabla v\| \|\nabla w\|,$$

we set $w = u - v$ and arrive at the estimate

$$\nu \|\nabla(u - v)\| \leq | \mathcal{F}_{\tau,f} | + \|\tau + \mathbb{I}\mathrm{q} - \nu \nabla v\|, \tag{6.2.2}$$

where q is an arbitrary function in $\widetilde{L}^2(\Omega)$ and τ is an arbitrary tensor-valued function in $\Sigma(\Omega)$. If $\tau \in H(\Omega, \operatorname{Div}) \cap \Sigma_s$, then

$$\int_\Omega (f \cdot w - \tau : \nabla w)\, dx = \int_\Omega (f + \operatorname{Div} \tau) \cdot w \, dx$$

and we find that

$$| \mathcal{F}_{\tau,f} | \leq C_{F\Omega} \|f + \operatorname{Div} \tau\|,$$

where $C_{F\Omega}$ is the constant in (6.1.10).

Now, the right-hand side (6.2.2) is presented by directly computable integrals, namely,

$$\nu \|\nabla(u - v)\| \leq \overline{\mathfrak{M}}_{\mathrm{ST1}}(v, \tau, \mathrm{q}) := \|\tau + \mathrm{q}\,\mathbb{I} - \nu \nabla v\| + C_{F\Omega} \|\operatorname{Div} \tau + f\|. \tag{6.2.3}$$

If $q \in H^1(\Omega)$, then a somewhat different form of the estimate follows by changing τ to η, where

$$\tau = \eta - \mathfrak{q}\,\mathbb{I}, \tag{6.2.4}$$

which gives

$$\nu\|\nabla(u-v)\| \leq \overline{\mathfrak{M}}_{\text{ST2}}(v,\eta,\mathfrak{q}) := \|\eta - \nu\nabla v\| + C_{F\Omega}\|\operatorname{Div}\eta + f - \nabla\mathfrak{q}\|. \tag{6.2.5}$$

Estimates (6.2.3) and (6.2.5) have a clear meaning. Estimate (6.2.3) shows that the upper bound of the error can be represented as the sum of two parts related to the decomposition of the Stokes system as

$$\begin{aligned} \sigma &= -\mathfrak{p}\,\mathbb{I} + \nu\nabla u, \\ -\operatorname{Div}\sigma &= f. \end{aligned}$$

Its right-hand side vanishes if and only the above relations are exactly satisfied. Since v is a solenoidal field satisfying the boundary condition, the right-hand side of the majorant is zero if and only if $v = u$. Similarly, (6.2.5) shows that the upper bound of the error can be represented as the sum of two parts related to the decomposition of the Stokes system as

$$\begin{aligned} \bar{\sigma} &= \nu\nabla u, \\ -\operatorname{Div}\bar{\sigma} &= f - \nabla\mathfrak{p}. \end{aligned}$$

We can also deduce other equivalent forms of (6.2.3) and (6.2.5). Squaring both parts of (6.2.3), we obtain an estimate the right-hand side of which is given by a quadratic functional:

$$\begin{aligned} \nu^2\|\nabla(u-v)\|^2 &\leq \overline{\mathfrak{M}}^2_{\text{ST1}}(v,y,\mathfrak{q},\beta) \\ &:= (1+\beta)\|\tau + \mathfrak{q}\,\mathbb{I} - \nu\nabla v\|^2 + \frac{(1+\beta)C^2_{F\Omega}}{\beta}\|\operatorname{Div}\tau + f\|^2, \end{aligned} \tag{6.2.6}$$

where β is an arbitrary positive number. Rearrange the first term on the right-hand side by (1.4.2)–(1.4.2). Since $\operatorname{div} v = \operatorname{tr}\nabla v = 0$, we have

$$\begin{aligned} \|\tau + \mathfrak{q}\,\mathbb{I} - \nu\nabla v\|^2 &= \|\frac{1}{d}(\operatorname{tr}\tau + qd)\,\mathbb{I} + \tau^{\mathrm{D}} - \nu\nabla v\|^2 \\ &= \int_\Omega \left(\frac{1}{d}(\operatorname{tr}\tau + d\mathfrak{q})^2 + |\tau^{\mathrm{D}} - \nu\nabla v|^2\right) dx. \end{aligned} \tag{6.2.7}$$

If τ is selected in such a way that

$$\{\operatorname{tr}\tau\}_\Omega = 0, \tag{6.2.8}$$

then we can set $\mathrm{q} = -\frac{1}{d}(\mathrm{tr}\tau)$ and obtain

$$\nu^2\|\nabla(u-v)\|^2 = (1+\beta)\|\tau^{\mathrm{D}} - \nu\nabla v\|^2 + \frac{(1+\beta)C_{F\Omega}^2}{\beta}\|\mathrm{Div}\,\tau + f\|^2. \quad (6.2.9)$$

Note that the right-hand side of (6.2.9) does not contain q. It vanishes if

$$\begin{aligned}\tau^{\mathrm{D}} - \nu\nabla v &= 0,\\ \mathrm{Div}\,\tau + f &= 0.\end{aligned}$$

Since $\mathrm{div}\, v = 0$, we have

$$\mathrm{tr}(\tau + \mathrm{q}\,\mathbb{I} - \nu(\nabla v)) = \mathrm{tr}\,\tau + d\mathrm{q} = 0,$$

and, therefore, the constitutive relation in terms of traces is also satisfied. Thus, we conclude that this majorant also vanishes if only if v, τ, and q coincide with the exact solutions.

Now consider the case, in which an approximate solution may not exactly satisfy the divergence-free condition. We mark such approximations by hats. Assume that $\widehat{v} \in u_0 + V_0$ and $\mathrm{div}\,\widehat{v}$ may be not equal to zero. In this case, the estimate of its deviation from u can be obtained by the following arguments.

First, by Corollary 6.4, we know that for $\widehat{v}$ one can find a function $w \in u_\phi + \overset{\circ}{\mathrm{S}}{}^1(\Omega)$ such that

$$\|\nabla(\widehat{v} - w)\| \leq \mathrm{c}_\Omega\|\mathrm{div}\,\widehat{v}\|. \quad (6.2.10)$$

Therefore, we have

$$\nu\|\nabla(u-\widehat{v})\| \leq \nu\|\nabla(u-v)\| + \nu\|\nabla(\widehat{v}-v)\|.$$

Use (6.2.3) to estimate the first norm on the right-hand side of this inequality. We arrive at the estimate

$$\begin{aligned}\nu\|\nabla(u-\widehat{v})\| &\leq \|\tau + \mathrm{q}\,\mathbb{I} - \nu\nabla v\| + C_{F\Omega}\|\mathrm{Div}\,\tau + f\| + \nu\|\nabla(\widehat{v}-v)\|\\ &\leq \|\tau + \mathrm{q}\,\mathbb{I} - \nu\nabla\widehat{v}\| + C_{F\Omega}\|\mathrm{Div}\,\tau + f\| + 2\nu\|\nabla(\widehat{v}-v)\|.\end{aligned}$$

In view of (6.2.10), we find that

$$\nu\|\nabla(u-\widehat{v})\| \leq \|\tau + \mathrm{q}\,\mathbb{I} - \nu\nabla\widehat{v}\| + C_{F\Omega}\|\mathrm{Div}\,\tau + f\| + 2\nu\mathrm{c}_\Omega\|\mathrm{div}\,\widehat{v}\|, \quad (6.2.11)$$

where $\tau \in H(\Omega, \mathrm{Div})$ and $\mathrm{q} \in \widetilde{L}^2(\Omega)$.

If $\{\mathrm{tr}\tau\}_\Omega = 0$, then we can set $\mathrm{q} = -\frac{1}{d}(\mathrm{tr}\tau)$ and obtain

$$\nu\,\|\nabla(u-v)\| = \|\tau^{\mathrm{D}} - \nu\,(\nabla v)^{\mathrm{D}}\| + C_{F\Omega}\|\mathrm{Div}\,\tau + f\| + 2\nu\mathrm{c}_\Omega\|\mathrm{div}\,\widehat{v}\|. \quad (6.2.12)$$

Thus, if the constants $C_{F\Omega}$ and c_Ω are known (or we know suitable upper bounds for them), then (6.2.11) provides a way for evaluating the deviation of $\widehat{v}$ from u. For this purpose, we should select certain finite-dimensional subspaces Σ_k and Q_k for the functions τ (or η) and q, respectively. The minimization of the right-hand side of (6.2.11) with respect to τ and q provides an estimate of the deviation, which will be the sharper the greater is the dimension of the subspaces used.

6.2.2 Estimates for pressure

Estimates of $\|\mathfrak{p} - \mathfrak{q}\|$ can also be derived with the help of Lemma 6.2. Since $(\mathfrak{p} - \mathfrak{q}) \in \widetilde{L}^2(\Omega)$, we know that

$$\operatorname{div} \widetilde{w} = \mathfrak{p} - \mathfrak{q} \quad \text{and} \quad \|\nabla \widetilde{w}\| \leq c_\Omega \|\mathfrak{p} - \mathfrak{q}\|$$

for a certain vector-valued function $\widetilde{w} \in V_0$. Hence,

$$\|\mathfrak{p} - \mathfrak{q}\|^2 = \int_\Omega \operatorname{div} \widetilde{w}(\mathfrak{p} - \mathfrak{q})\, dx.$$

Recall that the exact solution u satisfies (6.1.18) and, therefore,

$$\begin{aligned}\int_\Omega (\mathfrak{p} - \mathfrak{q}) \operatorname{div} \widetilde{w}\, dx &= \int_\Omega (\nu \nabla u : \nabla \widetilde{w} - f \cdot \widetilde{w} - \mathfrak{q} \operatorname{div} \widetilde{w})\, dx \\ &= \int_\Omega \nu \nabla (u - \widehat{v}) : \nabla \widetilde{w}\, dx \\ &\quad + \int_\Omega (\nu \nabla \widehat{v} : \nabla \widetilde{w} + f \cdot \widetilde{w} - \mathfrak{q} \operatorname{div} \widetilde{w})\, dx.\end{aligned}$$

We have

$$\int_\Omega \nu \nabla (u - \widehat{v}) : \nabla \widetilde{w}\, dx \leq c_\Omega \nu \|\nabla (u - \widehat{v})\| \, \|\mathfrak{p} - \mathfrak{q}\|$$

and

$$\begin{aligned}&\int_\Omega (\nu \nabla \widehat{v} : \nabla \widetilde{w} + f \cdot \widetilde{w} - \mathfrak{q} \operatorname{div} \widetilde{w})\, dx \\ &\quad = \int_\Omega (\nu \nabla \widehat{v} - \tau - \mathfrak{q}\, \mathbb{I}) : \nabla \widetilde{w}\, dx - \int_\Omega (\operatorname{Div} \tau + f) \cdot \widetilde{w}\, dx \\ &\quad \leq \Big(\|\nu \nabla \widehat{v} - \tau - \mathfrak{q}\, \mathbb{I}\| + \nu C_{F\Omega} \|\operatorname{Div} \tau + f\| \Big) c_\Omega \|\mathfrak{p} - \mathfrak{q}\|.\end{aligned}$$

Therefore,

$$\begin{aligned}\|\mathfrak{p} - \mathfrak{q}\| &\leq c_\Omega \Big(\nu \|\nabla (u - \widehat{v})\| + \|\nu \nabla \widehat{v} - \tau - \mathfrak{q}\, \mathbb{I}\| + C_{F\Omega} \|\operatorname{Div} \tau + f\| \Big) \\ &\leq 2 c_\Omega \Big(\|\nu \nabla \widehat{v} - \tau - \mathfrak{q}\, \mathbb{I}\| + C_{F\Omega} \|\operatorname{Div} \tau + f\| + \nu c_\Omega \|\operatorname{div} \widehat{v}\| \Big),\end{aligned}$$

and we arrive at the estimate

$$\frac{1}{2c_\Omega} \|\mathfrak{p} - \mathfrak{q}\| \leq \|\nu \nabla \widehat{v} - \tau - \mathfrak{q}\, \mathbb{I}\| + C_{F\Omega} \|\operatorname{Div} \tau + f\| + \nu c_\Omega \|\operatorname{div} \widehat{v}\|. \quad (6.2.13)$$

It is easy to note that the right-hand side of (6.2.13) consists of the same terms as the right-hand side of (6.2.3) and vanishes if and only if,

$$\widehat{v} = u, \quad \tau = \sigma, \quad \text{and} \quad \mathfrak{p} = \mathfrak{q}.$$

However, in this case, the dependence of the penalty multipliers on the constant c_Ω is stronger.

6.2.3 Estimates for stresses

Let $\tau \in \Sigma(\Omega)$ be an approximation of σ. We have

$$\|\tau - \sigma\| = \|\tau + \mathfrak{p}\mathbb{I} - \nu\nabla u\| \\ \leq \|\tau + \mathfrak{q}\,\mathbb{I} - \nu\nabla\widehat{v}\| + \nu\|\nabla(\widehat{v} - u)\| + \sqrt{d}\,\|\mathfrak{p} - \mathfrak{q}\|. \tag{6.2.14}$$

By (6.2.11) and (6.2.13) we conclude that

$$\|\tau - \sigma\| \leq 2(1 + \sqrt{d}\mathrm{c}_\Omega)\|\tau + \mathfrak{q}\,\mathbb{I} - \nu\nabla\widehat{v}\| + C_{F\Omega}(1 + 2\sqrt{d}\mathrm{c}_\Omega)\|\mathrm{Div}\,\tau + f\| \\ + 2\nu\mathrm{c}_\Omega(1 + \sqrt{d}\mathrm{c}_\Omega)\|\mathrm{div}\,\widehat{v}\|. \tag{6.2.15}$$

Since

$$\|\tau - \sigma\|^2_{\mathrm{Div}} = \|\tau - \sigma\|^2 + \|\mathrm{Div}\,\tau + f\|^2,$$

it is not difficult to estimate the deviation $\tau - \sigma$ in the norm of $H(\Omega, \mathrm{Div})$. However, the estimate has a more symmetric form if the deviation is expressed in terms of the norm

$$]\![\,\eta\,]\!\![_{\mathrm{Div}} := \|\eta\| + C_{F\Omega}\|\mathrm{Div}\,\eta\|.$$

In this case,

$$\bar{c}\;]\![\tau - \sigma]\!\![_{\mathrm{Div}} \leq \|\tau + \mathfrak{q}\,\mathbb{I} - \nu\nabla\widehat{v}\| + C_{F\Omega}\|\mathrm{Div}\,\tau + f\| + \nu\mathrm{c}_\Omega\|\mathrm{div}\,\widehat{v}\|, \tag{6.2.16}$$

where $\bar{c} = \dfrac{1}{2(1 + \sqrt{d}\mathrm{c}_\Omega)}$.

6.2.4 Estimates in combined norms

We can measure errors in terms of combined norms of the product space

$$W := (u_0 + \overset{\circ}{\mathrm{S}}{}^1(\Omega)) \times H(\Omega, \mathrm{Div}) \times \widetilde{L}^2(\Omega),$$

for which we introduce two equivalent norms

$$\|(v, \tau, \mathfrak{q})\|_W := \nu\|\nabla v\| + \|\tau\|_{\mathrm{Div}} + \sqrt{d}\,\|\mathfrak{q}\|,$$
$$]\![(v, \tau, \mathfrak{q})]\!\![_W := \nu\|\nabla v\| +]\![\tau]\!\![_{\mathrm{Div}g} + \sqrt{d}\,\|\mathfrak{q}\|.$$

It is easy to see that

$$\gamma_1\|(v, \tau, \mathfrak{q})\|_W \leq]\![(v, \tau, \mathfrak{q})]\!\![_W \leq \gamma_2\|(v, \tau, \mathfrak{q})\|_W, \tag{6.2.17}$$

where $\gamma_1 = \min\{1, C_{F\Omega}\}$ and $\gamma_2 = \max\{1, C_{F\Omega}\}$.

We can show that the majorant $\overline{\mathfrak{M}}_{\text{ST1}}(v, \tau, \mathfrak{q})$ is *equivalent to the error* in the combined norm $]\![(u - v, \sigma - \tau, \mathfrak{p} - \mathfrak{q})]\![_W$.

We have

$$
\begin{aligned}
]\![\sigma - t]\![_{\text{Div}} &=]\![\nu \nabla u - \tau - \mathfrak{p}\,\mathbb{I}]\![_{\text{Div}} \\
&\leq \nu \|\nabla(u - v)\| + \|\nu \nabla v - \tau - \mathfrak{p}\,\mathbb{I}\| + C_{F\Omega}\|\text{Div}\,\tau + f\| \\
&\leq \nu \|\nabla(u - v)\| + \|\nu \nabla v - \tau - \mathfrak{q}\,\mathbb{I}\| + \sqrt{d}\,\|\mathfrak{p} - \mathfrak{q}\| + C_{F\Omega}\|\text{Div}\,\tau + f\|.
\end{aligned}
$$

Therefore,

$$
\begin{aligned}
]\![(u - v, \sigma - \tau, \mathfrak{p} - \mathfrak{q})]\![_W &:= \nu \|\nabla(u - v)\| +]\![\sigma - \tau]\![_{\text{Div}} + \sqrt{d}\,\|\mathfrak{p} - \mathfrak{q}\| \\
&\leq 2\nu \|\nabla(u - v)\| + \|\nu \nabla v - \tau - \mathfrak{q}\,\mathbb{I}\| \\
&\quad + 2\sqrt{d}\,\|\mathfrak{p} - \mathfrak{q}\| + C_{F\Omega}\|\text{Div}\,\tau + f\|.
\end{aligned}
$$

Since $v \in u_0 + \overset{\circ}{\mathrm{S}}{}^1(\Omega)$, we can use (6.2.3) and (6.2.13) to estimate the terms $\|\nabla(u - v)\|$ and $\|\mathfrak{p} - \mathfrak{q}\|$. We obtain

$$
\begin{aligned}
]\![(u - v, \sigma - \tau, \mathfrak{p} - \mathfrak{q})]\![_W &\leq C_{\oplus}\big(\|\nu \nabla v - \tau - \mathfrak{q}\,\mathbb{I}\| + C_{F\Omega}\|\text{Div}\,\tau + f\|\big) \\
&= C_{\oplus}\overline{\mathfrak{M}}_{\text{ST1}}(v, \tau, \mathfrak{q}), \qquad (6.2.18)
\end{aligned}
$$

where $C_{\oplus} = 3 + 4\sqrt{d}\,c_{\Omega}$. On the other hand,

$$
\overline{\mathfrak{M}}_{\text{ST1}}(v, \tau, \mathfrak{q}) \leq \nu \|\nabla(v - u)\| +]\![\sigma - \tau]\![_{\text{Div}} + \sqrt{d}\,\|\mathfrak{p} - \mathfrak{q}\|. \qquad (6.2.19)
$$

Thus, we find that

$$
\overline{\mathfrak{M}}_{\text{ST1}}(v, \tau, q) \leq]\![(u - v, \sigma - \tau, \mathfrak{p} - \mathfrak{q})]\![_W \leq C_{\oplus}\overline{\mathfrak{M}}_{\text{ST1}}(v, \tau, \mathfrak{q}). \qquad (6.2.20)
$$

In view of (6.2.17), the majorant $\overline{\mathfrak{M}}_{\text{ST1}}(v, \tau, q)$ is also equivalent to the combined norm $\|(u - v, \sigma - \tau, \mathfrak{p} - \mathfrak{q})\|_W$.

Remark 6.5. It is easy to show that the majorant $\overline{\mathfrak{M}}_{\text{ST2}}(v, \eta, \mathfrak{q})$ is also equivalent to the combined error norm. Moreover, one can prove that on a wider set

$$
W := (u_0 + V_0) \times H(\Omega, \text{Div}) \times \widetilde{L}^2(\Omega),
$$

the majorant

$$
\overline{\mathfrak{M}}_{\widehat{ST1}}(\widehat{v}, \tau, q) := \|\tau + \mathfrak{q}\,\mathbb{I} - \nu \nabla \widehat{v}\| + C_{F\Omega}\|\text{Div}\,\tau + f\| + 2\nu c_{\Omega}\|\text{div}\,\widehat{v}\|
$$

is also equivalent to the combined error norm $]\![(u - v, \sigma - \tau, \mathfrak{p} - \mathfrak{q})]\![_W$.

6.2.5 Lower bounds of errors

A lower bound of $\|\nabla(u - v)\|$ is derived by the same arguments that we have used in Section 4.1. For $\widehat{v} \in u_0 + V_0$, we have

$$\begin{aligned}\frac{\nu}{2}\|\nabla(u-\widehat{v})\|^2 &= \int_\Omega \Big(\nu\nabla(u-\widehat{v}) : \nabla(u-\widehat{v}) - \frac{\nu}{2}\nabla(u-\widehat{v}) : \nabla(u-\widehat{v})\Big)\,dx \\ &\le \sup_{\widehat{w}\in V_0} \int_\Omega \Big(\nu\nabla(u-\widehat{v}) : \nabla\widehat{w} - \frac{\nu}{2}\nabla\widehat{w} : \nabla\widehat{w}\Big)\,dx \\ &\le \sup_{\tau\in\Sigma(\Omega)} \int_\Omega \Big(\nu\nabla(u-\widehat{v}) : \tau - \frac{\nu}{2}\tau : \tau\Big)\,dx = \frac{\nu}{2}\|\nabla(u-\widehat{v})\|^2.\end{aligned}$$

Thus,

$$\nu\|\nabla(u-\widehat{v})\|^2 = \sup_{\widehat{w}\in V_0} \int_\Omega (2\nu\nabla(u-\widehat{v}) : \nabla\widehat{w} - \nu\nabla\widehat{w} : \nabla\widehat{w})\,dx.$$

By (6.1.12), we reform the right-hand side of this relation and obtain

$$\nu\|\nabla(u-\widehat{v})\|^2 = \sup_{\widehat{w}\in V_0} \{-G(\widehat{v}, \widehat{w}, \mathfrak{p})\},$$

where G is a quadratic functional defined by the relation

$$G(\widehat{v}, \widehat{w}, \mathfrak{p}) = \int_\Omega \big(2\nu\nabla\widehat{v} : \nabla\widehat{w} + \nu|\nabla\widehat{w}|^2 - 2f\cdot\widehat{w} - 2\mathfrak{p}\,\mathrm{div}\,\widehat{w}\big)\,dx.$$

Thus, for any $\widehat{w} \in V_0$

$$\nu\|\nabla(u-\widehat{v})\|^2 \ge -G(\widehat{v}, \widehat{w}, \mathfrak{q}) - 2\|\mathfrak{p} - \mathfrak{q}\|\|\mathrm{div}\,\widehat{w}\|.$$

We estimate the last term by (6.2.13) and arrive at the estimate

$$\begin{aligned}\nu\|\nabla(u-\widehat{v})\|^2 &\ge \underline{\mathfrak{M}}_{\widehat{\mathrm{ST}}}(\widehat{v}, \widehat{w}, \tau) \\ &:= -G(\widehat{v}, \widehat{w}, \mathfrak{q}) - 4c_\Omega\Big(\|\nu\nabla\widehat{v} - \tau - \mathfrak{q}\,\mathbb{I}\| \\ &\qquad + C_{F\Omega}\|\mathrm{Div}\,\tau + f\| + \nu c_\Omega\|\mathrm{div}\,\widehat{v}\|\Big)\|\mathrm{div}\,\widehat{w}\|. \qquad (6.2.21)\end{aligned}$$

If the trial functions are taken from a narrower set $\overset{\circ}{\mathrm{S}}{}^1(\Omega)$, then the last-mentioned term in (6.2.21) vanishes, and we find that

$$\nu\|\nabla(u-\widehat{v})\|^2 \ge -G(\widehat{v}, w, \mathfrak{q}), \qquad \forall w \in \overset{\circ}{\mathrm{S}}{}^1(\Omega),\ \mathfrak{q} \in \widetilde{L}^2(\Omega). \quad (6.2.22)$$

Since

$$\int_\Omega \nu\nabla u : \nabla(u-\widehat{v})\,dx = \int_\Omega (f\cdot(u-\widehat{v}) + \mathfrak{p}\,\mathrm{div}\,(u-\widehat{v}))\,dx, \qquad (6.2.23)$$

we note that

$$
\begin{aligned}
&G(\widehat{v}, u-\widehat{v}, \mathfrak{p}) \\
&= \int_\Omega \big(2\nu\nabla\widehat{v} : \nabla(u-\widehat{v}) + \nu|\nabla(u-\widehat{v})|^2 - 2f\cdot(u-\widehat{v}) - 2\mathfrak{p}\operatorname{div}(u-\widehat{v})\big)\, dx \\
&= -\nu\|\nabla(\widehat{v}-u)\|^2. \qquad (6.2.24)
\end{aligned}
$$

Hence,

$$\nu\|\nabla(u-\widehat{v})\|^2 = \sup_{\substack{w\in\overset{\circ}{\mathrm{S}}{}^1(\Omega)\\ \mathfrak{q}\in\widetilde{L}^2(\Omega)}} \big\{-G(\widehat{v}, w, \mathfrak{q})\big\}. \tag{6.2.25}$$

6.2.6 Mixed boundary conditions

Now, we consider a more general statement of the Stokes problem (6.1.5) and (6.1.6) with mixed Dirichlet–Neumann boundary conditions

$$u = u_0 \qquad \text{on } \Gamma_1, \tag{6.2.26}$$

$$\sigma n = Fm \qquad \text{on } \Gamma_2, \tag{6.2.27}$$

where Γ_1 and Γ_2 of Γ are two measurable nonintersecting parts of the boundary, u_0 is a given function such that $\operatorname{div} u_0 = 0$, and $F \in L^2(\Gamma_2, \mathbb{R}^d)$. In this section, we define the space V_0 as follows:

$$V_0(\Omega) := \big\{v \in H^1(\Omega, \mathbb{R}^d) \mid v = 0 \quad \text{on } \Gamma_1\big\}.$$

Let $\mathrm{S}_0^1(\Omega)$ be a subspace of $V_0(\Omega)$ formed by solenoidal functions. A generalized solution is a function $u \in \mathrm{S}_0^1(\Omega)$ that satisfies the integral identity

$$\int_\Omega \nu\nabla u : \nabla w\, dx = \ell(w), \qquad \forall w \in \mathrm{S}_0^1(\Omega). \tag{6.2.28}$$

Here, $\ell : V_0(\Omega) \to \mathbb{R}$ is the functional

$$\ell(w) := \int_\Omega f\cdot w\, dx + \int_{\Gamma_2} F\cdot w\, ds.$$

It is easy to see that

$$|\ell(w)| \le C_\ell\|\nabla w\|, \qquad \forall w \in V_0(\Omega). \tag{6.2.29}$$

Note that C_ℓ depends on Ω and Γ_2 and on the constants in the respective Friedrichs and trace inequalities (for the functions vanishing on Γ_1). For any $\tau \in \Sigma(\Omega)$, the functional

$$\mathcal{F}_{\tau,\ell}(w) := \ell(w) - \int_\Omega \nu\tau : \nabla w\, dx$$

is linear and continuous on $V_0(\Omega)$. Its norm is defined by the relation

$$|\!|\!| \mathcal{F}_{\tau,\ell} |\!|\!| := \sup_{w \in V_0(\Omega)} \frac{|\mathcal{F}_{\tau,\ell}(w)|}{\|\nabla w\|} \leq C_\ell + \nu \|\tau\|. \tag{6.2.30}$$

The set $\mathcal{K}_{\tau,\ell} = \operatorname{Ker} \mathcal{F}_{\tau,\ell}$ contains tensor-valued functions that satisfy (in a generalized sense) the equilibrium equation

$$\operatorname{Div} \tau + f = 0 \qquad \text{in } \Omega \tag{6.2.31}$$

and the boundary condition

$$\tau n = F \qquad \text{on } \Gamma_2. \tag{6.2.32}$$

Theorem 6.6. *For any $v \in u_0 + \mathrm{S}_0^1(\Omega)$, $\mathfrak{q} \in \widetilde{L}^2(\Omega)$, and $\tau \in \Sigma(\Omega)$, the following estimate holds:*

$$\nu \|\nabla(u - v)\| \leq \|\tau + \mathfrak{q}\,\mathbb{I} - \nu \nabla v\| + |\!|\!| \mathcal{F}_{\tau,\ell} |\!|\!|. \tag{6.2.33}$$

If

$$\tau \in H_{\Gamma_2}(\Omega, \operatorname{Div}) := \left\{ \tau \in \Sigma \mid \operatorname{Div} \tau \in L^2(\Omega, \mathbb{R}^d),\ \tau n \in L^2(\Gamma_2, \mathbb{R}^d) \right\},$$

then the majorant for the Stokes problem with mixed boundary conditions is given by the relation

$$\begin{aligned} \nu \|\nabla(u - v)\| &\leq \overline{\mathfrak{M}}_{\mathrm{STm}}(v, \tau, \mathfrak{q}) \\ &:= \|\tau + \mathfrak{q}\,\mathbb{I} - \nu \nabla v\| + C_{F\Omega} \|\operatorname{Div} \tau + f\| \\ &\quad + C_{T\Gamma_2} \|F - \tau n\|_{\Gamma_2}, \end{aligned} \tag{6.2.34}$$

where $C_{F\Omega}$ and $C_{T\Gamma_2}$ are the constants in the inequalities

$$\|w\| \leq C_{F\Omega} \|\nabla w\|, \qquad \forall w \in V_0(\Omega), \tag{6.2.35}$$

$$\|w\|_{\Gamma_2} \leq C_{T\Gamma_2} \|\nabla w\|. \tag{6.2.36}$$

Proof. From (6.2.28) we conclude that for any $w \in \mathrm{S}_0^1(\Omega)$

$$\int_\Omega \nu \nabla(u - v) : \nabla w \, dx = \ell(w) - \int_\Omega \nu \nabla v : \nabla w \, dx.$$

Let $\tau \in \Sigma$. Then,

$$\begin{aligned} \int_\Omega \nu \nabla(u - v) : \nabla w \, dx &= \int_\Omega (\tau - \nu \nabla v) : \nabla w \, dx + \ell(w) - \int_\Omega \nu \tau : \nabla w \, dx \\ &\leq \left(\|\tau + \mathfrak{q}\,\mathbb{I} - \nu \nabla v\| + |\!|\!| \mathcal{F}_{\tau,\ell} |\!|\!| \right) \|\nabla w\|, \end{aligned}$$

where q is an arbitrary function in $\widetilde{L}^2(\Omega)$. By setting $w = u - v$ we arrive at the estimate (6.2.33).

Assume that $\tau \in H_{\Gamma_2}(\Omega, \mathrm{Div})$. Then,

$$\begin{aligned}\mathcal{F}_{\tau,\ell}(w) &= \int_\Omega (\mathrm{Div}\,\tau + f)\cdot w\,dx + \int_{\Gamma_2} (F - \tau n)\cdot w\,ds \\ &\le \|\mathrm{Div}\,\tau + f\|\|w\| + \|F - \tau n\|\|w\|_{\Gamma_2} \\ &\le \left(C_{F\Omega}\|\mathrm{Div}\,\tau + f\| + C_{T\Gamma_2}\|F - \tau n\|_{\Gamma_2}\right)\|\nabla w\|\end{aligned}$$

and, therefore,

$$|\,\mathcal{F}_{\tau,\ell}(w)\,| \le C_{F\Omega}\|\mathrm{Div}\,\tau + f\| + C_{T\Gamma_2}\|F - \tau n\|_{\Gamma_2}. \tag{6.2.37}$$

Now (6.2.34) follows from (6.2.33) and (6.2.37). □

The functional $\overline{\mathfrak{M}}_{\mathrm{STm}}(v, \tau, \mathsf{q})$ is *directly computable*, provided that the constants $C_{F\Omega}$ and $C_{T\Gamma_2}$ (or their upper bounds) are known. It vanishes if and only if

$$\tau = -\mathsf{q}\,\mathbb{I} + \nu\nabla v$$

and the relations $\mathrm{Div}\,\tau + f = 0$ in Ω and $\tau n = F$ on Γ_2 hold almost everywhere. Since v meets the Dirichlet boundary condition on Γ_1 and satisfies the relation $\mathrm{div}\,v = 0$, we conclude that in such a case $v = u$ and τ and q coincide with the exact stress and pressure fields, respectively.

Remark 6.7. A modification of the above a posteriori estimate is obtained if $\{\mathrm{tr}\tau\} = 0$ and we set $\mathsf{q} = -\frac{1}{d}\mathrm{tr}\tau$. Then (6.2.34) implies the estimate

$$\nu\,\|\nabla(u - v)\| \le \|\tau^{\mathrm{D}} - \nu\nabla v^{\mathrm{D}}\| + C_{F\Omega}\|\mathrm{Div}\,\tau + f\| + C_{T\Gamma_2}\|F - \tau n\|_{\Gamma_2}, \tag{6.2.38}$$

which that does not contain q. Note that if the right-hand side of (6.2.38) vanishes, then $\tau n = F$ on Ω and

$$\mathrm{Div}\,(\tau^{\mathrm{D}} + \frac{1}{d}\mathrm{tr}\tau\mathbb{I}) + f = 0 \quad \text{in } \Omega.$$

In addition, $\tau^{\mathrm{D}} - \nu(\nabla v)^{\mathrm{D}} = 0$ and $\mathsf{q} = -\frac{1}{d}\,\mathrm{tr}\,\tau$, so that

$$\tau = -q\,\mathbb{I} + \nu(\nabla v)^{\mathrm{D}}.$$

Since $v \in u_0 + \mathrm{S}^1_0(\Omega)$, we conclude that $\tau = -q\mathbb{I} + \nu\nabla v$ satisfies the equilibrium equation and, therefore, τ coincides with σ, v with u, and $\mathsf{q} = -\frac{1}{d}\mathrm{tr}\tau$ coincides with p (up to a constant).

A lower bound of the error can be derived by arguments similar to those used in Sections 3.2 and 4.1. It has the form

$$|||\nabla(u-v)|||^2 \geq 2\ell(w) - \int_\Omega \left(\nu|\nabla w|^2 + 2\nu\nabla v : \nabla w\right) dx, \qquad \forall w \in \mathrm{S}^1_0(\Omega).$$

To derive estimates for approximations in $u_0 + V_0(\Omega)$ we first prove an assertion below, which can be viewed as a generalization of Lemma 6.3.

Lemma 6.8. *Assume that*

$$v \in \widetilde{V}_0(\Omega) := \{v \in V_0(\Omega) \mid \{\operatorname{div} v\}_\Omega = 0\}.$$

Then, there exists $v_0 \in \mathrm{S}^1_0(\Omega)$ *such that*

$$\|\nabla(v - v_0)\| \leq c_\Omega \|\operatorname{div} v\|. \tag{6.2.39}$$

Proof. For any $a \in H^{1/2}(\Gamma, \mathbb{R}^d)$ satisfying the condition $\int_\Gamma a \cdot n\, ds = 0$ there exists a solution w_a of the Stokes problem

$$\begin{aligned} -\Delta w_a + \nabla \mathfrak{p} &= 0, \\ \operatorname{div} w_a &= 0 \qquad \text{in } \Omega, \\ w_a + a &= 0 \qquad \text{on } \Gamma. \end{aligned}$$

Let a be the trace of $v \in \widetilde{V}_0(\Omega)$ on Γ. Then, $w_a + v \in V_0$. By Lemma 6.2 we know that there exists $w_0 \in \overset{\circ}{\mathrm{S}}{}^1(\Omega)$ such that

$$\|\nabla(w_a + v) - \nabla w_0\| \leq c_\Omega \|\operatorname{div}(w_a + v)\| = c_\Omega \|\operatorname{div} v\|.$$

This estimate means that

$$\|\nabla v - \nabla(w_0 - w_a)\| \leq c_\Omega \|\operatorname{div} v\|,$$

where the function $v_0 = w_0 - w_a$ is solenoidal and $v_0 = 0$ on Γ_1. □

Theorem 6.9. *For any* $\widehat{v} \in u_0 + V_0(\Omega)$ *such that*

$$\{\operatorname{div} \widehat{v}\}_\Omega = 0, \tag{6.2.40}$$

$\mathfrak{q} \in \widetilde{L}^2(\Omega)$, *and* $\tau \in \Sigma(\Omega)$, *the following estimate holds:*

$$\nu\, \|\nabla(u - \widehat{v})\| \leq \|\tau + \mathfrak{q}\mathbb{I} - \nu\nabla\widehat{v}\| + |\!|\!| \mathcal{F}_{\tau,\ell} |\!|\!| + 2\nu c_\Omega \|\operatorname{div} \widehat{v}\|. \tag{6.2.41}$$

If $\tau \in H_{\Gamma_2}(\Omega, \mathrm{Div})$ *then*

$$\begin{aligned} \nu\, \|\nabla(u - \widehat{v})\| \leq{}& \|\tau + \mathfrak{q}\mathbb{I} - \nu\nabla(\widehat{v})\| + C_{F\Omega}\|\operatorname{Div} \tau + f\| \\ &+ C_{T\Gamma_2}\|\tau n - F\|_{\Gamma_2} + 2\nu c_\Omega \|\operatorname{div} \widehat{v}\|. \end{aligned} \tag{6.2.42}$$

Proof. Let $\widehat{w} := \widehat{v} - u_0$. This function belongs to $V_0(\Omega)$. In view of (6.2.40)

$$\{\operatorname{div} \widehat{w}\}_\Omega = \{\operatorname{div} \widehat{v} - \operatorname{div} u_0\}_\Omega = 0,$$

so that $\widehat{w} \in \widetilde{V}_0(\Omega)$. By Lemma 6.8, there exists a function $v_0 \in \mathrm{S}_0^1(\Omega)$ such that

$$\|\nabla(\widehat{w} - v_0)\| \leq \mathrm{c}_\Omega \|\operatorname{div} \widehat{v}\|. \tag{6.2.43}$$

We have

$$\begin{aligned} \nu \|\nabla(u - \widehat{v})\| &= \nu \|\nabla(u - \widehat{w} - u_0)\| \\ &\leq \nu \|\nabla(u - v_0 - u_0)\| + \nu \|\nabla(\widehat{w} - v_0)\| . \end{aligned} \tag{6.2.44}$$

Since $\operatorname{div}(v_0 + u_0) = 0$, we estimate the first norm by (6.2.33) and find that

$$\begin{aligned} \nu \|\nabla(u - \widehat{v})\| &\leq \|\tau + \mathfrak{q}\,\mathbb{I} - \nu\nabla(v_0 + u_0)\| + |\, \mathcal{F}_{\tau,\ell} \,| + \nu \|\nabla(\widehat{w} - v_0)\| \\ &\leq \|\tau + \mathfrak{q}\,\mathbb{I} - \nu\nabla\widehat{v}\| + |\, \mathcal{F}_{\tau,\ell} \,| + 2\nu \|\nabla(\widehat{w} - v_0)\| . \end{aligned}$$

By (6.2.43), we obtain (6.2.41). Estimate (6.2.42) follows from (6.2.37) and (6.2.41). □

Estimate (6.2.42) has the same principal structure as (6.2.4). The only difference is that a new term $C_{T\Gamma_2}\|\tau n - F\|_{\Gamma_2}$ arises. It serves as a penalty for a possible violation of the Neumann boundary condition.

Remark 6.10. If $\{\operatorname{tr}\tau\}_\Omega = 0$, then the pressure can be excluded, and we have the estimate

$$\begin{aligned} \nu \|\nabla(u - \widehat{v})\| \leq \left\|\tau^{\mathrm{D}} - \nu(\nabla\widehat{v})^{\mathrm{D}}\right\| &+ C_{F\Omega}\|\operatorname{Div} \tau + f\| \\ &+ C_{T\Gamma_2}\|\tau n - F\|_{\Gamma_2} + 2\nu\mathrm{c}_\Omega\|\operatorname{div} \widehat{v}\|. \end{aligned} \tag{6.2.45}$$

Now, our aim is to derive an upper bound of $\|\mathfrak{p} - \mathfrak{q}\|$.

Let $v^\dagger \in V_0(\Omega)$ be a vector-valued function such that $\operatorname{div} v^\dagger = 1$ in Ω. We note that there are many functions with such properties. Indeed, the nonhomogeneous Stokes problem

$$\begin{aligned} -\Delta v + \nabla\bar{\mathfrak{p}} &= 0, \\ \operatorname{div} v &= 1 \qquad \text{in } \Omega, \\ v &= 0 \qquad \text{on } \Gamma_1, \\ v &= a \qquad \text{on } \Gamma_2, \quad \int_{\Gamma_2} a \cdot n \, ds = |\Omega| \end{aligned}$$

has a solution (e.g., see [348]). The latter can be taken as $v^\dagger$.

Theorem 6.11. *Let* $\mathfrak{q} \in \widetilde{L}^2(\Omega)$ *be an approximation of the pressure field* $\mathfrak{p}$. *Then*

$$\frac{1}{2\mathrm{c}_\Omega^\dagger}\,\|\mathfrak{p}-\mathfrak{q}\| \leq \|\nu\nabla\widehat{v} - \tau - \mathfrak{q}\,\mathbb{I}\| + C_{F\Omega}\|\mathrm{Div}\,\tau + f\|$$
$$+ C_{T\Gamma_2}\|\tau n - F\|_{\Gamma_2} + \nu \mathrm{c}_\Omega\|\mathrm{div}\,\widehat{v}\|, \qquad (6.2.46)$$

where $\mathrm{c}_\Omega^\dagger = \mathrm{c}_\Omega + |\Omega|^{-1/2}\,\|\nabla v^\dagger\|$, $\widehat{v}$ *and* τ *are arbitrary functions in* $\widetilde{V}_0(\Omega)$ *and* $H_{\Gamma_2}(\Omega, \mathrm{Div})$, *respectively.*

Proof. Since

$$\mathfrak{p} - \mathfrak{q} - \{\mathfrak{p}-\mathfrak{q}\}_\Omega = \widetilde{\mathfrak{p}-\mathfrak{q}} \in \widetilde{L}^2(\Omega),$$

we apply Lemma 6.2 and conclude that there exists a function $w_0 \in V_0(\Omega)$ (note that w_0 vanishes on the boundary) such that

$$\mathrm{div}\, w_0 = \widetilde{\mathfrak{p}-\mathfrak{q}} \qquad (6.2.47)$$

and

$$\|\nabla w_0\| \;\leq\; \mathrm{c}_\Omega\|\widetilde{\mathfrak{p}-\mathfrak{q}}\|. \qquad (6.2.48)$$

It is easy to see that $w_0^\dagger := w_0 + \{\mathfrak{p}-\mathfrak{q}\}_\Omega v^\dagger \in V_0(\Omega)$,

$$\int_\Omega \mathrm{div}\, w_0^\dagger(\mathfrak{p}-\mathfrak{q})\,dx = \int_\Omega (\mathrm{div}\, w_0(\mathfrak{p}-\mathfrak{q}) + \mathrm{div}\, v^\dagger\{\mathfrak{p}-\mathfrak{q}\}_\Omega(\mathfrak{p}-\mathfrak{q}))\,dx$$
$$= \|\widetilde{\mathfrak{p}-\mathfrak{q}}\|^2 + |\Omega|\{\mathfrak{p}-\mathfrak{q}\}^2 = \|\mathfrak{p}-\mathfrak{q}\|^2, \qquad (6.2.49)$$

and

$$\|\nabla w_0^\dagger\| \leq \|\nabla w_0\| + \{\mathfrak{p}-\mathfrak{q}\}_\Omega\|\nabla v^\dagger\| \qquad (6.2.50)$$
$$\leq \mathrm{c}_\Omega\|\widetilde{\mathfrak{p}-\mathfrak{q}}\| + \{\mathfrak{p}-\mathfrak{q}\}_\Omega\|\nabla v^\dagger\| \leq \mathrm{c}_\Omega^\dagger\|\mathfrak{p}-\mathfrak{q}\|,$$

where $\mathrm{c}_\Omega^\dagger = \mathrm{c}_\Omega + |\Omega|^{-1/2}\|\nabla v^\dagger\|$.

Now, we use the integral identity

$$\int_\Omega \nu\nabla u : \nabla w_0^\dagger\,dx \;=\; \ell(w_0^\dagger) + \int_\Omega \mathfrak{p}\,\mathrm{div}\, w_0^\dagger\,dx \qquad (6.2.51)$$

and rearrange (6.2.49) as follows:

$$\|\mathfrak{p}-\mathfrak{q}\|^2 = \int_\Omega \mathrm{div}\, w_0^\dagger(\mathfrak{p}-\mathfrak{q})\,dx$$
$$= \int_\Omega \left(\nu\nabla u : \nabla w_0^\dagger - \mathfrak{q}\,\mathrm{div}\, w_0^\dagger\right)\,dx - \ell(w_0^\dagger)$$
$$= \int_\Omega \nu\nabla(u-\widehat{v}) : \nabla w_0^\dagger\,dx + \int_\Omega \left(\nu\nabla\widehat{v} : \nabla w_0^\dagger - \mathfrak{q}\,\mathrm{div}\, w_0^\dagger\right)\,dx - \ell(w_0^\dagger).$$

In view of (6.2.50),

$$\int_\Omega \nu\nabla(u-\widehat{v}) : \nabla w_0^\dagger\, dx \le \mathrm{c}_\Omega^\dagger \nu \,\|\nabla(u-\widehat{v})\|\, \|\mathfrak{p}-\mathfrak{q}\|$$

and

$$\begin{aligned}
&\int_\Omega \left(\nu\nabla\widehat{v} : \nabla w_0^\dagger - \mathfrak{q}\operatorname{div} w_0^\dagger\right) dx - \ell(w_0^\dagger) \\
&= \int_\Omega (\nu\nabla\widehat{v} - \tau - \mathfrak{q}\,\mathbb{I}) : \nabla w_0^\dagger\, dx - \int_\Omega (\operatorname{Div}\tau + f)\cdot w_0^\dagger\, dx + \int_{\Gamma_2} (\tau n - F)\cdot w_0^\dagger\, ds \\
&\le \left(\|\nu\nabla\widehat{v} - \tau - \mathfrak{q}\,\mathbb{I}\| + C_{F\Omega}\|\operatorname{Div}\tau + f\| + C_{T\Gamma_2}\|F - \tau n\|_{\Gamma_2}\right)\|\nabla w_0^\dagger\| \\
&\le \mathrm{c}_\Omega^\dagger\left(\|\nu\nabla\widehat{v} - \tau - \mathfrak{q}\,\mathbb{I}\| + C_{F\Omega}\|\operatorname{Div}\tau + f\| + C_{T\Gamma_2}\|F - \tau n\|_{\Gamma_2}\right)\|\mathfrak{p}-\mathfrak{q}\|.
\end{aligned}$$

Therefore,

$$\begin{aligned}
\|\mathfrak{p}-\mathfrak{q}\| \le \mathrm{c}_\Omega^\dagger\Big(&\nu\,\|\nabla(u-\widehat{v})\| + \|\nu\nabla\widehat{v} - \tau - \mathfrak{q}\,\mathbb{I}\| \\
&+ C_{F\Omega}\|\operatorname{Div}\tau + f\| + C_{T\Gamma_2}\|F - \tau n\|_{\Gamma_2}\Big).
\end{aligned}$$

Now, we apply (6.2.42) and obtain the estimate

$$\begin{aligned}
\|\mathfrak{p}-\mathfrak{q}\| \le 2\mathrm{c}_\Omega^\dagger\Big(&\|\tau + \mathfrak{q}\,\mathbb{I} - \nu\nabla(\widehat{v})\| + C_{F\Omega}\|\operatorname{Div}\tau + f\| \\
&+ C_{T\Gamma_2}\|\tau n - F\|_{\Gamma_2} + \nu\mathrm{c}_\Omega\|\operatorname{div}\widehat{v}\|\Big),
\end{aligned}$$

which is equivalent to (6.2.46). □

Remark 6.12. The constant $\mathrm{c}_\Omega^\dagger$ contains the norm of a subsidiary function $v^\dagger$, which must satisfy the condition $\operatorname{div} v^\dagger = 1$ and $v^\dagger = 0$ on Γ_1. Usually, such a function is not difficult to construct. For example, for polygonal domains $v^\dagger$ can be constructed with the help of Raviart–Thomas elements of the lowest order. It is desirable to have a function $v^\dagger$ such that $\|\nabla v^\dagger\|$ is as small as possible.

6.2.7 Problems for almost incompressible fluids

In models of almost incompressible fluids, the incompressibility condition is replaced by a term that contains the divergence with a large multiplier. The respective "penalized" version of the stationary Stokes problem with Dirichlet boundary conditions reads: Find $u_\delta \in u_0 + V_0$ (where $V_0 = \overset{\circ}{H}{}^1(\Omega, \mathbb{R}^d)$) satisfying the integral identity

$$\int_\Omega \left(\nu\nabla u_\delta : \nabla w + \frac{1}{\delta}\operatorname{div} u_\delta \operatorname{div} w\right) dx = \int_\Omega f\cdot w\, dx, \qquad w \in V_0(\Omega). \quad (6.2.52)$$

From (6.2.52) we find that

$$\nu\|\nabla u_\delta\| \le C_{F\Omega}\|f\|, \tag{6.2.53}$$

$$\|\operatorname{div} u_\delta\|^2 \le \delta\|f\|\|u_\delta\| \le C_{F\Omega}\delta\|f\|\|\nabla u_\delta\| \le \delta\frac{C_{F\Omega}^2}{\nu}\|f\|^2. \tag{6.2.54}$$

We can deduce an estimate of the difference between u and u_δ. For this purpose, we use (6.2.11) with $\widehat{v} = u_\delta$. Other functions in the right-hand side of (6.2.11), we define as follows:

$$\tau = \tau_\delta := \nu\nabla u_\delta + \frac{1}{\delta}\operatorname{div} u_\delta \mathbb{I} \quad \text{and} \quad q = p_\delta := -\frac{1}{\delta}\operatorname{div} u_\delta.$$

In view of (6.2.52),

$$\int_\Omega \tau_\delta : \nabla w\, dx = \int_\Omega f \cdot w\, dx, \qquad w \in V_0(\Omega), \tag{6.2.55}$$

so that $\operatorname{Div} \tau_\delta + f = 0$ almost everywhere in Ω. Moreover,

$$\tau_\delta + p_\delta \mathbb{I} - \nu\nabla u_\delta = 0.$$

Thus, by (6.2.11) we conclude that

$$\frac{1}{2c_\Omega}\|\nabla(u - u_\delta)\| \le \|\operatorname{div} u_\delta\|. \tag{6.2.56}$$

By (6.2.13)

$$\frac{1}{2\nu c_\Omega^2}\|p - p_\delta\| \le \|\operatorname{div} u_\delta\|. \tag{6.2.57}$$

We note that the difference between the exact solutions of the Stokes problem and its penalized counterpart is controlled by the L^2 norm of the divergence of the problem. Moreover, from (6.2.54), (6.2.56), and (6.2.57) it follows that

$$\|\nabla(u - u_\delta)\| \le 2c_\Omega C_{F\Omega}\sqrt{\frac{\delta}{\nu}}\|f\|, \tag{6.2.58}$$

$$\|p - p_\delta\| \le 2c_\Omega^2 C_{F\Omega}\sqrt{\nu\delta}\|f\|. \tag{6.2.59}$$

These estimates show that $u \to u_\delta$ and $p \to p_\delta$ (this fact is known, see, e.g., R. Temam [348]). Moreover, we find that the sequences converge with the rate no less than $\delta^{1/2}$.

A similar estimate can be obtained for approximations constructed by means of the Uzawa algorithm.

6.2.8 Problems with the condition $\operatorname{div} u = \phi$

In some cases, the following version of the Stokes problem is considered

$$-\operatorname{Div} \sigma = f - \nabla \mathfrak{p} \qquad \text{in } \Omega, \tag{6.2.60}$$

$$\sigma = \nu \nabla u \qquad \text{in } \Omega, \tag{6.2.61}$$

$$u = u_0 \qquad \text{on } \Gamma, \tag{6.2.62}$$

$$\operatorname{div} u = \phi \qquad \text{in } \Omega, \tag{6.2.63}$$

where ϕ is a given function in L^2 and u_0 satisfies the compatibility condition

$$\int_\Omega \phi \, dx = \int_\Gamma u_0 \cdot n \, ds. \tag{6.2.64}$$

Let $u_\phi \in u_0 + V_0$ be a function satisfying (6.2.63). We set

$$u = u_\phi + \bar{u}, \qquad \bar{u} \in \overset{\circ}{\mathrm{S}}{}^1(\Omega).$$

Then (6.2.60)–(6.2.63) is reduced to the Stokes problem with

$$\bar{f} = f + \nu \Delta u_\phi \in H^{-1},$$

instead of f. In other words, the problem is to find $\bar{u} \in \overset{\circ}{\mathrm{S}}{}^1(\Omega)$ such that

$$\int_\Omega \nu \nabla \bar{u} \cdot \nabla w \, dx = \int_\Omega (f \cdot w - \tau_\phi : \nabla w + \mathfrak{p} \operatorname{div} w) \, dx, \qquad \forall w \in V_0, \tag{6.2.65}$$

where $\tau_\phi = \nu \nabla u_\phi$.

Let u be approximated by $v = u_\phi + \bar{v}$, where $\bar{v} \in \overset{\circ}{\mathrm{S}}{}^1(\Omega)$. By (6.2.65), we find that for any $w \in \overset{\circ}{\mathrm{S}}{}^1(\Omega)$,

$$\begin{aligned} \int_\Omega \nu \nabla (\bar{u} - \bar{v}) \cdot \nabla w \, dx &= \int_\Omega (f \cdot w - \tau_\phi : \nabla w - \nu \nabla \bar{v} : \nabla w + \mathfrak{q} \operatorname{div} w) \, dx \\ &\leq \int_\Omega (f + \operatorname{Div} \tau) \cdot w \, dx + \int_\Omega (\tau - \tau_\phi - \nu \nabla \bar{v} + \mathfrak{q} \, \mathbb{I}) : \nabla w \, dx. \end{aligned} \tag{6.2.66}$$

Since $\bar{u} - \bar{v} \in \overset{\circ}{\mathrm{S}}{}^1(\Omega)$, we can set $w = \bar{u} - \bar{v}$. Then, (6.2.66) implies the estimate

$$\nu \|\nabla (\bar{u} - \bar{v})\| \leq \|\tau + \mathfrak{q} \, \mathbb{I} - \nu \nabla \bar{v} - \tau_\phi\| + C_{F\Omega} \|\operatorname{Div} \tau + f\|. \tag{6.2.67}$$

Since $u - v = \bar{u} - \bar{v}$, we have

$$\nu \|\nabla (u - v)\| \leq \|\tau + \mathfrak{q} \, \mathbb{I} - \nu \nabla v\| + C_{F\Omega} \|\operatorname{Div} \tau + f\|. \tag{6.2.68}$$

If $\widehat{v} = u_\phi + \bar{v}$ and $\bar{v} \in V_0$ (so that $\operatorname{div} \widehat{v} \neq \phi$), then (using the same arguments as before) we deduce an estimate analogous to (6.2.11):

$$\nu \|\nabla (u - \widehat{v})\| \leq \|\tau + \mathfrak{q} \, \mathbb{I} - \nu \nabla v\| + C_{F\Omega} \|\operatorname{Div} \tau + f\| + 2\nu c_\Omega \|\operatorname{div} \widehat{v} - \phi\|. \tag{6.2.69}$$

For $\phi = 0$ this estimate coincides with (6.2.11).

6.3 Generalized Stokes problem

Various generalized statements of the Stokes problem are motivated by semidiscrete formulations of evolutionary problems, in which $u(x,t)$ is replaced by a sequence of approximations $u^k(x)$ representing $u(x,t_k)$, where t_k, $k = 0,1,\dots,M$ are some selected values of the time-variable t. For example, the scheme

$$\begin{cases} \dfrac{u^k - u^{k-1}}{\delta_k} - \nu\Delta u^k + \operatorname{div}(u^{k-1} \otimes u^{k-1}) = f - \nabla \mathfrak{p}^k \text{ in } \Omega, \\ \operatorname{div} u^k = 0, \qquad \delta_k = t_k - t_{k-1}, \end{cases} \tag{6.3.1}$$

leads to a stationary problem for u_k. It differs from the Stokes problem by the presence of the term μu (where $\mu = \frac{1}{\delta_k}$). In other cases, additional terms in the basic equation (6.1.1) arise owing to some physical phenomena.

In this section, we first consider the system

$$\mu u - \nu \Delta u = f - \nabla \mathfrak{p} \qquad \text{in } \Omega, \tag{6.3.2}$$

$$\operatorname{div} u = 0 \qquad \text{in } \Omega. \tag{6.3.3}$$

As before, the function u additionally satisfies the prescribed boundary condition $u = u_0$ ($u_0 \in \overset{\circ}{\mathrm{S}}{}^1(\Omega)$). In general, both parameters $\nu > 0$ and $\mu \geq 0$ may be large or small. Henceforth, we assume that μ and ν are positive functions such that

$$\mu \in [\mu_\ominus, \mu_\oplus] \quad \text{and} \quad \nu \in [\nu_\ominus, \nu_\oplus].$$

Moreover, we assume that $\nu_\ominus = 1$ (this assumption does not lead to loss in generality, because after a proper scaling (6.3.2)–(6.3.3) can always be transformed to a system that satisfies it).

For our purposes, it is convenient to state the generalized Stokes problem in the so-called "three-field setting": Find u, $\mathfrak{p}$, and a tensor-valued function σ (stress) such that

$$-\operatorname{Div}\sigma + \mu u = f \qquad \text{in } \Omega, \tag{6.3.4}$$

$$\sigma = -\mathfrak{p}\,\mathbb{I} + \nu \nabla u \qquad \text{in } \Omega, \tag{6.3.5}$$

$$\operatorname{div} u = 0 \qquad \text{in } \Omega, \tag{6.3.6}$$

$$u = u_0 \qquad \text{on } \Gamma. \tag{6.3.7}$$

The velocity field $u \in u_0 + \overset{\circ}{\mathrm{S}}{}^1(\Omega)$ of (6.3.4)–(6.3.5) is defined by the integral identity

$$\int_\Omega (\nu \nabla u : \nabla w + \mu u \cdot w)\,dx = \int_\Omega f \cdot w\,dx, \qquad w \in \overset{\circ}{\mathrm{S}}{}^1(\Omega), \tag{6.3.8}$$

or by the variational problem for the functional

$$\inf_{w \in u_0 + \overset{\circ}{\mathrm{S}}{}^1(\Omega)} I(w) = I(u), \tag{6.3.9}$$

where

$$I(w) = \int_\Omega \Big(\frac{\nu}{2}|\nabla w|^2 + \frac{\mu}{2}|w|^2 - f \cdot w\Big)\, dx \tag{6.3.10}$$

is the corresponding *energy functional*. The existence and uniqueness of u is proved by the same arguments as for the Stokes problem.

Our goal is to deduce two-sided bounds of the error evaluated in terms of the norm

$$\|w\|_{\nu\mu}^2 := \int_\Omega (\nu|\nabla w|^2 + \mu|w|^2)\, dx,$$

which is the natural energy norm.

6.3.1 Estimates for solenoidal approximations

Let $v \in u_0 + \overset{\circ}{\mathrm{S}}{}^1(\Omega)$. Then (6.3.8) implies the relation

$$\int_\Omega (\nu\nabla(u - v) : \nabla w + \mu(u - v) \cdot w)\, dx$$
$$= \int_\Omega (f \cdot w - \mu v \cdot w - \nu\nabla v : \nabla w)\, dx, \tag{6.3.11}$$

which holds for any $w \in \overset{\circ}{\mathrm{S}}{}^1(\Omega)$. Let τ be a tensor-valued function in $H(\Omega, \mathrm{Div})$. Since w vanishes on the boundary, we rewrite (6.3.11) as follows:

$$\int_\Omega (\nu\nabla(u - v) : \nabla w + \mu(u - v) \cdot w)\, dx = \mathfrak{F}(w; v, \tau, \mathfrak{q}), \tag{6.3.12}$$

where $\mathfrak{F}(w; v, \tau, \mathfrak{q})$ is a linear functional with respect to w defined by the relation

$$\mathfrak{F}(w; v, \tau, \mathfrak{q}) := \int_\Omega ((f - \mu v + \mathrm{Div}\,\tau) \cdot w + (\tau + \mathfrak{q}\,\mathbb{I} - \nu\nabla v) : \nabla w)\, dx$$

and $\mathfrak{q} \in \widetilde{L}^2(\Omega)$. Henceforth, we denote

$$\mathrm{r}(v, \tau) := f - \mu v + \mathrm{Div}\,\tau \quad \text{and} \quad \mathrm{d}(v, \tau, \mathfrak{q}) := \tau + \mathfrak{q}\,\mathbb{I} - \nu\nabla v.$$

Represent the first term of $\mathfrak{F}(w; v, \tau, \mathfrak{q})$ in the form

$$\int_\Omega \mathrm{r}(v, \tau) \cdot w\, dx = \int_\Omega \frac{\alpha}{\sqrt{\mu}}\mathrm{r}(v, \tau) \cdot \sqrt{\mu}w\, dx + \int_\Omega (1 - \alpha)\mathrm{r}(v, \tau) \cdot w\, dx,$$

where $\alpha = \alpha(x)$ is a real function with values in $[0, 1]$. We have

$$\left|\int_\Omega \mathrm{r}(v, \tau) \cdot w\, dx\right| \le \left\|\frac{\alpha}{\sqrt{\mu}}\mathrm{r}(v, \tau)\right\| \|\sqrt{\mu}w\| + C_{\nu\Omega}\, \|(1 - \alpha)\mathrm{r}(v, \tau)\|\, \|\sqrt{\nu}\,\nabla w\|,$$

where $C_{\nu\Omega}$ is a constant in the Friedrichs type inequality

$$\|w\|^2 \leq C_{\nu\Omega}^2 \int_\Omega \nu(x)|\nabla w|^2\,dx, \qquad \forall\, w \in V_0(\Omega). \tag{6.3.13}$$

Since

$$\|w\| \leq C_{F\Omega}\|\nabla w\| \leq \frac{C_{F\Omega}}{\sqrt{\nu_\ominus}}\|\sqrt{\nu}\nabla w\|,$$

we can set $C_{\nu\Omega} = \frac{C_{F\Omega}}{\sqrt{\nu_\ominus}}$ (if $c_\ominus = 1$, then $C_{\nu\Omega} = C_{F\Omega}$). Next,

$$\left|\int_\Omega \mathrm{d}(v,\tau,\mathrm{q}) : \nabla w\,dx\right| \leq \left\|\frac{1}{\sqrt{\nu}}\mathrm{d}(v,\tau,\mathrm{q})\right\| \|\sqrt{\nu}\,\nabla w\|.$$

Thus, we obtain

$$\begin{aligned}
&|\mathfrak{F}(w;v,\tau,\mathrm{q})| \\
&\leq \left(C_{\nu\Omega}\,\|(1-\alpha)\mathrm{r}(v,\tau)\| + \left\|\frac{1}{\sqrt{\nu}}\mathrm{d}(v,\tau,\mathrm{q})\right\|\right)\|\sqrt{\nu}\nabla\,w\| + \left\|\frac{\alpha}{\sqrt{\mu}}\mathrm{r}(v,\tau)\right\|\,\|\sqrt{\mu}\,w\| \\
&\leq \left(\left(C_{\nu\Omega}\,\|(1{-}\alpha)\mathrm{r}(v,\tau)\| + \left\|\frac{\mathrm{d}(v,\tau,\mathrm{q})}{\sqrt{\nu}}\right\|\right)^2 + \left\|\frac{\alpha\mathrm{r}(v,\tau)}{\sqrt{\mu}}\right\|^2\right)^{1/2}\|w\|_{\nu\mu}.
\end{aligned} \tag{6.3.14}$$

Set $w = u - v$. Then, we arrive at the following estimate for the generalized Stokes problem:

$$\begin{aligned}
\|u-v\|_{\nu\mu}^2 \leq \overline{\mathfrak{M}}_{\alpha\,\mathrm{GST}}(v,\tau,\mathrm{q}) := \left\|\frac{\alpha}{\sqrt{\mu}}\mathrm{r}(v,\tau)\right\|^2 + \Big(C_{\nu\Omega}\,\|(1-\alpha)\mathrm{r}(v,\tau)\| \\
+ \left\|\frac{1}{\sqrt{\nu}}\mathrm{d}(v,\tau,\mathrm{q})\right\|\Big)^2.
\end{aligned} \tag{6.3.15}$$

This estimate is valid for any $\alpha \in [0,1]$, $\tau \in H(\Omega, \mathrm{Div})$, and $\mathrm{q} \in \widetilde{L}^2(\Omega)$. If $\mu = 0$ and $\nu = \text{const}$, then we set $\alpha = 0$ and arrive at (6.2.3).

Two particular forms of this estimate related to the choice $\alpha = 0$ and $\alpha = 1$ are as follows:

$$\|u-v\|_{\nu\mu} \leq C_{\nu\Omega}\,\|\mathrm{r}(v,\tau)\| + \left\|\frac{1}{\sqrt{\nu}}\mathrm{d}(v,\tau,\mathrm{q})\right\| =: \overline{\mathfrak{M}}_{0\,\mathrm{GST}}(v,\tau,\mathrm{q}) \tag{6.3.16}$$

and

$$\|u-v\|_{\nu\mu}^2 \leq \left\|\frac{1}{\sqrt{\nu}}\mathrm{d}(v,\tau,\mathrm{q})\right\|^2 + \left\|\frac{1}{\sqrt{\mu}}\,\mathrm{r}(v,\tau)\right\|^2 =: \overline{\mathfrak{M}}_{1\,\mathrm{GST}}(v,\tau,\mathrm{q}). \tag{6.3.17}$$

By (6.3.4) and (6.3.5) we find that

$$\left\| \frac{1}{\sqrt{\nu}}(\sigma + \mathfrak{p}\,\mathbb{I} - \nu\nabla v) \right\|^2 = \int_\Omega \nu |\nabla(u - v)|^2\, dx$$

and

$$\int_\Omega \frac{1}{\mu} |f - \mu v + \operatorname{Div} \sigma|^2\, dx = \int_\Omega \mu |u - v|^2\, dx.$$

Therefore,

$$\overline{\mathfrak{M}}_{1\,\mathrm{GST}}(v, \sigma, \mathfrak{p}) = \|u - v\|_{\nu\mu}$$

and the majorant $\overline{\mathfrak{M}}_{1\,\mathrm{GST}}(v, \tau, \mathfrak{q})$ can provide a sharp upper bound of the error if τ and $\mathfrak{q}$ are properly chosen. However, it has an essential drawback: if μ is small, then the second term involves a large multiplier, which makes the whole estimate sensitive to the residual $\mathrm{r}(v, \tau)$. In practice, this may lead to a considerable overestimation of the error.

The majorant $\overline{\mathfrak{M}}_{0\,\mathrm{GST}}(v, \tau, \mathfrak{q})$ does not contain large parameters, but we cannot prove that

$$\inf_{\tau, \mathfrak{q}} \overline{\mathfrak{M}}_{0\,\mathrm{GST}}(v, \tau, \mathfrak{q}) = \|u - v\|_{\nu\mu}\,.$$

In other words, there may be an inherent gap between the left- and right-hand sides of (6.3.17).

In order to avoid the above difficulties and to obtain an estimate, which possesses positive features of the above estimates without bad ones, we derive another upper bound for the deviation $u - v$. For this purpose, we apply the same method as was used in Chapter 4 for the reaction-diffusion problem.

Let us estimate the first term on the right-hand side of (6.3.14) and rewrite it in the form

$$\begin{aligned} \|u - v\|^2_{\nu\mu} \le {} & C^2_{\nu\Omega}(1 + \beta)\, \|(1 - \alpha)\mathrm{r}(v, \tau)\|^2 \\ & + \left\| \frac{\alpha}{\sqrt{\mu}} \mathrm{r}(v, \tau) \right\|^2 + \frac{1 + \beta}{\beta} \left\| \frac{1}{\sqrt{\nu}} \mathrm{d}(v, \tau, \mathfrak{q}) \right\|^2, \end{aligned}$$

where β is an arbitrary positive number. The minimum of the right-hand side with respect α is attained if

$$\alpha = H(\beta, C_{\nu\Omega}, \mu) := \frac{C^2_{\nu\Omega}\mu(1 + \beta)}{C^2_{\nu\Omega}\mu(1 + \beta) + 1} \in [0, 1),$$

which leads to another upper bound of the error for the generalized Stokes problem:

$$\begin{aligned} \|u - v\|^2_{\nu\mu} \le {} & \int_\Omega \frac{H(\beta, C_{\nu\Omega}, \mu)}{\mu} \mathrm{r}^2(v, \tau)\, dx + \frac{1 + \beta}{\beta} \left\| \frac{1}{\sqrt{\nu}} \mathrm{d}(v, \tau, \mathfrak{q}) \right\|^2 \\ =: {} & \overline{\mathfrak{M}}^{\,2}_{\beta\,\mathrm{GST}}(v, \tau, \mathfrak{q}). \end{aligned} \tag{6.3.18}$$

Since

$$\mathrm{r}(v,\sigma) = \mu(u-v), \quad \mathrm{d}(v,\sigma,\mathfrak{p}) = \sigma + \mathfrak{p}\,\mathbb{I} - \nu\nabla v = \nu\nabla(u-v),$$

we note that

$$\overline{\mathfrak{M}}^{\,2}_{\beta\,\mathrm{GST}}(v,\sigma,\mathfrak{p}) = \int_\Omega \left(\frac{C^2_{\nu\Omega}(1+\beta)}{C^2_{\nu\Omega}\mu(1+\beta)+1}\mu^2(v-u)^2 + \frac{1+\beta}{\beta}\nu|\nabla(v-u)|^2 \right) dx$$

and

$$\overline{\mathfrak{M}}^{\,2}_{\beta\,\mathrm{GST}}(v,\sigma,\mathfrak{p}) \to \|u-v\|^2_{\nu\mu} \quad \text{as } \beta \to +\infty.$$

Therefore, (6.3.18) has no gap between its left- and right-hand sides. At the same time, the structure of the first term of (6.3.18) is such that it is not sensitive to small values of μ.

If $\mu = 0$ and $\nu = \text{const}$, then (6.3.18) implies the estimate

$$\nu\|\nabla(u-v)\|^2 \le C^2_{\nu\Omega}(1+\beta)\,\|f + \operatorname{Div}\tau\|^2 + \frac{1+\beta}{\beta}\left\|\frac{\mathrm{d}(v,\tau,\mathfrak{q})}{\sqrt{\nu}}\right\|^2, \tag{6.3.19}$$

which gives an upper bound of the error for the generalized Stokes problem (cf. (6.2.3)).

If $\mu = \text{const}$ then we minimize $\overline{\mathfrak{M}}^{\,2}_{\beta\,\mathrm{GST}}(v,\tau,\mathfrak{q})$ with respect to β and set

$$\beta = \begin{cases} \dfrac{\mathrm{D}}{C_{\nu\Omega}}\dfrac{\mu C^2_{\nu\Omega}+1}{\mathrm{R}-\mu C_{\nu\Omega}\mathrm{D}} & \text{if } \mathrm{R} > \mu C_{\nu\Omega}\mathrm{D}, \\ +\infty & \text{if } \mathrm{R} \le \mu C_{\nu\Omega}\mathrm{D}, \end{cases} \tag{6.3.20}$$

where

$$\mathrm{D} := \left\|\frac{\mathrm{d}(v,\tau,\mathfrak{q})}{\sqrt{\nu}}\right\| \quad \text{and} \quad \mathrm{R} := \|\mathrm{r}(v,\tau)\|\,.$$

Then, we arrive at the estimate

$$\|u-v\|^2_{\nu\mu} \le \begin{cases} \dfrac{1}{1+\mu C^2_{\nu\Omega}}(C_{\nu\Omega}\mathrm{R}+\mathrm{D})^2 & \text{if } \mathrm{R} > \mu C_{\nu\Omega}\mathrm{D}, \\ \dfrac{1}{\mu}\mathrm{R}^2 + \mathrm{D}^2 & \text{if } \mathrm{R} \le \mu C_{\nu\Omega}\mathrm{D}. \end{cases} \tag{6.3.21}$$

Note that the second branch of (6.3.21) is bounded by the quantity $\mathrm{D}(C_{\nu\Omega}\mathrm{R}+\mathrm{D})$ and, therefore, (6.3.21) does not "blow up" if $\mu \to 0$.

6.3.2 Estimates for nonsolenoidal fields

Let $\widehat{v} \in u_0 + V_0$ be such that $\operatorname{div} \widehat{v} \neq 0$. To derive an upper bound of $\|u - \widehat{v}\|_{\nu\mu}$, we use the same arguments as for the Stokes problem.

By Lemma 6.2, we know that there exists $v_0 \in u_0 + \overset{\circ}{\mathrm{S}}{}^1(\Omega)$ such that

$$\|\nabla(\widehat{v} - v_0)\| \leq \mathrm{c}_\Omega \|\operatorname{div} \widehat{v}\|.$$

Therefore,

$$\|\widehat{v} - v_0\|_{\nu\mu}^2 \leq \kappa_\Omega^2 \|\operatorname{div} \widehat{v}\|^2, \qquad \kappa_\Omega^2 := \mathrm{c}_\Omega^2(\nu_\oplus + \mu_\oplus C_{F\Omega}^2), \tag{6.3.22}$$

and

$$\|u - \widehat{v}\|_{\nu\mu} \leq \|u - v\|_{\nu\mu} + \|v - \widehat{v}\|_{\nu\mu}, \qquad \forall v \in u_0 + \overset{\circ}{\mathrm{S}}{}^1(\Omega). \tag{6.3.23}$$

To estimate the first norm on the right-hand side of (6.3.23), we apply (6.3.18) (assuming for the sake of simplicity that μ is a constant) represented in the form

$$\|u - v\|_{\nu\mu}^2 \leq \frac{H(\beta, C_{\nu\Omega}, \mu)}{\mu} \|\mathrm{r}(v, \tau)\|^2 + \frac{1+\beta}{\beta} \left\| \frac{\mathrm{d}(v, \tau, \mathrm{q})}{\sqrt{\nu}} \right\|^2. \tag{6.3.24}$$

Let

$$\wp^2(\beta) = \max \left\{ H(\beta, C_{\nu\Omega}, \mu), \frac{1+\beta}{\beta} \right\}.$$

Since

$$\|\mathrm{r}(v, \tau)\|^2 \leq \|\mathrm{r}(\widehat{v}, \tau)\|^2 + 2\mu \|\mathrm{r}(\widehat{v}, \tau)\| \|v - \widehat{v}\| + \mu^2 \|v - \widehat{v}\|^2,$$

$$\left\| \frac{\mathrm{d}(v, \tau, \mathrm{q})}{\sqrt{\nu}} \right\|^2 \leq \left\| \frac{\mathrm{d}(\widehat{v}, \tau, \mathrm{q})}{\sqrt{\nu}} \right\|^2 + 2 \left\| \frac{\mathrm{d}(\widehat{v}, \tau, \mathrm{q})}{\sqrt{\nu}} \right\| \|\sqrt{\nu} \nabla(v - \widehat{v})\| + \|\sqrt{\nu} \nabla(v - \widehat{v})\|^2,$$

and

$$\begin{aligned} H(\beta, C_{\nu\Omega}, \mu) \|\mathrm{r}(\widehat{v}, \tau)\| \|v - \widehat{v}\| &+ \frac{1+\beta}{\beta} \left\| \frac{\mathrm{d}(\widehat{v}, \tau, \mathrm{q})}{\sqrt{\nu}} \right\| \|\sqrt{\nu} \nabla(v - \widehat{v})\| \\ &\leq \sqrt{\frac{H(\beta, C_{\nu\Omega}, \mu)}{\mu}} \|\mathrm{r}(\widehat{v}, \tau)\| \sqrt{H(\beta, C_{\nu\Omega}, \mu)} \|\sqrt{\mu}(v - \widehat{v})\| \\ &\quad + \frac{1+\beta}{\beta} \left\| \frac{\mathrm{d}(\widehat{v}, \tau, \mathrm{q})}{\sqrt{\nu}} \right\| \|\sqrt{\nu} \nabla(v - \widehat{v})\| \\ &\leq \wp(\beta) \, \overline{\mathfrak{M}}_{\beta\,\mathrm{GST}}(\widehat{v}, \tau, \mathrm{q}) \|v - \widehat{v}\|_{\nu\mu}, \end{aligned}$$

we find that

$$\|u - \widehat{v}\|_{\nu\mu} \le \left(\overline{\mathfrak{M}}^{\,2}_{\beta\,\mathrm{GST}}(\widehat{v}, \tau, \mathfrak{q}) + 2\wp(\beta)\overline{\mathfrak{M}}_{\beta\,\mathrm{GST}}(\widehat{v}, \tau, \mathfrak{q})\|v - \widehat{v}\|_{\nu\mu} + \wp^2(\beta)\|v - \widehat{v}\|^2_{\nu\mu}\right)^{1/2} + \|v - \widehat{v}\|_{\nu\mu}. \tag{6.3.25}$$

Set $v = v_0$ and apply (6.3.22). We obtain

$$\|u - \widehat{v}\|_{\nu\mu} \le \left(\overline{\mathfrak{M}}^{\,2}_{\beta\,\mathrm{GST}}(\widehat{v}, \tau, \mathfrak{q}) + 2\wp(\beta)\kappa_\Omega\overline{\mathfrak{M}}_{\beta\,\mathrm{GST}}(\widehat{v}, \tau, \mathfrak{q})\|\operatorname{div}\widehat{v}\| + \wp^2(\beta)\kappa^2_\Omega\|\operatorname{div}\widehat{v}\|^2\right)^{1/2} + \kappa_\Omega\|\operatorname{div}\widehat{v}\|. \tag{6.3.26}$$

With the help of positive parameters γ and δ, we represent (6.3.26) in quadratic form. Set $v = v_0$ and apply (6.3.22). We obtain

$$\|u - \widehat{v}\|^2_{\nu\mu} \le (1+\gamma)(1+\delta)\overline{\mathfrak{M}}^{\,2}_{\beta\,\mathrm{GST}}(\widehat{v}, \tau, \mathfrak{q}) + (1+\gamma)\left(\wp^2(\beta) + \frac{\wp^2(\beta)}{\delta} + \frac{1}{\gamma}\right)\kappa^2_\Omega\|\operatorname{div}\widehat{v}\|^2. \tag{6.3.27}$$

If $\widehat{v} \in u_0 + \overset{\circ}{\mathrm{S}}{}^1(\Omega)$, then $\operatorname{div}\widehat{v} = 0$. In this case, we set $\gamma = \delta = 0$ and see that (6.3.27) is converted to (6.3.24).

6.3.3 Estimates for the pressure field

Let $\mathfrak{q} \in \widetilde{L}^2(\Omega)$ be an approximation of the pressure field $\mathfrak{p}$. Then, there exists a function $\bar{w} \in V_0$ such that

$$\operatorname{div}\bar{w} = \mathfrak{p} - \mathfrak{q}, \quad \text{and} \quad \|\nabla\bar{w}\| \le \mathrm{c}_\Omega\|\mathfrak{p} - \mathfrak{q}\|. \tag{6.3.28}$$

Therefore,

$$\|\bar{w}\|_{\nu\mu} \le \kappa_\Omega\|\mathfrak{p} - \mathfrak{q}\|. \tag{6.3.29}$$

In view of (6.3.28), we note that

$$\|\mathfrak{p} - \mathfrak{q}\|^2 = \int_\Omega \operatorname{div}\bar{w}(\mathfrak{p} - \mathfrak{q})\,dx.$$

By the relation

$$\int_\Omega \operatorname{div}\bar{w}\,\mathfrak{p}\,dx = \int_\Omega (\nu\nabla u : \nabla\bar{w} + \mu u \cdot \bar{w} - f \cdot \bar{w})\,dx,$$

we obtain

$$\begin{aligned}\|\mathfrak{p}-\mathfrak{q}\|^2 &= \int_\Omega (\nu\nabla(u-v) : \nabla\bar{w} + \mu(u-v)\cdot\bar{w})\,dx \\ &\quad + \int_\Omega (\nu\nabla v : \nabla\bar{w} + \mu v\bar{w} - \mathfrak{q}\,\mathrm{div}\,\bar{w} - f\cdot\bar{w})\,dx \\ &= \int_\Omega (\nu\nabla(u-v) : \nabla\bar{w} + \mu(u-v)\cdot\bar{w})\,dx - \mathfrak{F}(\bar{w}; v, \tau, \mathfrak{q}),\end{aligned} \quad (6.3.30)$$

where v is an arbitrary function in $u_0 + \overset{\circ}{\mathrm{S}}{}^1(\Omega)$.

We have

$$\int_\Omega (\nu\nabla(u-v) : \nabla\bar{w} + \mu(u-v)\cdot\bar{w})\,dx \leq \|u-v\|_{\nu\mu}\|\bar{w}\|_{\nu\mu}.$$

By (6.3.14),

$$|\mathfrak{F}(\bar{w}; v, \tau, \mathfrak{q})| \\ \leq \left(\left(C_{\nu\Omega}\,\|(1-\alpha)\mathrm{r}(v,\tau)\| + \left\|\frac{\mathrm{d}(v,\tau,\mathfrak{q})}{\sqrt{\nu}}\right\|\right)^2 + \left\|\frac{\alpha\mathrm{r}(v,\tau)}{\sqrt{\mu}}\right\|^2\right)^{1/2} \|\bar{w}\|_{\nu\mu}.$$

Therefore, from (6.3.30) we find that

$$\|\mathfrak{p}-\mathfrak{q}\| \leq \kappa_\Omega\|u-v\|_{\nu\mu} \\ + \kappa_\Omega\sqrt{\left(C_{\nu\Omega}\,\|(1-\alpha)\mathrm{r}(v,\tau)\| + \left\|\frac{\mathrm{d}(v,\tau,\mathfrak{q})}{\sqrt{\nu}}\right\|\right)^2 + \left\|\frac{\alpha\mathrm{r}(v,\tau)}{\sqrt{\mu}}\right\|^2}.$$

In view of (6.3.15), we obtain

$$\frac{1}{4\kappa_\Omega^2}\|\mathfrak{p}-\mathfrak{q}\|^2 \leq \left(C_{\nu\Omega}\,\|(1-\alpha)\mathrm{r}(v,\tau)\| + \left\|\frac{\mathrm{d}(v,\tau,\mathfrak{q})}{\sqrt{\nu}}\right\|\right)^2 + \left\|\frac{\alpha\mathrm{r}(v,\tau)}{\sqrt{\mu}}\right\|^2. \quad (6.3.31)$$

In other words, the upper bound is given by the same expression as in (6.3.15) but with multiplier $2\kappa_\Omega$. Since $\alpha(x)$ is in our disposal, we can select it, e.g., as $\alpha(x) = \min\{\mu(x), 1\}$, which makes the estimate robust with respect to small values of μ.

If we set $\alpha(x) = 1$, then (6.3.31) reads

$$\frac{1}{4\kappa_\Omega^2}\|\mathfrak{p}-\mathfrak{q}\|^2 \leq \left\|\frac{\mathrm{d}(v,\tau,\mathfrak{q})}{\sqrt{\nu}}\right\|^2 + \left\|\frac{\mathrm{r}(v,\tau)}{\sqrt{\mu}}\right\|^2. \quad (6.3.32)$$

If $\alpha(x) = 0$, then we have another estimate:

$$\frac{1}{2\kappa_\Omega}\|\mathfrak{p}-\mathfrak{q}\| \leq C_{\nu\Omega}\,\|\mathrm{r}(v,\tau)\| + \left\|\frac{\mathrm{d}(v,\tau,\mathfrak{q})}{\sqrt{\nu}}\right\|. \quad (6.3.33)$$

Let $\mu = 0$ and $\nu = \text{const}$. In this case, $\kappa_\Omega = \sqrt{\nu}\mathrm{c}_\Omega$ and (6.3.33) is converted into

$$\frac{1}{2\sqrt{\nu}\mathrm{c}_\Omega}\|\mathfrak{p} - \mathfrak{q}\| \leq \left\| \frac{\mathrm{d}(v, \tau, \mathfrak{q})}{\sqrt{\nu}} \right\| + C_{\nu\Omega}\|\mathrm{r}(v, \tau, \mathfrak{q})\|. \tag{6.3.34}$$

Since $C_{\nu\Omega} = \frac{C_{F\Omega}}{\sqrt{\nu}}$ and $\operatorname{div} v = 0$, this estimate coincides with (6.2.13) derived for the Stokes problem.

6.3.4 Error minorant

In the case considered, lower bounds can be derived by variational arguments based on (6.1.28). Assume that $v \in u_0 + \overset{\circ}{\mathrm{S}}{}^1(\Omega)$. For any $w \in \overset{\circ}{\mathrm{S}}{}^1(\Omega)$, we have

$$\begin{aligned}
\frac{1}{2}\|u - v\|^2_{\nu\mu} &= I(v) - I(u) \geq I(v) - I(v + w) \\
&= \int_\Omega \left(-\frac{\nu}{2}|\nabla w|^2 - \frac{\mu}{2}|w|^2 - \nu\nabla v : \nabla w - \mu v \cdot w + f \cdot w\right) dx.
\end{aligned}$$

It is easy to see that

$$\begin{aligned}
&\frac{1}{2}\|u - v\|^2_{\nu\mu} \\
&= \sup_{w \in \overset{\circ}{\mathrm{S}}{}^1(\Omega)} \int_\Omega \left(-\frac{\nu}{2}|\nabla w|^2 - \frac{\mu}{2}|w|^2 - \nu\nabla v : \nabla w - \mu v \cdot w + f \cdot w\right) dx.
\end{aligned} \tag{6.3.35}$$

To prove this, it suffices to take $w = u - v$ and use the relation

$$\int_\Omega \nu(\nabla u : \nabla (u - v) + \mu u \cdot (u - v))\, dx = \int_\Omega f \cdot (u - v)\, dx.$$

Therefore, the maximization of the right-hand side of (6.3.35) with respect to a certain finite-dimensional subspace of $\overset{\circ}{\mathrm{S}}{}^1(\Omega)$ gives a computable lower bound of the error norm.

6.3.5 Models with polymerization

Another version of the generalized Stokes problem is related to models of fluids with polymerization (e.g., see J. Bonvin, M. Picasso, and R. Stenberg [62]). It can be represented in the form

$$\operatorname{Div}\sigma + f = 0 \quad \text{in } \Omega, \tag{6.3.36}$$

$$\operatorname{div} u = 0 \quad \text{in } \Omega, \tag{6.3.37}$$

$$\sigma = \sigma_0 - \mathfrak{p}\,\mathbb{I} + \nu\nabla u \quad \text{in } \Omega, \tag{6.3.38}$$

$$u = 0 \quad \text{on } \Gamma, \tag{6.3.39}$$

where σ_0 is a given tensor-valued function such that $\operatorname{tr}\sigma_0 = 0$. Note that (6.3.37) decomposes σ into spherical and deviatoric parts, respectively.

The generalized solution u of the system (6.3.36)–(6.3.39) is a function in $\overset{\circ}{\mathrm{S}}{}^1(\Omega)$ satisfying the integral identity

$$\int_\Omega (\nu\nabla u : \nabla w + \sigma_0 : \nabla w)\, dx = \int_\Omega f \cdot w\, dx, \qquad w \in \overset{\circ}{\mathrm{S}}{}^1(\Omega). \tag{6.3.40}$$

For an approximation $v \in \overset{\circ}{\mathrm{S}}{}^1(\Omega)$, we have

$$\begin{aligned}\int_\Omega \nu\nabla(u - v) : \nabla w\, dx &= \int_\Omega (f \cdot w - \sigma_0 : \nabla w - \nu\nabla v : \nabla w)\, dx \\ &= \int_\Omega \big((f + \operatorname{Div}\tau)\cdot w + (\tau + \mathrm{q}\,\mathbb{I} - \sigma_0 - \nu\nabla v) : \nabla w\big)\, dx, \end{aligned}\tag{6.3.41}$$

where $\tau \in H(\Omega, \mathrm{Div})$ and $\mathrm{q} \in \widetilde{L}^2(\Omega)$.

From (6.3.41) it follows that (if ν is a constant)

$$\nu\|\nabla(u - v)\| \leq \|\tau + \mathrm{q}\,\mathbb{I} - \sigma_0 - \nu\nabla v\| + C_{F\Omega}\|f + \operatorname{Div}\tau\|. \tag{6.3.42}$$

If an approximation $\widehat{v}$ belongs to a wider set V_0, then we apply the same arguments as for the Stokes problem and deduce the estimate

$$\nu\|\nabla(u - \widehat{v})\| \leq \|\tau + \mathrm{q}\,\mathbb{I} - \sigma_0 - \nu\nabla v\| + C_{F\Omega}\|f + \operatorname{Div}\tau\| + 2\nu c_\Omega\|\operatorname{div}\widehat{v}\|. \tag{6.3.43}$$

6.3.6 Models with rotation

In certain models, the Navier–Stokes problem is considered in a rotating coordinate system. Then, additional terms arise in the equation of motion and we write the whole system in the form (see J. P. Vanyo [353])

$$\partial_t u + \operatorname{div}(u \times u) - \operatorname{Div}\sigma = f - 2\boldsymbol{\varpi} \times v - \boldsymbol{\varpi} \times (\boldsymbol{\varpi} \times r), \tag{6.3.44}$$

$$\sigma = -\mathrm{p}\,\mathbb{I} + \nu\nabla u, \tag{6.3.45}$$

$$\operatorname{div} u = 0, \tag{6.3.46}$$

$$u = u_0 \qquad \text{on } \Gamma. \tag{6.3.47}$$

In (6.3.44), the term $2\boldsymbol{\varpi} \times v$ is due to the Coriolis force and the term $\boldsymbol{\varpi} \times (\boldsymbol{\varpi} \times r)$ is related to the centrifugal force (the latter term is usually appended to the source function and disappears from the equation). The vector $\boldsymbol{\varpi}$ is oriented along the axis x_3, and its value depends on the rotation velocity. Mathematical properties of such models were studied by a number of authors (e.g., see A. Babin, A. Mahalov, and B. Nicolaenko [18, 19] and the literature cited in those papers).

A linearized version of (6.3.44)–(6.3.47) can be viewed as a certain generalization of the Stokes problem. It is defined by the relations

$$-\operatorname{Div}\sigma + \mu u = f - \boldsymbol{\varpi} \times u, \tag{6.3.48}$$

$$\sigma = -\mathrm{p}\,\mathbb{I} + \nu\nabla u, \tag{6.3.49}$$

$$\operatorname{div} u = 0, \tag{6.3.50}$$

$$u = u_0 \qquad \text{on } \Gamma. \tag{6.3.51}$$

This system of equations arises if the problem (6.3.44)–(6.3.47) is solved by semi-discrete approximations (then $\mu > 0$ comes from an approximation of the term $\partial_t u$).

A generalized solution of the problem (6.3.48)–(6.3.51) is defined by the integral relation

$$\int_\Omega (\nu\nabla u : \nabla w + \mu u \cdot w + (\boldsymbol{\varpi} \times u) \cdot w)\, dx = \int_\Omega f \cdot w\, dx, \tag{6.3.52}$$

which holds for any $w \in \overset{\circ}{\mathrm{S}}{}^1(\Omega)$. As in the models considered before, this integral relation generates an estimate of the difference between u and any $v \in u_0 + \overset{\circ}{\mathrm{S}}{}^1(\Omega)$. We have

$$\begin{aligned}
&\int_\Omega (\nu\nabla(u-v) : \nabla w + \mu(u-v)\cdot w + (\boldsymbol{\varpi} \times (u-v))\cdot w)\, dx \\
&\quad = \int_\Omega (f\cdot w - \nu\nabla v : \nabla w - \mu v \cdot w - (\boldsymbol{\varpi} \times v)\cdot w)\; dx \\
&\quad = \int_\Omega (f - \mu v - (\boldsymbol{\varpi} \times v))\cdot w\, dx + \int_\Omega (\mathrm{q}\,\mathbb{I} - \nu\nabla v) : \nabla w\, dx,
\end{aligned}$$

where $\mathrm{q} \in \widetilde{L}^2(\Omega)$. Set $w = u - v$ and note that

$$(\boldsymbol{\varpi} \times w)\cdot w = 0. \tag{6.3.53}$$

Let $\tau \in H(\Omega, \operatorname{Div})$ and

$$\begin{aligned}
\mathrm{r}(v,\tau) &:= f - \mu v - \boldsymbol{\varpi} \times v + \operatorname{Div}\tau, \\
\mathrm{d}(v,\tau,\mathrm{q}) &:= \tau + \mathrm{q}\,\mathbb{I} - \nu\nabla v.
\end{aligned}$$

By arguments used at the beginning of this section, we obtain the estimate

$$\begin{aligned}
\|u - v\|^2_{\nu\mu} &\le \overline{\mathfrak{M}}_{\alpha\boldsymbol{\varpi}}(v,\tau,\mathrm{q}) \\
&:= \left(\left\| \frac{\mathrm{d}(v,\tau,\mathrm{q})}{\sqrt{\nu}} \right\| + C_{\nu\Omega}\, \|(1-\alpha)\mathrm{r}(v,\tau)\| \right)^2 + \left\| \frac{\alpha\mathrm{r}(v,\tau)}{\sqrt{\mu}} \right\|^2,
\end{aligned} \tag{6.3.54}$$

which is valid for any $\alpha \in [0,1]$. Estimates for nonsolenoidal velocity fields and for approximations of the pressure function can also be derived quite analogously (see [159] for details).

6.4 The Oseen problem

The Oseen problem is often regarded as a linearization of the Navier–Stokes problem at a neighborhood of a constant velocity field $\mathrm{a} \in \overset{\circ}{\mathrm{S}}{}^1(\Omega)$, which leads to the system

$$u_t - \nu\Delta u + \operatorname{div}(\mathrm{a} \otimes u) = f - \nabla \mathfrak{p} \qquad \text{in } \Omega, \tag{6.4.1}$$

$$u(x,0) = \widehat{u}(x), \tag{6.4.2}$$

$$\operatorname{div} u = 0, \tag{6.4.3}$$

$$u = u_0 \qquad \text{on } \Gamma. \tag{6.4.4}$$

In the stationary case, we formulate the problem as follows: Find u, $\mathfrak{p}$, and σ such that

$$-\operatorname{Div}\sigma + \operatorname{Div}(\mathrm{a} \otimes u) = f - \nabla \mathfrak{p} \qquad \text{in } \Omega, \tag{6.4.5}$$

$$\sigma = \nu\nabla u, \tag{6.4.6}$$

$$u = u_0 \qquad \text{on } \Gamma, \tag{6.4.7}$$

$$\operatorname{div} u = 0, \tag{6.4.8}$$

where it is assumed that $\operatorname{div} u_0 = 0$. A generalized solution of the above system is defined as a function $u \in u_0 + \overset{\circ}{\mathrm{S}}{}^1(\Omega)$ that satisfies the integral identity

$$\int_\Omega (\nu\nabla u : \nabla w - (\mathrm{a} \otimes u) : \nabla w)\, dx = \int_\Omega f \cdot w\, dx, \qquad w \in \overset{\circ}{\mathrm{S}}{}^1(\Omega). \tag{6.4.9}$$

We assume that the problem data are such that u exists and is unique.

Estimates for the velocity field. (6.4.9) implies computable bounds of errors for solenoidal approximations of u. Let $v \in \overset{\circ}{\mathrm{S}}{}^1(\Omega)$. We rearrange (6.4.9) into the form

$$\int_\Omega (\nu\nabla(u - v) : \nabla w - (a \otimes (u - v)) : \nabla w)\, dx$$
$$= \int_\Omega (f \cdot w - \nu\nabla v : \nabla w + (a \otimes v) : \nabla w)\, dx, \qquad w \in \overset{\circ}{\mathrm{S}}{}^1(\Omega). \tag{6.4.10}$$

Note that

$$\int_\Omega (\mathrm{a} \otimes w) : \nabla w\, dx = -\int_\Omega \operatorname{Div}(\mathrm{a} \otimes w) \cdot w\, dx = -\int_\Omega (\mathrm{a} \cdot \nabla w) \cdot w\, dx$$
$$= -\int_\Omega \mathrm{a} \cdot ((\nabla w) w)\, dx = -\frac{1}{2}\int_\Omega \mathrm{a} \cdot \nabla(|w|^2)\, dx = 0.$$

Take a symmetric tensor-valued function $\tau \in H(\Omega, \operatorname{Div})$ and rewrite (6.4.10) as follows:

$$\int_\Omega \nu\nabla(u - v) : \nabla w\, dx$$
$$= \int_\Omega \Big((f + \operatorname{Div}\tau) \cdot w + (\tau + \mathbb{I}\mathfrak{q} + \mathrm{a} \otimes v - \nu\nabla v) : \nabla w \Big)\, dx, \tag{6.4.11}$$

where $q \in \widetilde{L}^2(\Omega)$. By setting $w = u - v$, we note that (6.4.11) leads to the inequality

$$\begin{aligned} \nu\|\nabla(u-v)\| &\leq \overline{\mathfrak{M}}_{\rm OS}(v,\tau,\mathfrak{q}) \\ &:= \|\tau + \mathfrak{q}\,\mathbb{I} + \mathrm{a}\otimes v - \nu\nabla v\| + C_{F\Omega}\|f + \operatorname{Div}\tau\|. \qquad (6.4.12) \end{aligned}$$

Remark 6.13. If u is approximated by the function $\widehat{v} \in u_0 + V_0$, then the corresponding estimate for the deviation norm can be derived in exactly the same way as for the Stokes problem. In this case, the majorant includes an additional term that penalizes possible violations of the incompressibility condition. This estimate has the following form:

$$\nu\|\nabla(u-\widehat{v})\| \leq \|\tau + \mathfrak{q}\,\mathbb{I} + \mathrm{a}\otimes v - \nu\nabla\widehat{v}\| + C_{F\Omega}\|f + \operatorname{Div}\tau\| + \overline{c}_\Omega\|\operatorname{div}\widehat{v}\|, \quad (6.4.13)$$

where $\overline{c}_\Omega$ depends on c_Ω, $C_{F\Omega}$, $\|\mathrm{a}\|$, and ν.

Remark 6.14. We note that

$$\overline{\mathfrak{M}}_{\rm OS}(v,\sigma,\mathfrak{p}) \leq \nu\|\nabla(u-v)\| + \|\mathrm{a}\|\,\|u-v\| \leq (\nu + \|\mathrm{a}\|C_{F\Omega})\|\nabla(u-v)\|.$$

Therefore, the minimization of $\overline{\mathfrak{M}}_{\rm OS}(v,\tau,\mathfrak{q})$ with respect to τ and q gives an upper bound, which is equivalent to the energy error norm.

Estimates for the pressure field. Assume that $\mathfrak{q} \in \widetilde{L}^2(\Omega)$ is an approximation of the pressure field $\mathfrak{p}$. We take $\widetilde{w} \in V_0$ as in Section 6.2.2, i.e., $\operatorname{div}\widetilde{w} = \mathfrak{p} - \mathfrak{q}$ and $\|\nabla\widetilde{w}\| \leq \mathrm{c}_\Omega\|\mathfrak{p}-\mathfrak{q}\|$. Since

$$\begin{aligned} \int_\Omega (\mathfrak{p}-\mathfrak{q})\operatorname{div}\widetilde{w}\,dx &= \int_\Omega (\nu\nabla u : \nabla\widetilde{w} - f\widetilde{w} - (\mathrm{a}\otimes u) : \nabla\widetilde{w} - \mathfrak{q}\operatorname{div}\widetilde{w})\,dx \\ &\leq \int_\Omega (\nu\nabla(u-v) : \nabla\widetilde{w} - (\mathrm{a}\otimes(u-v)) : \nabla\widetilde{w})\,dx \\ &\quad + \int_\Omega (\nu\nabla v : \nabla\widetilde{w} - f\widetilde{w} - (\mathrm{a}\otimes v) : \nabla\widetilde{w} - \mathfrak{q}\operatorname{div}\widetilde{w})\,dx, \end{aligned}$$

we find that

$$\begin{aligned} \|\mathfrak{p}-\mathfrak{q}\|^2 &\leq (\nu + C_{F\Omega}\|\mathrm{a}\|)\|\nabla(u-v)\|\,\|\nabla\widetilde{w}\| \\ &\quad + \|\operatorname{Div}\tau + f\|C_{F\Omega}\|\nabla\widetilde{w}\| + \|\tau + \mathfrak{q}\,\mathbb{I} + \mathrm{a}\otimes v - \nu\nabla v\|\,\|\nabla\widetilde{w}\|. \end{aligned}$$

This relation implies the estimate

$$\begin{aligned} \|\mathfrak{p}-\mathfrak{q}\| \leq \mathrm{c}_\Omega\big((\nu + C_{F\Omega}\|\mathrm{a}\|)\|\nabla(u-v)\| &+ C_{F\Omega}\|\operatorname{Div}\tau + f\| \\ &+ \|\tau + \mathfrak{q}\,\mathbb{I} + \mathrm{a}\otimes v - \nu\nabla v\|\big), \qquad (6.4.14) \end{aligned}$$

where $\|\nabla(u-v)\|$ is estimated by (6.4.12).

Remark 6.15. For the product space

$$W := (u_0 + \overset{\circ}{\mathrm{S}}{}^1(\Omega)) \times H(\Omega, \mathrm{Div}) \times \widetilde{L}^2(\Omega),$$

we introduce two equivalent norms

$$\|(v, \tau, \mathfrak{q})\|_W := \nu\|\nabla v\| + \|\tau\|_{\mathrm{Div}} + \sqrt{d}\,\|\mathfrak{q}\|,$$

$$][(v, \tau, \mathfrak{q})][_W := \nu\|\nabla v\| +][\tau][_{\mathrm{Div}g} + \sqrt{d}\,\|\mathfrak{q}\|.$$

By the same arguments as in Section 6.2.4 one can show that the majorant $\overline{\mathfrak{M}}_{\mathrm{OS}}(v, \tau, \mathfrak{q})$ is equivalent to the error in the above-defined combined norms.

Generalized form of the Oseen problem. As in the case of the Stokes problem, a generalized form of the Oseen problem arises if semidiscrete approximations of the Navier–Stokes problem are used. For example, the scheme

$$\begin{cases} \frac{u^k - u^{k-1}}{\delta_k} - \nu\Delta u^k + \mathrm{div}\,(u^{k-1} \otimes u^k) = f - \nabla \mathfrak{p}^k \text{ in } \Omega, \\ \mathrm{div}\, u^k = 0, \end{cases} \tag{6.4.15}$$

leads to a stationary Oseen problem (for u^k): Find $u \in u_0 + \overset{\circ}{\mathrm{S}}{}^1(\Omega)$ that satisfies the integral identity

$$\int_\Omega (\nu\nabla u : \nabla w + \mu u \cdot w - (\mathrm{a} \otimes u) : \nabla w)\, dx = \int_\Omega f \cdot w\, dx, \quad w \in \overset{\circ}{\mathrm{S}}{}^1(\Omega). \tag{6.4.16}$$

Estimates for such problems can be obtained without any serious difficulties by repeating arguments we used in Section 6.3 for the generalized Stokes problems. In this case, the estimates have the same form as, e.g., (6.3.15) and (6.3.31) but with $\mathrm{d}(v, \tau, \mathfrak{q}) := \tau + \mathfrak{q}\,\mathbb{I} + \mathrm{a} \otimes v - \nu\nabla v$.

6.5 Stationary Navier–Stokes problem for $d = 2$

The Navier–Stokes equation

$$u_t - \nu\Delta u + \mathrm{Div}\,(u \otimes u) = f - \nabla \mathfrak{p} \tag{6.5.1}$$

is the most known model in the theory of viscous incompressible fluids. From the mathematical point of view, the Navier–Stokes equation has yet to be completely understood.[1] It is known that for sufficiently regular initial data it has a rather weak *Leray–Hopf* solution (e.g, see G. Galdi [146], where the reader will find a consequent

[1]The problem to prove or counter the existence of a unique smooth solution in $(0, T] \times \mathbb{R}^3$ is formulated as one of the Millennium Prize Problems stated by the Clay Mathematical Institute.

exposition of the mathematical theory related to Navier–Stokes equation). So far, all existence and uniqueness results of a stronger type have been conditional. In view of these difficulties, it is not surprising that at present we have no reliable a posteriori estimates for this class of problems. Below, we consider one special case generated by the stationary Navier–Stokes equation in a bounded Lipschitz domain $\Omega \subset \mathbb{R}^2$ for which existence of a unique solution is established (see O. Ladyzhenskaya [210]). We consider the problem

$$-\mathrm{Div}\,(\nu \nabla u) + \mathrm{Div}\,(u \otimes u) = f - \nabla \mathfrak{p} \qquad \text{in } \Omega, \tag{6.5.2}$$

$$\mathrm{div}\, u = 0, \tag{6.5.3}$$

$$u = u_0 \qquad \text{on } \Gamma. \tag{6.5.4}$$

A generalized solution $u \in u_0 + \overset{\circ}{\mathrm{S}}{}^1(\Omega)$ is defined by the integral identity

$$\int_\Omega (\nu \nabla u : \nabla w - (u \otimes u) : \nabla w)\, dx = \int_\Omega f \cdot w dx \quad w \in \overset{\circ}{\mathrm{S}}{}^1(\Omega). \tag{6.5.5}$$

For any solenoidal u and $w \in \overset{\circ}{\mathrm{S}}{}^1(\Omega)$, we have

$$\int_\Omega (u \otimes w) : \nabla w\, dx = -\frac{1}{2} \int_\Omega u \cdot \nabla(|w|^2)\, dx = 0. \tag{6.5.6}$$

In particular, $\int_\Omega (u \otimes u) : \nabla u\, dx = 0$, and the relation

$$\int_\Omega (\nu \nabla u : \nabla u - (u \otimes u) : \nabla u)\, dx = \int_\Omega f \cdot u\, dx$$

furnishes the energy estimate $\nu \|\nabla u\| \leq C_{F\Omega} \|f\|$.

Now, we use (6.5.5), for the derivation of an a posteriori estimate. Assume that $v \in \overset{\circ}{\mathrm{S}}{}^1(\Omega)$ is an approximation of u. Then,

$$\int_\Omega (\nu \nabla (u - v) : \nabla w - (u \otimes (u - v)) : \nabla w)\, dx$$
$$= \int_\Omega (f \cdot w - \nu \nabla v : \nabla w + (u \otimes v) : \nabla w)\, dx, \qquad \forall w \in \overset{\circ}{\mathrm{S}}{}^1(\Omega). \tag{6.5.7}$$

Set $w = u - v$. In view of (6.5.6), the second term in the left-hand side of (6.5.7) vanishes. Also, we note that

$$(u \otimes v) : \nabla(u - v) = (v \otimes v) : \nabla(u - v) + ((u - v) \otimes v) : \nabla(u - v)$$

and

$$\begin{aligned}\int_\Omega ((u - v) \otimes v) : \nabla(u - v)\, dx &= \int_\Omega \mathrm{Div}\,((u - v) \otimes v) \cdot (u - v)\, dx \\ &= -\int_\Omega ((u - v) \cdot \nabla v) \cdot (u - v)\, dx \\ &= -\int_\Omega \nabla v : ((u - v) \otimes (u - v))\, dx. \end{aligned} \tag{6.5.8}$$

From (6.5.8), it follows that

$$\nu\|\nabla(u-v)\|^2 + \int_\Omega \nabla v : \big((u-v)\otimes(u-v)\big)\,dx$$
$$\leq \int_\Omega (f\cdot(u-v) - \nu\nabla v : \nabla(u-v) + (v\otimes v) : \nabla(u-v))\;dx. \quad (6.5.9)$$

For a symmetric $\tau \in H(\Omega, \mathrm{Div})$, we represent the right-hand side as follows:

$$\int_\Omega (f - \mathrm{Div}\,\tau + \mathrm{Div}\,(v\otimes v))\cdot(u-v)\,dx + \int_\Omega (\tau + \mathrm{q}\,\mathbb{I} - \nu\nabla v) : \nabla(u-v)\,dx.$$

Note that

$$\int_\Omega \nabla v : ((u-v)\otimes(u-v))\,dx \leq \int_\Omega |\nabla v||(u-v)|^2\,dx \leq \|\nabla v\|\|(u-v)\|^2_{4,\Omega}$$
$$\leq \mu\|\nabla v\|\|\nabla(u-v)\|^2, \quad (6.5.10)$$

where μ is a constant in the inequality $\|(u-v)\|_{4,\Omega} \leq \mu\|\nabla(u-v)\|$ (which holds in view of embedding theorems).

A computable upper bound of $\|\nabla(u-v)\|$ follows from (6.5.10) if we assume that

$$\bar{\nu} = \nu - \mu\|\nabla v\| > 0. \quad (6.5.11)$$

It should be remarked that this assumption is very demanding (for approximate solutions of problems with high velocities it does not hold). By (6.5.9)–(6.5.11), we conclude that

$$\bar{\nu}\|\nabla(u-v)\| \leq \|\tau + \mathrm{q}\,\mathbb{I} - \nu\nabla v\| + C_{F\Omega}\|f - \mathrm{Div}\,\tau + \mathrm{Div}\,(v\otimes v)\| \quad (6.5.12)$$

has the same structure as the estimates derived for the Stokes and Oseen problems. However, this is a conditional estimate valid only for sufficiently slow flows.

Finally, we note that in view of the relation

$$\int_\Omega \nabla(u-v) : \big((u-v)\otimes(u-v)\big)\,dx = 0,$$

(6.5.9) can be represented in the form

$$\nu\|\nabla(u-v)\|^2 + \int_\Omega \nabla u : \big((u-v)\otimes(u-v)\big)\,dx$$
$$= \int_\Omega (f - \mathrm{Div}\,\tau + \mathrm{Div}\,(v\otimes v))\cdot(u-v)\,dx$$
$$+ \int_\Omega (\tau + \mathrm{q}\,\mathbb{I} - \nu\nabla v) : \nabla(u-v)\,dx. \quad (6.5.13)$$

If

$$\widetilde{\nu} = \nu - \mu\|\nabla u\| > 0, \tag{6.5.14}$$

then we find that

$$\widetilde{\nu}\|\nabla(u - v)\| \leq \|\tau + \mathfrak{q}\,\mathbb{I} - \nu\nabla v\| + C_{F\Omega}\|f - \operatorname{Div}\tau + \operatorname{Div}(v \otimes v)\|, \tag{6.5.15}$$

which implies the uniqueness of u. Indeed, assume that $\bar{u}$ and $\bar{p}$ is a pair of solutions satisfying (6.5.2)–(6.5.4) that differs from u and $\mathfrak{p}$. In this case, we set $v = \bar{u}$, $\mathfrak{q} = \bar{\mathfrak{p}}$, and $\tau = -\bar{\mathfrak{p}}\,\mathbb{I} + \nu\nabla\bar{u}$. By (6.5.15), we conclude that

$$\widetilde{\nu}\|\nabla(u - \bar{u})\| = 0$$

and, consequently, $u = \bar{u}$. In view of the energy estimate, (6.5.14) is satisfied if

$$\mu\|\nabla u\| \leq \frac{\mu C_{F\Omega}}{\nu}\|f\| \leq \nu. \tag{6.5.16}$$

Hence, if f is sufficiently small (namely, if $\|f\| \leq \nu^2\mu^{-1}C_{F\Omega}^{-1}$), then a solution u to the stationary Navier–Stokes problem is unique (this fact is well known, see, e.g., [210]).

6.6 Notes for the chapter

Numerical methods for viscous flow problems are represented in many publications. The reader will find a systematic discussion of the topic in, e.g., M. Feistauer [131], V. Girault and P. A. Raviart [152], R. Glowinski and O. Pironneau [157], R. Rannacher [271], R. Rannacher and S. Turek [274], R. Temam [348], and S. Turek [352].

In the last decades, adaptive methods and a posteriori error indicators for approximate solutions of viscous flow problems attracted serious attention of many researches who used methods different from those considered in this chapter. We cannot give here a consequent overview of these results and confine ourselves to a short discussion of some publications that set out the main approaches. The reader will find more literature references in the papers cited.

Residual type a posteriori methods for viscous flow problems are considered in the book by R. Verfürth [356]. A posteriori analysis of approximations computed with the help of the backward Euler scheme is given in C. Bernardi and R. Verfürth [55]. A posteriori error estimators for finite-element approximations of the Stokes problem were obtained in R. E. Bank and B. D. Welfert [43] and R. Verfürth [354]. Error indicators for the Navier–Stokes equations in stream function and vorticity statement are discussed in M. Amara, M. Ben Younes, and C. Bernardi [13]. In D. Kay and D. Silvester [193], various a posteriori estimators are investigated for stabilized mixed approximations of the Stokes problem. Adaptive methods and a posteriori estimates in computational fluid dynamics are exposed, e.g., in M. Ainsworth and J. T. Oden [8], J. G. Heywood and R. Rannacher [172], C. Johnson, R. Rannacher, and M. Boman [189], J. T. Oden, W. Wu and M. Ainsworth [255], J. T. Oden, L. Demkowicz, T. Strouboulis and P. Devloo [252], R. Rannacher [271], T. Strouboulis and J. T. Oden [345]. A posteriori error estimators for some quasi-Newtonian fluids were considered in C. Padra [258] and in A. Bermúdez, R. Durán and R. Rodríguez [53] for combined fluid-solid systems. J. Wang and X. Ye [368] investigated error indicators based on the superconvergence of finite-element approximations for Stokes and Navier–Stokes equations.

For the stationary Stokes problem with Dirichlet boundary conditions, a posteriori estimates of the functional type were originally derived by the author with the help of a variational method in [288]. In [293], it was shown that for the Stokes problem the same estimates follow from a pertinent integral identity. The material exposed in Section 6.2 follow the lines of this paper. Later the variational technique was applied to some classes of generalized Newtonian fluids (see M. Bildhauer, M. Fuchs, and S. Repin [57], M. Fuchs and S. Repin [139], and in the author's papers [289, 292, 293]). Estimates of the same type were derived (by a nonvariational method) for generalizations of the Stokes problem in S. Repin and R. Stenberg [316]. Estimates for flow models with polymerization discussed in Section 6.3.5 were derived in [315]. Estimates for the Oseen problem considered in Section 6.4 were obtained in [293]. Within the framework of a posteriori analysis, problems with rotation were considered in E. Gorshkova, A. Mahalov, P. Neittaanmáki, and S. Repin [159] and in the PhD thesis of E. Gorshkova [158], which also contains results of numerical tests and a comparative study of different error indication methods for approximate solution of viscous flow problems.

7 Generalizations

7.1 Linear elliptic problem

First, we consider the following general form of a linear elliptic problem: Find $u \in u_0 + V_0$ such that

$$(\mathcal{A}\Lambda u, \Lambda w) + \langle \ell, w \rangle = 0, \qquad \forall w \in V_0. \tag{7.1.1}$$

Here V_0 is a closed subspace of a reflexive Banach space V, Λ is a linear bounded operator acting from V to a Hilbert space U with scalar product $(\cdot,\cdot)$, $\ell \in V_0^*$, and $\mathcal{A} \in \mathcal{L}(U,U)$ is a self-adjoint operator. In this section, $\|\cdot\|$ stands for the norm in U. We assume that the operators Λ and $\mathcal{A}$ satisfy the relations

$$c_1^2\|y\|^2 \leq (\mathcal{A}y, y) \leq c_2^2\|y\|^2, \qquad \forall y \in U, \tag{7.1.2}$$

and

$$\|\Lambda w\| \geq c_3\|w\|_V, \qquad \forall w \in V_0, \tag{7.1.3}$$

with positive constants c_1, c_2, and c_3. In most applied problems, the functional ℓ is representable in the form $\langle \ell, w \rangle = (f, w)_{\mathcal{V}} + (g, \Lambda w)$, where $f \in \mathcal{V}$, $g \in U$, and $\mathcal{V}$ is a Hilbert space such that $V \subset \mathcal{V} \subset V_0^*$. Such a functional is well defined and finite on the elements of V. Henceforth, we assume that ℓ satisfies this condition.

For our analysis, it is convenient to introduce two additional spaces. The quantity $(\mathcal{A}y, y)^{1/2}$ determines a new norm $|||y|||$, which is equivalent to the original norm $\|y\| = (y, y)^{1/2}$. Another equivalent norm is

$$|||y|||_* = (\mathcal{A}^{-1}y, y)^{1/2},$$

where $\mathcal{A}^{-1}$ is the operator inverse to $\mathcal{A}$. The spaces Y and Y^* contain elements of U equipped with the norms $|||\cdot|||$ and $|||\cdot|||_*$, respectively.

Here and later on, $\Lambda^* : U \to V_0^*$ denotes the operator conjugate to Λ in the sense that

$$(y, \Lambda w) = \langle \Lambda^* y, w \rangle, \quad w \in V_0, \tag{7.1.4}$$

and $\langle w^*, w \rangle$ is the value of the functional $w^* \in V_0^*$ at $w \in V_0$. Below we derive functional a posteriori estimates for the problem (7.1.1) with the help of two different methods.

7.1.1 The variational method

In the variational method, we rely on the variational statement of the problem: Find $u \in u_0 + V_0$ such that

$$J(u) = \inf_{w \in u_0 + V_0} J(w), \quad \text{where} \quad J(w) := \frac{1}{2}|||\Lambda w|||^2 + \langle \ell, w \rangle.$$

We call this problem *primal* or Problem $\mathcal{P}$.

It is easy to show that it is coercive on $u_0 + V_0$. For any $w \in V_0$, we have

$$\begin{aligned} J(w) &\geq \frac{c_1^2}{2}\|\Lambda(w+u_0)\|^2 - \langle \ell, u_0 \rangle - |\ell|\, |||\Lambda(w)||| \\ &\geq \frac{c_1^2}{2}\|\Lambda w\|^2 + \frac{c_1^2}{2}\|\Lambda u_0\|^2 - \langle \ell, u_0 \rangle - c_2 |\ell|\, \|\Lambda w\| - c_1^2(\Lambda w, \Lambda u_0)_H, \end{aligned}$$

where $|\ell| < +\infty$ is a positive quantity defined by the relation

$$|\ell| := \sup_{w \in V_0} \frac{\langle \ell, w \rangle}{|||\Lambda w|||}.$$

In view of (7.1.3), this norm is equivalent to the standard norm $\sup_{w \in V_0} \frac{\langle \ell, w \rangle}{\|w\|_V}$. It is clear that $J(w) \to +\infty$ as $\|\Lambda w\| \to +\infty$, which together with (7.1.3) proves the coercivity of J on $u_0 + V_0$. Since J is also continuous (on V) and strictly convex, we conclude that the minimizer u exists and is unique.

The variational method of deriving an a posteriori error estimate attracts the so-called dual problem, which we introduce below.

Let $(u_0 + V_0) \times Y^* \to \mathbb{R}$ be the Lagrangian

$$L(v, y) = (y, \Lambda v) - \frac{1}{2}|||y|||_*^2 + \langle \ell, v \rangle.$$

It is easy to see that

$$\sup_{y \in U} L(v, y) = J(v).$$

Define the functional

$$I^*(y) = \inf_{v \in u_0 + V_0} L(v, y) = \begin{cases} (y, \Lambda u_0) - \frac{1}{2}|||y|||_*^2 + \langle \ell, u_0 \rangle, & y \in Q_\ell, \\ -\infty, & y \notin Q_\ell, \end{cases}$$

where

$$Q_\ell := \left\{ y \in U \;\middle|\; (y, \Lambda w) + \langle \ell, w \rangle = 0, \quad \forall w \in V_0 \right\}.$$

The functional I^* generates the following (*dual*) Problem $\mathcal{P}^*$: Find $p \in Q_\ell$ such that

$$I^*(p) = \sup_{y \in Q_\ell} I^*(y) := \sup \mathcal{P}^*.$$

The functionals J and $(-I^*)$ are convex and coercive on u_0+V_0 and Y^*, respectively. The sets u_0+V_0 and Q_ℓ are closed affine manifolds. Therefore (e.g., see [121]), there exist $u \in u_0+V_0$ and $p \in Q_\ell$ such that

$$J(u) = \inf \mathcal{P}, \qquad I^*(p) = \sup \mathcal{P}^*. \tag{7.1.5}$$

The minimizer u satisfies (7.1.1). For any $\epsilon \in \mathbb{R}$ the maximizer p satisfies the relation

$$(p+\epsilon\eta, \Lambda u_0) - \frac{1}{2}|||p+\epsilon\eta|||_*^2 \leq (p, \Lambda u_0) - \frac{1}{2}|||p|||_*^2, \qquad \forall \eta \in Q_0^*,$$

which implies the necessary condition

$$(\Lambda u_0 - \mathcal{A}^{-1}p, \eta) = 0, \qquad \forall \eta \in Q_0^*, \tag{7.1.6}$$

where

$$Q_0^* := \left\{ y \in U \;\middle|\; (y, \Lambda w) = 0, \;\; \forall w \in V_0 \right\}.$$

Note that $\mathcal{A}\Lambda u \in Q_\ell$, so that we obtain

$$I^*(\mathcal{A}\Lambda u) = (\mathcal{A}\Lambda u, \Lambda u_0) - \frac{1}{2}|||\mathcal{A}\Lambda u|||_*^2 + \langle \ell, u_0 \rangle \leq \sup \mathcal{P}^*. \tag{7.1.7}$$

Since

$$|||\mathcal{A}\Lambda u|||_*^2 = (\mathcal{A}^{-1}\mathcal{A}\Lambda u, \mathcal{A}\Lambda u) = |||\Lambda u|||^2$$

and

$$(\mathcal{A}\Lambda u, \Lambda(u-u_0)) + \langle \ell, (u-u_0) \rangle = 0,$$

we find that

$$\sup \mathcal{P}^* \geq I^*(\mathcal{A}\Lambda u) = J(u) = \inf \mathcal{P}. \tag{7.1.8}$$

On the other hand, we know that $\inf \mathcal{P} \geq \sup \mathcal{P}^*$. Therefore,

$$\sup \mathcal{P}^* = \inf \mathcal{P}. \tag{7.1.9}$$

From (7.1.9) it follows that

$$(p, \Lambda u) - \frac{1}{2}|||p|||_*^2 + \langle \ell, u \rangle = \frac{1}{2}|||\Lambda u|||^2 + \langle \ell, u \rangle,$$

which is equivalent to the relation

$$\frac{1}{2}|||\Lambda u|||^2 + \frac{1}{2}|||p|||_*^2 - (p, \Lambda u) = \frac{1}{2}|||\Lambda u - \mathcal{A}^{-1}p|||^2 = 0 \tag{7.1.10}$$

or simply to

$$p = \mathcal{A}\Lambda u. \tag{7.1.11}$$

This is the *duality relation* for the pair (u, p).

Let $v \in u_0+V_0$ and $y \in U$ be some approximations of u and p, respectively. First, we establish the following basic result.

Theorem 7.1. *For any $v \in u_0 + V_0$ and $q \in Q_\ell$, we have*

$$|||\Lambda(v-u)|||^2 + |||q-p|||_*^2 = 2\,(J(v) - I^*(q)), \tag{7.1.12}$$

$$|||\Lambda(v-u)|||^2 + |||q-p|||_*^2 = 2\,D(\Lambda v, q), \tag{7.1.13}$$

where

$$D(\Lambda v, q) := \frac{1}{2}|||\Lambda v|||^2 + \frac{1}{2}|||q|||_*^2 - (q, \Lambda v) = \frac{1}{2}|||\mathcal{A}\Lambda v - q|||_*^2.$$

Proof. In view of (7.1.1) and (7.1.6), we have

$$\begin{aligned}\frac{1}{2}|||\Lambda(v-u)|||^2 &= J(v) - J(u) + (\mathcal{A}\Lambda u, \Lambda(v-u)) + \langle \ell, v-u\rangle \\ &= J(v) - J(u),\end{aligned}$$

and

$$\begin{aligned}\frac{1}{2}|||q-p|||_*^2 &= I^*(p) - I^*(q) + (\Lambda u_0 - \mathcal{A}^{-1}p, p-q) \\ &= I^*(p) - I^*(q).\end{aligned}$$

Since $J(u) = I^*(p)$, these relations imply (7.1.12). For $q \in Q_\ell$ we have

$$\begin{aligned}J(v) - I^*(q) &= \frac{1}{2}|||\Lambda v|||^2 + \langle \ell, v-u_0\rangle - (q, \Lambda u_0) + \frac{1}{2}|||q|||_*^2 \\ &= \frac{1}{2}|||\Lambda v|||^2 + \frac{1}{2}|||q|||_*^2 - (q, \Lambda v) \\ &= D(\Lambda v, q),\end{aligned}$$

so that (7.1.13) follows from (7.1.12). □

Remark 7.2. We note that (7.1.12) and (7.1.13) can be viewed as generalizations of the Mikhlin and Prager–Synge estimates (2.3.1) and (2.2.3), respectively.

By (7.1.13), we conclude that

$$|||\Lambda(v-u)||| \le |||\mathcal{A}\Lambda v - q|||_*, \qquad \forall q \in Q_\ell.$$

Let $y \in U$, then

$$|||\Lambda(v-u)||| \le |||\mathcal{A}\Lambda v - y|||_* + \inf_{q \in Q_\ell} |||q-y|||_*. \tag{7.1.14}$$

Thus, we need a computable estimate of the distance between y and the set Q_ℓ. For this purpose, we introduce the space

$$Q^* := \{y \in U \mid \Lambda^* y \in \mathcal{V}\}$$

endowed with the norm

$$\|y\|_{Q^*} := |\!|\!|y|\!|\!|_* + \|\Lambda^* y\|_{\mathcal{V}}.$$

If $y \in Q^*$, then

$$(y, \Lambda w) = (\Lambda^* y, w)_{\mathcal{V}}, \qquad \forall w \in V_0.$$

We estimate the distance to the set Q_ℓ by the following lemma.

Lemma 7.3. *For any $y \in U$,*

$$\inf_{q \in Q_\ell} |\!|\!|y - q|\!|\!|_* \leq |\Lambda^* y + \ell| := \sup_{w \in V_0} \frac{\langle \ell + \Lambda^* y, w \rangle}{|\!|\!|\Lambda w|\!|\!|}. \tag{7.1.15}$$

If $\langle \ell, w \rangle = (f, w)_{\mathcal{V}}$, then for any $y \in Q^*$,

$$\inf_{q \in Q_\ell} |\!|\!|y - q|\!|\!|_* \leq \mathrm{C}\|\Lambda^* y + f\|_{\mathcal{V}}, \tag{7.1.16}$$

where $\mathrm{C} = \dfrac{c}{c_1}$ and c is the constant in the inequality

$$\|w\|_{\mathcal{V}} \leq c\,\|\Lambda w\|, \qquad \forall w \in V_0. \tag{7.1.17}$$

Proof. Consider the problem

$$(\mathcal{A}\Lambda w_\ell, \Lambda w) = -\langle \ell + \Lambda^* y, w \rangle. \tag{7.1.18}$$

Represent the right-hand side in the form $-\langle \ell, w \rangle - (y, \Lambda w)$. Then, (7.1.18) implies the relation

$$|\!|\!|\Lambda w_\ell|\!|\!| \leq \sup_{w \in V_0} \frac{|(y, \Lambda w) + \langle \ell, w \rangle|}{|\!|\!|\Lambda w|\!|\!|} = |\Lambda^* y + \ell|.$$

Also, (7.1.18) has the form

$$((\mathcal{A}\Lambda w_\ell + y), \Lambda w) + \langle \ell, w \rangle = 0, \qquad \forall w \in V_0,$$

which means that $q_\ell := \mathcal{A}\Lambda w_\ell + y \in Q_\ell$. Therefore,

$$\inf_{q \in Q_\ell} |\!|\!|y - q|\!|\!|_* \leq |\!|\!|y - q_\ell|\!|\!|_* = |\!|\!|\mathcal{A}\Lambda w_\ell|\!|\!|_* = |\!|\!|\Lambda w_\ell|\!|\!| \leq |\Lambda^* y + \ell|.$$

If $\langle \ell, w \rangle = (f, w)_{\mathcal{V}}$ and $y \in Q^*$, then

$$|\Lambda^* y + \ell| = \sup_{w \in V_0} \frac{\langle \Lambda^* y + \ell, w \rangle}{|\!|\!|\Lambda w|\!|\!|} = \sup_{w \in V_0} \frac{(\Lambda^* y + f, w)_{\mathcal{V}}}{|\!|\!|\Lambda w|\!|\!|}$$

$$\leq \sup_{w \in V_0} \frac{\|\Lambda^* y + f\|_{\mathcal{V}} \|w\|_{\mathcal{V}}}{|\!|\!|\Lambda w|\!|\!|} \leq \|\Lambda^* y + f\|_{\mathcal{V}}\, c_1^{-1} \sup_{w \in V_0} \frac{\|w\|_{\mathcal{V}}}{\|\Lambda w\|}.$$

Since V_0 is continuously embedded into $\mathcal{V}$ and (7.1.3) holds, we conclude that (7.1.17) also holds with a constant $c > 0$ independent of w. Then

$$|\Lambda^* y + \ell| \leq \frac{c}{c_1} \|\Lambda^* y + f\|_{\mathcal{V}},$$

and we arrive at (7.1.16). □

Now, (7.1.14) implies two estimates (see also [277, 282, 286]):

$$|||\Lambda(v-u)||| \leq |||\mathcal{A}\Lambda v - y|||_* + |\Lambda^* y + \ell|, \tag{7.1.19}$$

$$|||\Lambda(v-u)||| \leq |||\mathcal{A}\Lambda v - y|||_* + \mathrm{C}\, \|f + \Lambda^* y\|_{\mathcal{V}}. \tag{7.1.20}$$

We recall that in (7.1.19) $y \in U$ and in (7.1.20) $y \in Q^*$.

Henceforth, we denote the right-hand side of (7.1.20) by $\overline{\mathfrak{M}}_\Lambda(v, y)$.

Remark 7.4. It is easy to see that

$$|||\Lambda(u-v)||| = \inf_{y \in Q^*} \Big\{ |||\mathcal{A}\Lambda v - y|||_* + \mathrm{C}\|\ell + \Lambda^* y\| \Big\}.$$

To prove this fact, it suffices to set $y = p$. Hence, the upper bound given by $\overline{\mathfrak{M}}_\Lambda$ has no gap.

From (7.1.20), it also follows that for any $\beta > 0$,

$$\begin{aligned} |||\Lambda(v-u)|||^2 &\leq (1+\beta)|||\Lambda v - y|||_*^2 + \Big(1 + \frac{1}{\beta}\Big)\mathrm{C}^2 \| f + \Lambda^* y\|_{\mathcal{V}}^2 \\ &=: \overline{\mathfrak{M}}^2_\Lambda(v, y, \beta). \end{aligned} \tag{7.1.21}$$

This estimate provides a majorant having the form of a quadratic functional.

Error reduction property. Theorem 7.1 has a simple consequence related to the so-called *error reduction property*. In practice, it is often of interest to predict how significantly an approximation error would decrease on account of a certain improvement of a mesh. Assume that a coarse mesh $\mathcal{T}_h$ is replaced by a refined one $\mathcal{T}_{h_{\mathrm{ref}}}$. For problems associated with quadratic functionals (similar to J), the value of the error reduction is easy to compute. Indeed, for any $v \in u_0 + V_0$,

$$\frac{1}{2}|||\Lambda(u-v)|||^2 = J(v) - I^*(p) = J(v) - J(u).$$

Therefore,

$$\frac{1}{2} e_h^2 := \frac{1}{2}|||\Lambda(u-u_h)|||^2 = J(u_h) - J(u),$$

$$\frac{1}{2} e_{h_{\mathrm{ref}}}^2 := \frac{1}{2}|||\Lambda(u-u_{h_{\mathrm{ref}}})|||^2 = J(u_{h_{\mathrm{ref}}}) - J(u).$$

From the above, a simple *error reduction relation* follows:

$$e^2_{h_{\mathrm{ref}}} = e^2_h - 2\big(J(u_h) - J(u_{h_{\mathrm{ref}}})\big). \tag{7.1.22}$$

We note that the reduction of the approximation error is equal to the difference of the values of corresponding functionals. If an upper bound of the approximation error related to the coarser mesh $\mathcal{T}_h$ is known, i.e.,

$$|||\Lambda(u - u_h)|||^2 \le \overline{M},$$

then for the refined mesh we have the estimate

$$|||\Lambda(u - u_{h_{\mathrm{ref}}})|||^2 \le \overline{M} - 2\big(J(u_h) - J(u_{h_{\mathrm{ref}}})\big). \tag{7.1.23}$$

If $\overline{M}$ is a sharp upper bound of the error on $\mathcal{T}_h$, then (7.1.23) gives a realistic value of the error on $\mathcal{T}_{h_{\mathrm{ref}}}$. Since the right-hand side of (7.1.23) is easily computable, getting the upper bound can be performed with minimal expenditures.

From (7.1.23), it also follows that a mesh-adaptation strategy of refining the mesh should minimize local contributions in the energy functional.

7.1.2 The method of integral identities

Upper bound. Let $v \in u_0 + V_0$ be an approximation of u. Then, the a posteriori estimate can be derived directly from (7.1.1). Indeed,

$$(\mathcal{A}\Lambda(u - v), \Lambda w) = -\langle \ell + \Lambda^* y, w\rangle + (y - \mathcal{A}\Lambda v, \Lambda w). \tag{7.1.24}$$

Since

$$(\mathcal{A}\Lambda v - y, \Lambda w) \le |||\mathcal{A}\Lambda v - y|||_* |||\Lambda w|||$$

and

$$\langle \ell + \Lambda^* y, w\rangle \le |\ell + \Lambda^* y|\, |||\Lambda w|||,$$

we find that

$$(\mathcal{A}\Lambda(u - v), \Lambda w) \le \big(|||\mathcal{A}\Lambda v - y|||_* + |\ell + \Lambda^* y|\big)|||\Lambda w|||.$$

Setting $w = u - v$, we arrive at (7.1.19).

If $\langle \ell, w\rangle = (f, w)_{\mathcal{V}}$ and $y \in Q^*$, then

$$\begin{aligned}\langle \ell + \Lambda^* y, u - v\rangle &= (\ell + \Lambda^* y, u - v)_{\mathcal{V}} \le \|\ell + \Lambda^* y\|_{\mathcal{V}} \|u - v\|_{\mathcal{V}} \\ &\le c\|\ell + \Lambda^* y\|_{\mathcal{V}} \|\Lambda(u - v)\| \le C\|\ell + \Lambda^* y\|_{\mathcal{V}} |||\Lambda(u - v)|||\end{aligned}$$

and (7.1.24) implies (7.1.20).

If $\langle \ell, w\rangle = (f, w)_V + (g, \Lambda w)$ and $y \in Q^*$, then

$$\langle \ell + \Lambda^* y, w\rangle = \langle \ell, w\rangle + \langle \Lambda^* y, w\rangle = (f + \Lambda^* y, w)_V + (g, \Lambda w).$$

From (7.1.24), we obtain

$$(\mathcal{A}\Lambda(u - v), \Lambda w) = -(f + \Lambda^* y, w)_V + (y - \mathcal{A}\Lambda v - g, \Lambda w). \quad (7.1.25)$$

For $w = u - v$, (7.1.25) implies the estimate

$$|||\Lambda(u - v)||| \le |||g + \mathcal{A}\Lambda v - y|||_* + \mathrm{C}\|f + \Lambda^* y\|_V. \quad (7.1.26)$$

If g is efficiently approximated by $\widetilde{g} \in Q^*$, then another estimate can be used:

$$|||\Lambda(u - v)||| \le |||\mathcal{A}\Lambda v - y|||_* + \mathrm{C}\|f + \Lambda^*(y + \widetilde{g})\|_V + \|g - \widetilde{g}\|. \quad (7.1.27)$$

Lower bound. A lower bound of $|||\Lambda(u - v)|||$ can be derived in the following way. Note that

$$\begin{aligned}\sup_{w \in V_0} \left\{(\mathcal{A}\Lambda(u - v), \Lambda w) - \frac{1}{2}(\mathcal{A}\Lambda w, \Lambda w)\right\} &\le \sup_{\tau \in Y} \left\{(\mathcal{A}\Lambda(u - v), \tau) - \frac{1}{2}(\mathcal{A}\tau, \tau)\right\} \\ &= \frac{1}{2}|||\Lambda(u - v)|||^2.\end{aligned}$$

However,

$$\begin{aligned}&\sup_{w \in V_0} \left\{(\mathcal{A}\Lambda(u - v), \Lambda w) - \frac{1}{2}(\mathcal{A}\Lambda w, \Lambda w)\right\} \\ &\quad \ge (\mathcal{A}\Lambda(u - v), \Lambda(u - v)) - \frac{1}{2}(\mathcal{A}\Lambda(u - v), \Lambda(u - v)) = \frac{1}{2}|||\Lambda(u - v)|||^2.\end{aligned}$$

Thus, we conclude that

$$\begin{aligned}\frac{1}{2}|||\Lambda(u - v)|||^2 &= \sup_{w \in V_0} \left\{(\mathcal{A}\Lambda(u - v), \Lambda w) - \frac{1}{2}(\mathcal{A}\Lambda w, \Lambda w)\right\} \\ &\ge \sup_{w \in V_0} \left\{-\frac{1}{2}(\mathcal{A}\Lambda w, \Lambda w) - (\mathcal{A}\Lambda v, \Lambda w) - \langle \ell, w\rangle\right\}. \quad (7.1.28)\end{aligned}$$

It is easy to see that the lower bound given by the left-hand side of the above estimate is sharp (set $w = u - v$).

Modifications of the estimates. Let $\vartheta \in V_0$ be such that $\Lambda^* \mathcal{A}\Lambda\vartheta \in \mathcal{V}$. Then, the product $\langle \Lambda^* \mathcal{A}\Lambda\vartheta, w\rangle$ is represented by the scalar product of $\mathcal{V}$ and we have

$$(\Lambda^* \mathcal{A}\Lambda\vartheta, v - u)_{\mathcal{V}} = -(\mathcal{A}\Lambda\vartheta, \Lambda u) + (\mathcal{A}\Lambda\vartheta, \Lambda v) = \langle \ell, \vartheta\rangle + (\mathcal{A}\Lambda\vartheta, \Lambda v).$$

Then, for $\langle \ell, w\rangle = (f, w)_{\mathcal{V}}$ and $y \in Q^*$ we express (7.1.24) in the form

$$\begin{aligned} |||\Lambda(u - v)|||^2 = (f + \Lambda^* y - \Lambda^* \mathcal{A}\Lambda\vartheta, v - u)_{\mathcal{V}} \\ + (y - \mathcal{A}\Lambda v, \Lambda(u - v)) + \mathcal{F}_v(\vartheta), \end{aligned} \tag{7.1.29}$$

where $\mathcal{F}_v(\vartheta) := \langle \ell, \vartheta\rangle + (\mathcal{A}\Lambda\vartheta, \Lambda v)_{\mathcal{V}}$. Hence, we obtain

$$\begin{aligned} |||\Lambda(u - v)|||^2 \le \frac{2\gamma}{2\gamma - 1}\mathcal{F}_v(\vartheta) + \frac{\gamma^2}{2\gamma - 1}\Big(|||\mathcal{A}\Lambda v - y|||_* \\ + \mathrm{C}\|f + \Lambda^*(y - \mathcal{A}\Lambda\vartheta)\|_{\mathcal{V}}\Big)^2. \end{aligned} \tag{7.1.30}$$

If ϑ solves the problem

$$(\mathcal{A}\Lambda\vartheta, \Lambda w) = (f + \Lambda^* y, w)_{\mathcal{V}}, \qquad \forall w \in V_0,$$

then (7.1.29) implies that

$$|||\Lambda(u - v)|||^2 \le \frac{2\gamma}{2\gamma - 1}\mathcal{F}_v(\vartheta) + \frac{\gamma^2}{2\gamma - 1}|||\mathcal{A}\Lambda v - y|||_*^2. \tag{7.1.31}$$

Another modification is based on a generalization of the idea that was used in Section 3.5.3. Consider the functional spaces U, $\mathcal{V}$, V, and Y contain functions defined in a domain Ω, which is decomposed into a collection of subdomains Ω_i, $i = 1, 2, \ldots, N$. By $\mathcal{V}(\Omega_i)$ we denote the restriction of $\mathcal{V}$ associated with Ω_i and assume that the respective scalar product is additive with respect to the above decomposition. Let

$$K_i := \{w \in \mathcal{V}(\Omega_i) \mid \Lambda w = 0 \text{ in } \Omega_i\}$$

and $\{w\}_{\Omega_i} \in K_i$ denote the orthogonal projection of w, i.e.,

$$\|w - \{w\}_{\Omega_i}\|_{\mathcal{V}(\Omega_i)} = \inf_{w_0 \in K_i} \|w - w_0\|_{\mathcal{V}(\Omega_i)}.$$

In this case,

$$(w - \{w\}_{\Omega_i}, w_0)_{\mathcal{V}(\Omega_i)} = 0, \qquad \forall\, w_0 \in K_i. \tag{7.1.32}$$

Assume that for any Ω_i,

$$\|w - \{w\}_{\Omega_i}\|_{\mathcal{V}(\Omega_i)} \le C_{\Omega_i}\|\Lambda w\|_{U(\Omega_i)}, \tag{7.1.33}$$

where C_{Ω_i} is a positive constant depending only on Ω_i (this estimate is a generalization of the Poincaré inequality). Since $\{w\}_{\Omega_i} \in K_i$, we have

$$(\mathrm{r}(y), w)_{\mathcal{V}(\Omega_i)} = (\mathrm{r}(y) - \{\mathrm{r}(y)\}_{\Omega_i}, w)_{\mathcal{V}(\Omega_i)} + (\{\mathrm{r}(y)\}_{\Omega_i}, w)_{\mathcal{V}(\Omega_i)}$$
$$= (\mathrm{r}(y) - \{\mathrm{r}(y)\}_{\Omega_i}, w - \{w\}_{\Omega_i})_{\mathcal{V}(\Omega_i)} + (\{\mathrm{r}(y)\}_{\Omega_i}, w)_{\mathcal{V}(\Omega_i)}, \quad \forall w \in \mathcal{V}(\Omega_i),$$

where $\mathrm{r}(y) := f + \Lambda^* y$. Then

$$\begin{aligned}(\mathrm{r}(y), w)_{\mathcal{V}} &= \sum_{i=1}^{N} (\mathrm{r}(y), w)_{\mathcal{V}(\Omega_i)} \\ &= \sum_{i=1}^{N} \big((\mathrm{r}(y) - \{\mathrm{r}(y)\}_{\Omega_i}, w - \{w\}_{\Omega_i})_{\mathcal{V}(\Omega_i)} + (\{\mathrm{r}(y)\}_{\Omega_i}, w)_{\mathcal{V}(\Omega_i)} \big) \\ &\leq \left(\sum_{i=1}^{N} C^2_{\Omega_i} \|\mathrm{r}(y) - \{\mathrm{r}(y)\}\|^2_{\Omega_i} \right)^{1/2} \left(\sum_{i=1}^{N} \|\Lambda w\|^2_{U(\Omega_i)} \right)^{1/2} \\ &\quad + \left(\sum_{i=1}^{N} \|\{\mathrm{r}(y)\}_{\Omega_i}\|^2_{\mathcal{V}(\Omega_i)} \right)^{1/2} \left(\sum_{i=1}^{N} \|w\|^2_{\mathcal{V}(\Omega_i)} \right)^{1/2},\end{aligned}$$

and we arrive at the estimate

$$\begin{aligned}|(\mathrm{r}(y), w)_{\mathcal{V}}| &\leq \left(\sum_{i=1}^{N} C^2_{\Omega_i} \|\mathrm{r}(y) - \{\mathrm{r}(y)\}\|^2_{\Omega_i} \right)^{1/2} \|\Lambda w\| \\ &\quad + c \left(\sum_{i=1}^{N} \|\{\mathrm{r}(y)\}_{\Omega_i}\|^2_{\mathcal{V}(\Omega_i)} \right)^{1/2} \|\Lambda w\|. \end{aligned} \tag{7.1.34}$$

By (7.1.24), we have

$$(\mathcal{A}\Lambda(u - v), \Lambda w) \leq |(\mathrm{r}(y), w)_{\mathcal{V}}| + |||y - \mathcal{A}\Lambda v|||_* |||\Lambda w|||. \tag{7.1.35}$$

Set here $w = u - v$ and use (7.1.34). Then, we obtain the estimate

$$\begin{aligned}|||\Lambda(u - v)||| &\leq |||y - \mathcal{A}\Lambda v|||_* + \frac{1}{c_1} \left(\sum_{i=1}^{N} C^2_{\Omega_i} \|\mathrm{r}(y) - \{\mathrm{r}(y)\}_{\Omega_i}\|^2_{\Omega_i} \right)^{1/2} \\ &\quad + \mathrm{C} \left(\sum_{i=1}^{N} \|\{\mathrm{r}(y)\}_{\Omega_i}\|^2_{\mathcal{V}(\Omega_i)} \right)^{1/2}. \end{aligned} \tag{7.1.36}$$

If the residuals are post-processed in such a way that $\{\mathrm{r}(y)\}_{\Omega_i} = 0$ on any Ω_i, then (7.1.36) takes a simplified form:

$$|||\Lambda(u-v)||| \le |||y - \mathcal{A}\Lambda v|||_* + \frac{1}{c_1}\left(\sum_{i=1}^{N} C^2_{\Omega_i}\|\mathrm{r}(y) - \{\mathrm{r}(y)\}_{\Omega_i}\|^2_{\Omega_i}\right)^{1/2}, \tag{7.1.37}$$

which can be viewed as a generalization of (3.5.20).

Remark 7.5. If Λw is defined by the operator of small strains $\varepsilon(w)$, then $\{w\}_{\Omega_i}$ denotes the orthogonal projection to the space $\mathrm{R}(\Omega_i)$ of rigid deflections.

7.1.3 Error estimates for the dual variable

Consider y as an approximation of p (which is the exact solution of the dual problem). To obtain an upper bound of $|||p - y|||_*$, we use the relation

$$\begin{aligned} |||p - y|||_* &\le |||y - \mathcal{A}\Lambda v|||_* + |||\Lambda(v-u)||| \\ &\le 2|||y - \mathcal{A}\Lambda v|||_* + |\ell + \Lambda^* y|. \end{aligned} \tag{7.1.38}$$

If $\ell = f \in \mathcal{V}$ and $y \in Q^*$, then

$$\begin{aligned} |||p - y|||_* &\le |||y - \mathcal{A}\Lambda v|||_* + |||\Lambda(v-u)||| \\ &\le 2|||y - \mathcal{A}\Lambda v|||_* + \mathrm{C}\|\Lambda^* y + f\|_{\mathcal{V}}. \end{aligned} \tag{7.1.39}$$

The norm $\|y\|_{Q^*}$ is another measure that characterizes the error $p - y$. Since

$$\|\Lambda^*(p-y)\| = \|\Lambda^* y + f\|,$$

we note that

$$\|p - y\|_{Q^*} \le 2|||y - \mathcal{A}\Lambda v|||_* + (\mathrm{C}+1)\|\Lambda^* y + f\|_{\mathcal{V}}. \tag{7.1.40}$$

7.1.4 Two-sided estimates for combined norms

If a pair (u, p) is viewed as a solution, then it is natural to measure the corresponding error in terms of combined (primal-dual) norms of the product space $W := V \times Q^*$, for which we introduce the norm

$$\|(v, y)\|_W := |||\Lambda v||| + \|y\|_{Q^*}.$$

We show that the majorant $\overline{\mathfrak{M}}_\Lambda(v, y)$ is *equivalent to the error* in the combined norm $\|(v, y)\|_W$. We have

$$\begin{aligned} \|(u - v, p - y)\|_W &:= |||\Lambda(u-v)||| + |||p - y|||_{Q^*} \\ &\le 3|||\mathcal{A}\Lambda v - y||| + (2\mathrm{C}+1)\|\Lambda^* y + f\| \le c_\oplus \overline{\mathfrak{M}}_\Lambda(v, y), \end{aligned} \tag{7.1.41}$$

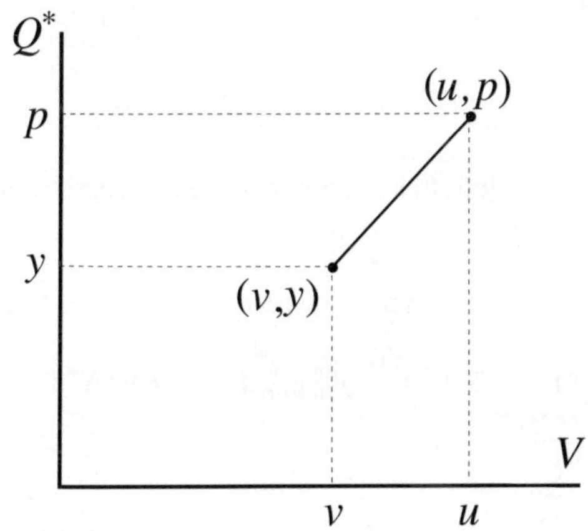

Figure 7.1.1 Error in terms of the combined primal-dual norm.

where $c_{\oplus} = \max\left\{3, 2 + \frac{1}{\mathrm{C}}\right\}$. On the other hand,

$$\begin{aligned}\overline{\mathfrak{M}}_{\Lambda}(v, y) &\leq |||\Lambda(v-u)||| + |||p - y|||_* + \mathrm{C}\|\Lambda^* y + f\| \\ &\leq \max\{1, \mathrm{C}\}\,\|(u-v, p-y)\|_W .\end{aligned} \tag{7.1.42}$$

Thus, we note that the following two-sided estimate holds:

$$c_{\ominus}\overline{\mathfrak{M}}_{\Lambda}(v, y) \leq \|(u-v, p-y)\|_W \leq c_{\oplus}\overline{\mathfrak{M}}_{\Lambda}(v, y), \tag{7.1.43}$$

where $c_{\ominus} = \frac{1}{\max\{1,\mathrm{C}\}}$.

Hence, the efficiency index of the majorant (with respect to the combined error norm) is estimated from above as follows:

$$I_{\mathrm{eff}} \leq \frac{c_{\oplus}}{c_{\ominus}} = \max\{1, \mathrm{C}\} \max\left\{3, 2 + \tfrac{1}{\mathrm{C}}\right\}. \tag{7.1.44}$$

Therefore, $\overline{\mathfrak{M}}_{\Lambda}$ is an *efficient* and *reliable* measure of the error in the combined norm $\|(u-v, p-y)\|_W$.

Remark 7.6. Since

$$\overline{\mathfrak{M}}_{\Lambda}(v, p) = \|\Lambda(u-v)\| \qquad \text{and} \qquad \overline{\mathfrak{M}}_{\Lambda}(u, y) = |||p - y|||_* + \|\Lambda^*(p-y)\|,$$

we note that

$$\|(u-v, p-y)\|_W := \overline{\mathfrak{M}}_{\Lambda}(v, p) + \overline{\mathfrak{M}}_{\Lambda}(u, y). \tag{7.1.45}$$

Remark 7.7. If the error is measured in a different (but equivalent) norm

$$\|(v, y)\|_W^{(1)} := |||\Lambda v||| + |||y|||_* + \mathrm{C}\|\Lambda^* y\|,$$

then

$$\overline{\mathfrak{M}}_{\Lambda}(v,y) \le \|(u-v,p-y)\|_{W}^{(1)} \le 3\,\overline{\mathfrak{M}}_{\Lambda}(v,y). \tag{7.1.46}$$

Thus, the majorant is equivalent to such a norm and the corresponding efficiency index does not depend on c.

We can give another interpretation of the results discussed. Introduce the functionals

$$\overline{\mathbb{M}}_{\Lambda}(v,y) = 3|||\mathcal{A}\Lambda v - y|||_* + (1+2c)\,\|\Lambda^* y + \ell\|$$

and

$$\underline{\mathbb{M}}_{\Lambda}(v,y) := |||\mathcal{A}\Lambda v - y|||_* + \|\Lambda^* y + \ell\|,$$

which consist of the same terms as does the majorant but with different weights. We have proved that

$$\underline{\mathbb{M}}_{\Lambda}(v,y) \le \|(u-v,p-y)\|_{W} \le \overline{\mathbb{M}}_{\Lambda}(v,y). \tag{7.1.47}$$

In other words, if the terms $|||\mathcal{A}\Lambda v - y|||_*$ and $\|\Lambda^* y + \ell\|$ are supplied with proper weights, then their sum furnishes two-sided bounds of the error in the combined norm.

Comments. The functionals $\overline{\mathfrak{M}}_{\Lambda}(v,y)$, $\overline{\mathbb{M}}_{\Lambda}(v,y)$ and $\underline{\mathbb{M}}_{\Lambda}(v,y)$ provide natural error estimation tools for approximations computed with the help of the *mixed* finite element method. Mixed approximations of boundary value problems are given by a pair of functions (u_h, p_h), which can be substituted in the corresponding majorant directly or after a certain post-processing procedure. A priori and a posteriori estimates for mixed approximations were investigated by many authors. An elaborated theory of mixed finite element methods can be found in the books by F. Brezzi and M. Fortin [79], D. Braess [67], J. E. Roberts and J.-M. Thomas [325]. Various a posteriori error estimators for mixed finite element methods were studied by many authors (e.g., see A. Alonso [12], I. Babuška and G. N. Gatica [26], G. F. Carey and A. I. Pehlivanov [85], C. Carstensen [86, 88], C. Carstensen and G. Dolzmann [91], C. Carstensen and R. H. W. Hoppe [94], B. I. Wohlmuth and R. H. W. Hoppe [372]). In G. Gatica [149], the reader will find a proof of the efficiency of residual-based a posteriori error estimators for mixed approximations. A posteriori estimates for approximations based on Raviart–Thomas elements are represented by D. Braess and R. Verfürth [70], and estimates based on superconvergence phenomenon are analyzed by J. H. Brandts [74]. A posteriori error estimators for mixed approximations of problems in linear elasticity theory were investigated by M. Lonsing and R. Verfürth [223] and for mixed approximations of the Stokes problem by X. Cheng, W. Han, and H. Huang [104]. In B. Engelmann, R. H. W. Hoppe, B. Wohlmuth, Yu. Kuznetsov, Yu. Iliash, and Yu. Vasilevskii [122], adaptive methods for hybrid finite element approximations are considered.

Functional a posteriori estimates for mixed approximations have been studied in S. Repin, S. Sauter, and A. Smolianski [314, 313], where they were derived for linear elliptic problems by variational techniques.

These estimates are applicable to mixed approximations of all types. In particular, they applicable to the so-called "dual mixed approximations", which use nonconforming (with respect to the energy space V) approximations of u and "cell-equilibrated" approximations for p (which integrally satisfy the relation $\Lambda^* p_h + f = 0$ on each cell/element). Approximations of such a type gained high popularity, because they preserve relations coming from the physical (conservation) law and obtain a high flexibility. The latter property was used to adapt them to approximations on highly distorted meshes (e.g., see Yu. Kuznetsov and S. Repin [206, 205]). A posteriori estimates for the dual-mixed approximations follow from the above-presented estimates, provided that the function v is post-processed (e.g., smoothed) in such a way that $v \in u_0 + V_0$.

7.2 Elliptic problems with lower terms

Now we consider the problem: Find $u \in u_0 + V_0$ such that

$$(\mathcal{A}\Lambda u, \Lambda w) + (Tu, w)_H + \langle \ell, w \rangle = 0, \qquad w \in V_0, \tag{7.2.1}$$

where $T : \mathcal{V} \to \mathcal{V}$ is a linear continuous operator satisfying the relation

$$\mu_1^2 \|w\|_{\mathcal{V}}^2 \leq (Tw, w)_{\mathcal{V}} \leq \mu_2^2 \|w\|_{\mathcal{V}}^2. \tag{7.2.2}$$

The function u minimizes the functional

$$J(w) = \frac{1}{2}\Big((\mathcal{A}\Lambda w, \Lambda w) + (Tw, w)_{\mathcal{V}}\Big) + \langle \ell, w \rangle$$

on the set $u_0 + V_0$. Under the assumptions made, $J(w)$ is coercive on $u_0 + V_0$ and, therefore, the minimizer u exists.

For a function $v \in u_0 + V_0$, we have

$$(\mathcal{A}\Lambda(u - v), \Lambda w) + (T(u - v), w)_{\mathcal{V}} = -\langle \ell, w \rangle - (\mathcal{A}\Lambda v, \Lambda w) - (Tv, w)_{\mathcal{V}}. \tag{7.2.3}$$

Let $y \in U$ and $w = u - v$. Then, (7.2.3) infers the relation

$$|[u - v|]^2 = -\langle \ell + \Lambda^* y, u - v \rangle - (\mathcal{A}\Lambda v - y, \Lambda(u - v)) - (Tv, u - v)_{\mathcal{V}}. \tag{7.2.4}$$

Here,

$$|[u - v|]^2 := (\mathcal{A}\Lambda(u - v), \Lambda(u - v)) + (T(u - v), u - v)_{\mathcal{V}}$$

is the energy error norm associated with the problem.

If ℓ is defined by an element $f \in \mathcal{V}$ and $y \in Q^*$, then (7.2.4) is rearranged as follows:

$$|[u - v]|^2 = (f + \Lambda^* y + Tv, v - u)_{\mathcal{V}} + (\mathcal{A}\Lambda v - y, \Lambda(v - u)). \tag{7.2.5}$$

We have

$$(f + \Lambda^* y + Tv, u - v)_{\mathcal{V}} \leq \|\mathrm{r}(v,y)\|_{\mathcal{V}} \|u - v\|_{\mathcal{V}}$$
$$\leq (T^{-1}\mathrm{r}(v,y), \mathrm{r}(v,y))_{\mathcal{V}}^{1/2} (T(u-v), u-v)_{\mathcal{V}}^{1/2},$$
$$(\mathcal{A}\Lambda v - y, \Lambda(u-v)) \leq |||\mathcal{A}\Lambda v - y|||_* |||\Lambda(u-v)|||,$$

where $\mathrm{r}(v,y) := f + \Lambda^* y + Tv$ With the help of these estimates and (7.2.5), we obtain

$$|[u - v]|^2 \leq (T^{-1}\mathrm{r}(v,y), \mathrm{r}(v,y))_{\mathcal{V}} + |||\mathcal{A}\Lambda v - y|||_*^2. \tag{7.2.6}$$

This estimate is sharp. Indeed, let $y = \mathcal{A}\Lambda u$. By (7.3.1), we find that

$$(Tu, w)_{\mathcal{V}} = -(f, w)_{\mathcal{V}} - (y, \Lambda w), \qquad w \in V_0,$$

which means that

$$(T(u - v), w)_{\mathcal{V}} = -(f + \Lambda^* y + Tv, w)_{\mathcal{V}}, \qquad w \in V_0,$$

and, therefore, $\mathrm{r}(v, y) = T(v - u)$. Hence,

$$(T^{-1}\mathrm{r}(v,y), \mathrm{r}(v,y))_{\mathcal{V}} = (T(u - v), u - v)_{\mathcal{V}}.$$

Also,

$$|||\mathcal{A}\Lambda v - y|||_* = |||\mathcal{A}\Lambda(v - u)|||_* = |||\Lambda(v - u)|||$$

and the right-hand side of (7.2.6) is equal to the error norm $|[u - v]|^2$.

Remark 7.8. Since the problem has a variational statement, we can derive a posteriori estimates using the variational method (see [276, 277, 282]), which gives the same result.

Another estimate follows from (7.2.4) if we apply (7.1.17) and estimate the first term on the right-hand side of (7.2.5) as follows:

$$(\Lambda^* y + f + Tv, u - v)_{\mathcal{V}} \leq \mathrm{C}\|\Lambda^* y + f + Tv\|_{\mathcal{V}} \|\Lambda(u - v)\|.$$

In this case, we arrive at the estimate

$$|||\Lambda(u - v)||| \leq \|\mathcal{A}\Lambda v - y\|_* + \mathrm{C}\|\Lambda^* y + f + Tv\|_{\mathcal{V}}. \tag{7.2.7}$$

Remark 7.9. Also, we can derive a hybrid estimate if we introduce a function $\alpha(x) \in [0, 1]$ and split the first term on the right-hand side of (7.2.5) (cf. Section 4.2). Then, we obtain the estimate

$$|[\Lambda(u - v)]|^2 \leq (\|\mathcal{A}\Lambda v - y\|_* + \mathrm{C}\|(1 - \alpha)\mathrm{r}(v,y)\|_{\mathcal{V}})^2$$
$$+ (\alpha T^{-1}\mathrm{r}(v,y), \mathrm{r}(v,y))_{\mathcal{V}}. \tag{7.2.8}$$

For $\alpha = 0$ and $\alpha = 1$, (7.2.8) implies the estimates (7.2.7) and (7.2.6), respectively.

7.3 Problems with solutions defined in subspaces

7.3.1 Abstract problem

The above-discussed method can be extended to problems the solutions of which belong to a certain subspace of the basic energy space.

Define another pair of mutually conjugate linear operators $B : V_0 \to H$ and $B^* : H \to V_0^*$, where H is a Hilbert space endowed with scalar product $(\cdot,\cdot)_H$. The spaces and operators introduced are conveniently represented by the following diagram:

$$\begin{array}{ccccc} H & \overset{B}{\longleftarrow} & V_0 & \overset{\Lambda}{\longrightarrow} & U \quad (Y, Y^*) \\ & & \updownarrow & & \\ H & \overset{B^*}{\longrightarrow} & V_0^* & \overset{\Lambda^*}{\longleftarrow} & U. \end{array}$$

Consider the problem: Find $p \in H$ and $u \in V_{0,B}$ satisfying the relation

$$(\mathcal{A}\Lambda u, \Lambda w) + \langle \ell - B^* p, w\rangle = 0, \qquad w \in V_0, \tag{7.3.1}$$

where

$$V_{0,B} := \{v \in V_0 \mid Bv = 0\}$$

is a subspace defined by the kernel of B.

For $q \in H$, we have

$$\langle B^* q, w\rangle = (q, Bw)_H = 0, \qquad \forall w \in V_{0,B}. \tag{7.3.2}$$

Therefore, we can also define u by the relation

$$(\mathcal{A}\Lambda u, \Lambda w) + \langle \ell, w\rangle = 0, \qquad w \in V_{0,B}. \tag{7.3.3}$$

The solvability of this problem requires special properties of the operator B, which are analogous to the LBB condition (see Lemma 7.10). We assume that the necessary conditions are satisfied and (7.3.1) has a solution $u \in V_{0,B}$. Our goal is to present a general scheme that delivers guaranteed estimates of the deviation from u.

7.3.2 Estimate for approximations lying in the subspace

Let $v \in V_{0,B}$. From (7.3.2), it follows that for any $w \in V_{0,B}$ and $y \in U$ the identity

$$(\mathcal{A}\Lambda(u - v), \Lambda w) = (y - \mathcal{A}\Lambda v, \Lambda w) - \langle \ell + \Lambda^* y, w\rangle \tag{7.3.4}$$

holds. In view of (7.3.2), we have

$$\sup_{w \in V_{0,B}} \frac{|\langle \ell + \Lambda^* y, w\rangle|}{|||\Lambda w|||} = \sup_{w \in V_{0,B}} \frac{|\langle \ell + \Lambda^* y - B^* q, w\rangle|}{|||\Lambda w|||}$$
$$\leq \sup_{w \in V_0} \frac{|\langle \ell + \Lambda^* y - B^* q, w\rangle|}{|||\Lambda w|||} =: |\ell + \Lambda^* y - B^* q|.$$

Thus (7.3.4) with $w = u - v$ leads to the estimate

$$|||\Lambda(u-v)||| \leq |||y - \mathcal{A}\Lambda v|||_* + |\ell + \Lambda^* y - B^* q|, \tag{7.3.5}$$

where $q \in H$. If ℓ is defined by $f \in \mathcal{V}$ and $\Lambda^* y$ and $B^* q$ also belong to $\mathcal{V}$, then

$$\langle \ell + \Lambda^* y - B^* q, w\rangle = (f + \Lambda^* y - B^* q, w)_{\mathcal{V}} \leq \|f + \Lambda^* y - B^* q\|_{\mathcal{V}} \|w\|_{\mathcal{V}},$$

and we find that

$$\sup_{w \in V_0} \frac{|\langle \ell + \Lambda^* y - B^* q, w\rangle|}{|||\Lambda w|||} \leq c \sup_{w \in V_0} \frac{\|f + \Lambda^* y - B^* q\|_{\mathcal{V}} \|\Lambda w\|}{|||\Lambda w|||}.$$

In this case,

$$|\ell + \Lambda^* y - B^* q| \leq \mathrm{C}\|f + \Lambda^* y - B^* q\|_{\mathcal{V}},$$

and we arrive at the estimate

$$|||\Lambda(u-v)||| \leq |||y - \mathcal{A}\Lambda v|||_* + \mathrm{C}\,\|f + \Lambda^* y - B^* q\|_{\mathcal{V}}. \tag{7.3.6}$$

Another form of the estimate arises if there exists a linear continuous operator $T : H \to U$ such that

$$\langle B^* q, w\rangle = (Tq, \Lambda w), \qquad \forall w \in V_0.$$

Then, (7.3.4) is rearranged as

$$(\mathcal{A}\Lambda(u-v), \Lambda w) = (y - \mathcal{A}\Lambda v - Tq, \Lambda w) - \langle \ell + \Lambda^* y, w\rangle, \tag{7.3.7}$$

and instead of (7.3.5) we obtain

$$|||\Lambda(u-v)||| \leq |||y - \mathcal{A}\Lambda v - Tq|||_* + |\ell + \Lambda^* y|. \tag{7.3.8}$$

If $\ell = f$ and $\Lambda^* y$ belongs to $\mathcal{V}$, then (7.3.8) implies the estimate

$$|||\Lambda(u-v)||| \leq |||y - \mathcal{A}\Lambda v - Tq|||_* + \mathrm{C}\|f + \Lambda^* y\|_{\mathcal{V}}, \tag{7.3.9}$$

which is a counterpart of (7.3.5).

7.3.3 Estimate for approximations lying in the energy space

For functions that do not belong to the subspace $V_{0,B}$, the deviations from u can be estimated, provided that we can prove Lemma 7.10 (which is analogous to Lemma 6.2).

Lemma 7.10. *For any $g \in \operatorname{Im} B$, there exists $v_g \in V_0$ such that*

$$Bv_g = g \quad \text{and} \quad \|v_g\|_V \leq \kappa \|g\|_H, \tag{7.3.10}$$

where $\kappa > 0$ does not depend on g.

In general, this lemma can be proved with the help of *closed range lemma* (e.g., see K. Yosida [374]), which is valid for operators with closed range (we recall that the operator B has a closed range if for any sequence $\{v_k\} \in V$ such that Bv_k converges in H there exists $v \in V$ such that $Bv_k \to Bv$).

Corollary 7.11. *Set $g = B\widehat{v}$, where $\widehat{v}$ is an element of V_0. Then, we can find an element $v_g \in V_0$ satisfying the relation $B(\widehat{v} - v_g) = 0$ and such that*

$$\|v_g\|_V \leq \kappa \|B\widehat{v}\|. \tag{7.3.11}$$

Hence, the function $w_0 = (\widehat{v} - v_g)$ belong to $V_{0,B}$ and satisfies the relation

$$\|\widehat{v} - w_0\|_V \leq \kappa \|B\widehat{v}\|.$$

Since Λ is a bounded operator, we have the inequality

$$\|\Lambda(\widehat{v} - w_0)\| \leq \bar{\kappa} \|B\widehat{v}\|, \tag{7.3.12}$$

where the constant $\bar{\kappa}$ does not depend on $\widehat{v}$.

Let u be compared with a function $\widehat{v} \in V_0$. Take a function $w_0 \in V_{0,B}$ that satisfies (7.3.12). Then,

$$\begin{aligned} |||\Lambda(u - \widehat{v})||| &\leq |||\Lambda(u - w_0)||| + |||\Lambda(\widehat{v} - w_0)||| \\ &\leq |||\mathcal{A}\Lambda w_0 - y|||_* + |\ell + \Lambda^* y - B^* q| + |||\Lambda(\widehat{v} - w_0)|||. \end{aligned}$$

Insert $\widehat{v}$ into the first term and use the triangle inequality. We have

$$\begin{aligned} |||\Lambda(u - \widehat{v})||| &\leq |||\mathcal{A}\Lambda(\widehat{v} - w_0)|||_* + |||\mathcal{A}\Lambda\widehat{v} - y|||_* \\ &\quad + |\ell + \Lambda^* y - B^* q| + |||\Lambda(\widehat{v} - w_0)||| \\ &= 2|||\Lambda(\widehat{v} - w_0)||| + |||\mathcal{A}\Lambda w_0 - y|||_* + |\ell + \Lambda^* y - B^* q| \\ &\leq 2c_2\|\Lambda(\widehat{v} - w_0)\| + |||\mathcal{A}\Lambda w_0 - y|||_* + |\ell + \Lambda^* y - B^* q|. \end{aligned} \tag{7.3.13}$$

Hence,

$$|||\Lambda(u - \widehat{v})||| \leq 2c_2\bar{\kappa}\|B\widehat{v}\| + |||\mathcal{A}\Lambda\widehat{v} - y|||_* + |\ell + \Lambda^* y - B^* q|. \tag{7.3.14}$$

If ℓ is defined by $f \in \mathcal{V}$ and $\Lambda^* y$ and $B^* q$ also belong to $\mathcal{V}$, then (7.3.14) implies the estimate

$$|||\Lambda(u - \widehat{v})||| \leq 2c_2\bar{\kappa}\|B\widehat{v}\| + |||\mathcal{A}\Lambda\widehat{v} - y|||_* + \mathrm{C}\|f + \Lambda^* y - B^* q\|_{\mathcal{V}}. \tag{7.3.15}$$

Remark 7.12. We note that the problem considered can be expressed as a system

$$\langle \Lambda^* \sigma + \ell - B^* p, w \rangle = 0, \qquad w \in V_0, \tag{7.3.16}$$

$$\sigma = \mathcal{A} \Lambda u, \tag{7.3.17}$$

$$Bv = 0. \tag{7.3.18}$$

Each term on the right-hand side of (7.3.14) and (7.3.15) is a certain penalty for a possible violation of one of the relations of this system.

Remark 7.13. For the Stokes problem, we have

$$\Lambda v = \varepsilon(v), \quad Bv = -\operatorname{div} v, \quad \text{and} \quad A = \nu \mathbb{I}.$$

It is easy to see that in this case, $c_1 = c_2 = \nu$,

$$|||\mathcal{A}\Lambda\widehat{v} - y|||_* = \frac{1}{\sqrt{\nu}} \|\nu\varepsilon(v) - y\|, \quad \text{and} \quad |||\Lambda(u - \widehat{v})||| = \sqrt{\nu}\|\Lambda(u - \widehat{v})\|$$

and we find that the estimate (7.3.14) coincides with (6.2.17).

7.4 Derivation of a posteriori estimates from saddle point relations

In this section we again consider the problem (7.1.1) and show that *a posteriori estimates can be obtained by simple transformations of the saddle point relations.*

First, we recall that the problem (7.1.1) can be stated as a minimax problem: Find $u \in u_0 + V_0$ and $p \in U$ such that

$$L(u, q) \le L(u, p) \le L(v, p), \qquad \forall v \in u_0 + V_0, \ q \in U, \tag{7.4.1}$$

where

$$L(w, q) = (\Lambda w, q) - \frac{1}{2}(\mathcal{A}^{-1} q, q) + \langle \ell, w \rangle.$$

The corresponding system that defines the saddle point (u, p) follows from (7.4.1). Indeed,

$$L(u, p + \epsilon q) \le L(u, p), \qquad \forall q \in U,$$

where ϵ is an arbitrary positive number and

$$L(u, p) \le L(u + w, p), \qquad \forall w \in V_0.$$

Therefore, (u, p) must satisfy the relations

$$(\Lambda u, q) - (\mathcal{A}^{-1} p, q) = 0, \qquad \forall q \in U, \tag{7.4.2}$$

$$(p, \Lambda w) + \langle \ell, w \rangle = 0, \qquad \forall w \in V_0, \tag{7.4.3}$$

or (equivalent) relations

$$(\mathcal{A}\Lambda u, q) - (p, q) = 0, \qquad \forall q \in U, \tag{7.4.4}$$

$$\langle \Lambda^* p, w \rangle + \langle \ell, w \rangle = 0, \qquad \forall w \in V_0. \tag{7.4.5}$$

Let the pair $(v, y) \in (u_0 + V_0) \times U$ be an approximation of the saddle point. Estimates of the deviations $u - v$ and $p - y$ directly follow from (7.4.4) and (7.4.5). Indeed, for $u - v$ and $p - y$ we have

$$(\mathcal{A}\Lambda(u - v) - (p - y), q) = (y - \mathcal{A}\Lambda v, q), \qquad \forall q \in U, \tag{7.4.6}$$

$$\langle \Lambda^*(p - y), w \rangle = -(\Lambda w, y) - \langle \ell, w \rangle, \quad w \in V_0. \tag{7.4.7}$$

Here, we set $q = \Lambda(u - v)$ and $w = u - v$. Then,

$$|||\Lambda(u - v)|||^2 - (p - y, \Lambda(u - v)) = (y - \mathcal{A}\Lambda v, \Lambda(u - v)), \tag{7.4.8}$$

$$\langle \Lambda^*(p - y), u - v \rangle = \langle \ell + \Lambda^* y, v - u \rangle. \tag{7.4.9}$$

We sum (7.4.8) and (7.4.9) and obtain

$$|||\Lambda(u - v)|||^2 = (\mathcal{A}\Lambda v - y, \Lambda(v - u)) + \langle \ell + \Lambda^* y, v - u \rangle,$$

which leads to (7.1.19) and (7.1.20).

Another *modus operandi* applied to (7.4.2) and (7.4.3) implies estimates for the dual variable. We put $q = p - y$ and $w = v - u$, and we have

$$(\Lambda(u - v), p - y) + (\Lambda v - \mathcal{A}^{-1} y, p - y) = (\mathcal{A}^{-1}(p - y), p - y), \tag{7.4.10}$$

$$(p - y, \Lambda(v - u)) + (y, \Lambda(v - u)) + \langle \ell, v - u \rangle = 0. \tag{7.4.11}$$

From (7.4.10) and (7.4.11) it follows that

$$\begin{aligned} |||y - p|||_*^2 &= (\Lambda v - \mathcal{A}^{-1} y, y - p) + \langle \ell + \Lambda^* y, v - u \rangle \\ &\le |||y - \mathcal{A}\Lambda v|||_* |||y - p|||_* + (y - p, \Lambda(v - u)) \\ &\le |||y - \mathcal{A}\Lambda v|||_* |||y - p|||_* + |||\Lambda(v - u)|||\,|||y - p|||_*. \end{aligned}$$

Hence, we arrive at the estimate

$$|||p - y|||_* \le |||y - \mathcal{A}\Lambda v|||_* + |||\Lambda(v - u)||| \tag{7.4.12}$$

and all estimates (see Section 7.1.3) that follow from it.

8 Nonlinear problems

In this chapter, we discuss a posteriori estimates for certain classes of nonlinear problems. As in previous chapters, we derive a posteriori estimates by two different methods. The first method is based on variational techniques and the second one operates with integral type relations that define generalized solutions. Both methods can be applied to nonlinear problems but we pay the main attention to the second (nonvariational) method. Some results obtained with the help of variational techniques are also discussed but without detailed proofs. The reader can find them in [61, 57, 58, 82, 139, 244, 276, 281, 277, 283, 319, 318, 320] and other publications.

8.1 Variational inequalities

Variational inequalities form an important class of nonlinear problems, which often arise in mechanics and physics (e.g., see G. Duvaut and J.-L. Lions [120], A. Friedman [133], and R. Glowinski [153]).

Let V be a reflexive Banach space, $a : V \times V \to \mathbb{R}$ be a bilinear V-elliptic form, and $j : V \to \mathbb{R}$ be a given convex continuous functional. Consider the following problem: Find $u \in K$ such that the inequality

$$a(u, w-u) + j(w) - j(u) \geq \langle \ell, w-u \rangle \tag{8.1.1}$$

holds for any $w \in K$, where K is a convex closed subset of V_0 and $\ell \in V_0^*$,

$$a(v, w) := \int_\Omega A\nabla v \cdot \nabla w \, dx,$$

Ω is a bounded domain in $\mathbb{R}^2$ with Lipschitz continuous boundary, $A = \{a_{ij}\}$ is a symmetric matrix satisfying the conditions (4.1.4). It is well known (e.g., see [120, 153]) that (8.1.1) is equivalent to the variational problem: Find $u \in K$ such that

$$J(u) = \inf_{w \in K} J(w), \tag{8.1.2}$$

$$J(w) = \frac{1}{2}a(w, w) + j(w) - \langle \ell, w \rangle.$$

The existence of a minimizer to this problem follows from the coercivity of a on V. In the literature, variational inequalities the nonlinear features of which arise owing to the set K are often called inequalities of the "first kind". If a nonlinearity is caused by the presence of a (nondifferentiable) functional j, then the inequality is assigned to the "second kind".

Exact solutions of variational inequalities may have a complicated structure. Typically, they contain unknown free boundaries the location and structure of which are a priori unknown and investigation of their properties is an important part of the a priori analysis of variational inequalities (e.g., see [133]).

In this section, we show that computable error bounds for elliptic variational inequalities of both kinds follow from the corresponding variational inequality.

8.1.1 Variational inequalities of the first kind

We begin with a classical problem related to the variational inequalities of the first kind. We set $j \equiv 0$ and define K with the help of pointwise restrictions:

$$K := \{v \in V_0 := \overset{\circ}{H}{}^1(\Omega) \mid \phi(x) \leq v(x) \leq \psi(x) \text{ a.e. in } \Omega\},$$

where ϕ and $\psi \in H^2(\Omega)$ are two given functions such that

$$\begin{aligned} \phi(x) \leq 0 \quad \text{and} \quad \psi(x) \geq 0 \qquad & \text{on } \Gamma, \\ \phi(x) \leq \psi(x) \qquad & \text{in } \Omega. \end{aligned}$$

In this case, (8.1.1) reads: Find $u \in K$ satisfying the inequality

$$a(u, w - u) \geq \langle \ell, w - u \rangle \tag{8.1.3}$$

Henceforth, we assume that ℓ is defined by an integrable function, i.e.,

$$\langle \ell, w \rangle = \int_\Omega f w \, dx$$

and Ω is a bounded domain with Lipschitz continuous boundary.

Problem (8.1.3) is the classical obstacle problem (e.g., see [133, 153]). Under the assumptions made, the solution u exists and is unique. In general, Ω is divided into three sets, where u is determined either by the differential equation or by obstacles. They are as follows:

$$\begin{aligned} \Omega^u_\psi &:= \{x \in \Omega \mid u(x) = \psi(x)\}, \\ \Omega^u_\phi &:= \{x \in \Omega \mid u(x) = \phi(x)\}, \\ \Omega^u_0 &:= \{x \in \Omega \mid \phi(x) < u(x) < \psi(x)\}. \end{aligned}$$

The sets Ω^u_ψ and Ω^u_ϕ are *upper* and *lower coincidence sets* and Ω^u_0 is an open set where a solution satisfies the differential equation. Thus, we see that this problem involves free boundaries, which are a priori unknown.

In [82, 283], the variational statement (8.1.3) was used to derive a guaranteed upper bound of the error

$$|||\nabla(u - v)||| := (a(u - v, u - v))^{1/2}.$$

Below we derive the same estimate by transformations of (8.1.3). Assume that $v \in K$ is a function compared with u. Then,

$$a(u-v, u-v) \leq \int_\Omega (f(u-v) - A\nabla v \cdot \nabla(u-v))\,dx. \tag{8.1.4}$$

For any vector-valued function y in the space $H(\Omega, \mathrm{div})$, we have

$$\int_\Omega (w \operatorname{div} y + y \cdot \nabla w)\,dx = 0, \qquad \forall w \in V_0.$$

Therefore, (8.1.4) is transformed as follows:

$$\begin{aligned} a(u-v, u-v) &\leq \int_\Omega (f + \operatorname{div} y)(u-v)\,dx + \int_\Omega (y - A\nabla v) \cdot \nabla(u-v)\,dx \\ &= \int_\Omega (y - A\nabla v) \cdot \nabla(u-v)\,dx + \int_{\Omega_\psi^v} (f + \operatorname{div} y)(u-v)\,dx \\ &\quad + \int_{\Omega_\phi^v} (f + \operatorname{div} y)(u-v)\,dx \\ &\quad + \int_{\Omega_0^v} (f + \operatorname{div} y)(u-v)\,dx, \end{aligned} \tag{8.1.5}$$

where

$$\begin{aligned} \Omega_\psi^v &:= \{x \in \Omega \mid v(x) = \psi(x)\}, \\ \Omega_\phi^v &:= \{x \in \Omega \mid v(x) = \phi(x)\}, \\ \Omega_0^v &:= \{x \in \Omega \mid \phi(x) < v(x) < \psi(x)\}. \end{aligned}$$

Introduce the function

$$\lessdot f + \operatorname{div} y \gtrdot_v := \begin{cases} (f + \operatorname{div} y)_- & \text{a.e. in } \Omega_\psi^v, \\ f + \operatorname{div} y & \text{a.e. in } \Omega_0^v, \\ (f + \operatorname{div} y)_+ & \text{a.e. in } \Omega_\phi^v, \end{cases}$$

where $(g)_+ = \max\{0, g\}$ and $(g)_- = \min\{0, g\}$ denote the positive and negative parts of g, respectively. Note that

$$\lessdot f + \operatorname{div} y \gtrdot_v \in L_2(\Omega).$$

Further analysis rests upon the relation

$$\int_\Omega (f + \operatorname{div} y)(u-v)\,dx \leq \int_\Omega \lessdot f + \operatorname{div} y \gtrdot_v (u-v)\,dx, \tag{8.1.6}$$

from which it follows that the first term on the right-hand side of (8.1.5) is bounded from above by the quantity $\| \lessdot f + \operatorname{div} y \gtrdot_v \| \|v - u\|$. Since

$$\int_\Omega (A\nabla v - y) \cdot \nabla w \, dx \leq |||A\nabla v - y|||_* |||\nabla w|||,$$

we arrive at the estimate (cf. [283])

$$|||\nabla(u - v)||| \leq \overline{\mathfrak{M}}_{\text{OBS}}(v, y) := |||A\nabla v - y|||_* + \mathrm{C}\| \lessdot f + \operatorname{div} y \gtrdot_v \|, \quad (8.1.7)$$

where C is the constant in the inequality $\|w\| \leq \mathrm{C}|||\nabla w|||$.

We note that the error majorant $\overline{\mathfrak{M}}_{\text{OBS}}(v, y)$ derived for the obstacle problem consists of two terms. The first term penalizes the error in the relation

$$A\nabla v = y. \quad (8.1.8)$$

Another term of $\overline{\mathfrak{M}}_{\text{OBS}}(v, y)$ penalizes the "improper" behavior of $\operatorname{div} y + f$ on the sets Ω_0^v, Ω_ψ^v, and Ω_ϕ^v, respectively.

It is easy to see that the majorant is a nonnegative functional, which vanishes if and only if (8.1.8) holds and

$$\lessdot \operatorname{div} y + f \gtrdot_v = 0 \qquad \text{for a.e. } x \in \Omega. \quad (8.1.9)$$

The latter relation means that

$$\operatorname{div} y(x) + f(x) \leq 0 \quad \text{a.e. in } \Omega_\phi^v, \quad (8.1.10)$$

$$\operatorname{div} y(x) + f(x) = 0 \quad \text{a.e. in } \Omega_0^v, \quad (8.1.11)$$

$$\operatorname{div} y(x) + f(x) \geq 0 \quad \text{a.e. in } \Omega_\psi^v. \quad (8.1.12)$$

If (8.1.8) and (8.1.10)–(8.1.12) hold, then for any $w \in K$, we have

$$\int_\Omega A\nabla v \cdot \nabla(w - v) \, dx - \int_\Omega f(w - v) \, dx = \int_\Omega (\operatorname{div} y + f)(v - w) \, dx$$

$$= \int_{\Omega_\phi^v} (\operatorname{div} y + f)(\phi - w) \, dx + \int_{\Omega_0^v} (\operatorname{div} y + f)(v - w) \, dx$$

$$+ \int_{\Omega_\psi^v} (\operatorname{div} y + f)(\psi - w) \, dx \geq 0.$$

Thus, v satisfies the variational inequality

$$a(v, w - v) \geq \int_\Omega f(w - v) \, dx, \qquad \forall w \in K,$$

which coincides with (8.1.3). We have proved the following assertion.

Theorem 8.1. *For any $v \in V_0$ and $y \in H(\Omega, \mathrm{div})$, the estimate (8.1.7) provides a guaranteed upper bound of $|||\nabla(u-v)|||$. Moreover, if $\overline{\mathfrak{M}}_{\mathrm{OBS}}(v,y) = 0$, then $v = u$ and $y = A\nabla u$.*

Remark 8.2. Theorem 8.1 shows that problem (8.1.7) admits a new variational statement, namely: Find $u \in K$ and $p \in H(\Omega, \mathrm{div})$ such that

$$\overline{\mathfrak{M}}_{\mathrm{OBS}}(u,p) = \inf_{\substack{v \in V_0, \\ y \in H(\Omega,\mathrm{div})}} \overline{\mathfrak{M}}_{\mathrm{OBS}}(v,y). \tag{8.1.13}$$

Unlike the variational statement (8.1.2), the value of the exact lower bound for this problem is known (it is zero).

A more sophisticated estimate can be derived if (8.1.5) is given in the form

$$\begin{aligned} a(u-v,u-v) \leq {} & \int_\Omega (f + \mathrm{div}\, y + \lambda_1 - \lambda_2)(u-v)\, dx \\ & + \int_\Omega (y - A\nabla v)\cdot\nabla(u-v)\, dx + \int_\Omega (\lambda_1(v-u) + \lambda_2(u-v))\, dx, \end{aligned}$$

where $\lambda_i(x)$, $i = 1, 2$, are arbitrary nonnegative functions. Since

$$\int_\Omega \lambda_1(v-u)\, dx \leq \int_\Omega \lambda_1(v-\phi)\, dx$$

and

$$\int_\Omega \lambda_2(u-v)\, dx \leq \int_\Omega \lambda_2(\psi-v)\, dx,$$

from (8.1.5) it follows that

$$\begin{aligned} |||\nabla(u-v)|||^2 \leq {} & \Big(\mathrm{C}\|f + \mathrm{div}\, y + \lambda_1 - \lambda_2\| + |||y - A\nabla v|||_*\Big)|||\nabla(u-v)||| \\ & + \int_\Omega (\lambda_1(v-\phi) + \lambda_2(\psi - v))\, dx. \end{aligned} \tag{8.1.14}$$

By (8.1.14), we deduce an advanced error majorant:

$$\begin{aligned} |||\nabla(u-v)||| \leq {} & M(v,y,\lambda_1,\lambda_2) \\ & + \sqrt{\int_\Omega (\lambda_1(v-\phi) + \lambda_2(\psi-v))\, dx + M^2(v,y,\lambda_1,\lambda_2)}, \end{aligned} \tag{8.1.15}$$

where

$$M(v,y,\lambda_1,\lambda_2) := \frac{1}{2}\Big(|||y - A\nabla v|||_* + \mathrm{C}\|f + \mathrm{div}\, y + \lambda_1 - \lambda_2\|\Big).$$

Note that (8.1.7) is a special case of (8.1.15). Indeed, set $\lambda_1 = \bar{\lambda}_1$ and $\lambda_2 = \bar{\lambda}_2$, where

$$\bar{\lambda}_1 = 0 \qquad \text{on } \left(\Omega \setminus \Omega_\phi^v\right) \cup \left(\Omega_\phi^v \cap \{x \mid \operatorname{div} y(x) + f(x) \geq 0\}\right),$$
$$\bar{\lambda}_1(x) = |\operatorname{div} y(x) + f(x)| \quad \text{on } \Omega_\phi^v \cap \{x \mid \operatorname{div} y(x) + f(x) < 0\}$$

and

$$\bar{\lambda}_2 = 0 \qquad \text{on } \left(\Omega \setminus \Omega_\psi^v\right) \cup \left(\Omega_\psi^v(v) \cap \{x \mid \operatorname{div} y(x) + f(x) \leq 0\}\right),$$
$$\bar{\lambda}_2(x) = |\operatorname{div} y(x) + f(x)| \quad \text{on } \Omega_\psi^v \cap \{x \mid \operatorname{div} y(x) + f(x) > 0\}.$$

Then,

$$\int_\Omega (\bar{\lambda}_1(v - \phi) + \bar{\lambda}_2(\psi - v))\, dx = 0$$

and (8.1.15) takes the form

$$|||\nabla(u - v)||| \leq 2M(v, y, \bar{\lambda}_1, \bar{\lambda}_2). \tag{8.1.16}$$

Note that

$$\begin{aligned}
&\|f + \operatorname{div} y + \bar{\lambda}_1 - \bar{\lambda}_2\|^2 \\
&= \int_{\Omega_\phi^v} |f + \operatorname{div} y + \bar{\lambda}_1|^2\, dx + \int_{\Omega_0^v} |f + \operatorname{div} y|^2\, dx + \int_{\Omega_\psi^v} |f + \operatorname{div} y - \bar{\lambda}_2|^2\, dx \\
&= \int_{\Omega_\phi^v} (f + \operatorname{div} y)_+^2\, dx + \int_{\Omega_0^v} |f + \operatorname{div} y|^2\, dx + \int_{\Omega_\psi^v} (f + \operatorname{div} y)_-^2\, dx \\
&= \| \lessdot \operatorname{div} y + f \gtrdot_v \|^2.
\end{aligned}$$

For this reason, $2M(v, y, \bar{\lambda}_1, \bar{\lambda}_2)$ coincides with $\overline{\mathfrak{M}}_{\mathrm{OBS}}(v, y)$.

From (8.1.14) it also follows that

$$\begin{aligned}
\left(1 - \frac{\alpha}{2}\right) |||\nabla(u - v)|||^2 \leq{}& \frac{1}{2\alpha}\left(\mathrm{C}\|f + \operatorname{div} y + \lambda_1 - \lambda_2\| + |||y - A\nabla v|||_*\right)^2 \\
&+ \int_\Omega (\lambda_1(v - \phi) + \lambda_2(\psi - v))\, dx,
\end{aligned} \tag{8.1.17}$$

where $\alpha \in (0, 2)$. Using Young's inequality, we rewrite (8.1.17) in the form

$$\begin{aligned}
(2 - \alpha)\, |||\nabla(u - v)|||^2 \leq{}& \frac{1 + \beta}{\alpha} \mathrm{C}^2 \|f + \operatorname{div} y + \lambda_1 - \lambda_2\|^2 + \frac{1 + \beta}{\alpha\beta} |||y - A\nabla v|||_*^2 \\
&+ 2\int_\Omega (\lambda_1(v - \phi) + \lambda_2(\psi - v))\, dx,
\end{aligned} \tag{8.1.18}$$

where β is an arbitrary positive number. Let

$$c_{\alpha\beta} := \frac{1+\beta}{\alpha}\mathrm{C}^2 \quad \text{and} \quad \mathrm{r}(y) := \operatorname{div} y + f.$$

Now, we are aimed at selecting λ_1 and λ_2 in such a way that the value of the right-hand side of (8.1.18) attains its minimal value. Unconstrained minimization leads to the relations

$$\int_\Omega \left(\mathrm{r}(y) + \lambda_1 - \lambda_2 + \frac{v-\phi}{c_{\alpha\beta}}\right) dx = 0 \tag{8.1.19}$$

and

$$\int_\Omega \left(\mathrm{r}(y) + \lambda_1 - \lambda_2 + \frac{v-\psi}{c_{\alpha\beta}}\right) dx = 0, \tag{8.1.20}$$

which cannot be satisfied simultaneously. Therefore, either $\lambda_1 = 0$ or $\lambda_2 = 0$ (or both of them are equal to zero). Assume that $\lambda_1 = 0$. Then, we should set

$$\lambda_2 = \mathrm{r}(y) + \frac{v-\psi}{c_{\alpha\beta}}, \tag{8.1.21}$$

provided that $\lambda_2 > 0$. Let

$$\Phi(\mathrm{r}(y), v) := c_{\alpha\beta}(\mathrm{r}(y) + \lambda_1 - \lambda_2)^2 + 2(\lambda_1(v-\phi) + \lambda_2(\psi - v))$$

denote the integrand of two terms that contain λ_1 and λ_2. By (8.1.21), we obtain

$$\begin{aligned}\Phi(\mathrm{r}(y), v) &= c_{\alpha\beta}(\mathrm{r}(y) - \lambda_2)^2 + 2\lambda_2(\psi - v) \\ &= 2\mathrm{r}(y)(\psi - v) - \frac{(\psi - v)^2}{c_{\alpha\beta}} \qquad \text{if } \mathrm{r}(y) > \frac{\psi - v}{c_{\alpha\beta}}.\end{aligned} \tag{8.1.22}$$

Another option is to take $\lambda_2 = 0$ and

$$\lambda_1 = \frac{\phi - v}{c_{\alpha\beta}} - \mathrm{r}(y), \tag{8.1.23}$$

provided that $\lambda_1 > 0$. Then,

$$\begin{aligned}\Phi(\mathrm{r}(y), v) &= c_{\alpha\beta}(\mathrm{r}(y) + \lambda_1)^2 + 2\lambda_1(v - \phi) \\ &= 2\mathrm{r}(y)(\phi - v) - \frac{(\phi - v)^2}{c_{\alpha\beta}} \qquad \text{if } \frac{\phi - v}{c_{\alpha\beta}} > \mathrm{r}(y).\end{aligned} \tag{8.1.24}$$

If both λ_1 and λ_2 are equal to zero, then

$$\Phi(\mathrm{r}(\mathrm{y}), v) = c_{\alpha\beta}\mathrm{r}^2(y) \qquad \text{if } \frac{\phi - v}{c_{\alpha\beta}} < \mathrm{r}(y) < \frac{\psi - v}{c_{\alpha\beta}}. \tag{8.1.25}$$

Now, (8.1.18) takes the form

$$(2-\alpha)\,|||\nabla(u-v)|||^2 \le \frac{1+\beta}{\alpha\beta}|||y-A\nabla v|||_*^2 + \int_\Omega \Phi(\mathrm{r}(y),v)\,dx, \quad (8.1.26)$$

where $\Phi(\mathrm{r}(y),v)$ is defined in accordance with (8.1.22), (8.1.24), and (8.1.25).

Remark 8.3. Estimates (8.1.7) and (8.1.26) were derived in [298] by the method, which we have discussed above. In [283], the estimate (8.1.7) was derived by the variational method. Set $\alpha = 1$. Then $c_{\alpha\beta} = c_\beta := (1+\beta)\mathrm{C}^2$, and we arrive at the estimate

$$|||\nabla(u-v)|||^2 \le \frac{1+\beta}{\beta}|||y-A\nabla v|||_*^2 + \int_\Omega \Phi(\mathrm{r}(y),v)\,dx, \quad (8.1.27)$$

which was also derived in [283].

8.1.2 Variational inequalities of the second kind

Another group of variational inequalities is related to problems with nondifferentiable functionals. In this case, a solution u is defined by the inequality

$$a(u,w-u) + j(w) - j(u) \ge \langle \ell, w-u\rangle, \quad (8.1.28)$$

which holds for any $w \in V_0$. As for the first kind inequalities, the corresponding a posteriori estimate can be derived directly from (8.1.28). Let v be a function compared with u. Set $w = v$ and rearrange (8.1.28) as follows:

$$a(u-v,v-u) + j(v) - j(u) \ge \langle \ell, v-u\rangle - a(v,v-u). \quad (8.1.29)$$

Since j is a convex functional, we know that (cf. (1.4.43))

$$j(u) - j(v) \ge \langle v^*, u-v\rangle, \qquad \forall v^* \in \partial j(v),$$

where $\partial j(v) \in V_0^*$ denotes the subdifferential of j at v. Therefore, we rewrite (8.1.29) in the form

$$a(u-v,u-v) \le \langle v^*, v-u\rangle + \langle \ell, u-v\rangle + a(v,v-u).$$

Note that for $y \in L^2(\Omega,\mathbb{R}^d)$ (the divergence of which is an element of V_0^*), we have

$$\langle \operatorname{div} y, w\rangle + (y,\nabla w) = 0, \qquad \forall w \in V_0.$$

By this identity, we obtain

$$a(u-v,u-v) \le \langle v^* - \ell + \operatorname{div} y, v-u\rangle + \int_\Omega (A\nabla v - y)\cdot\nabla(v-u)\,dx, \quad (8.1.30)$$

which implies the estimate

$$|||\nabla(u-v)||| \leq |\,v^* - \ell + \operatorname{div} y\,| + |||A\nabla v - y|||_*, \tag{8.1.31}$$

where $|\,\xi^*\,| := \sup_{w \in V_0} \frac{\langle \xi^*, w\rangle}{|||\nabla w|||}$.

Assume that $y \in H(\Omega, \operatorname{div})$ and $\partial j(v)$ contains an element that can be identified with an $L^2(\Omega)$-function. Then, (8.1.31) is represented in the form

$$|||\nabla(u-v)||| \leq \mathrm{C}\|v^* - f + \operatorname{div} y\| + |||A\nabla v - y|||_*, \tag{8.1.32}$$

where v^* is such an element. In particular, if j is Gâteaux differentiable and $j'(v) \in L^2(\Omega)$ denotes the Gâteaux derivative, then we express (8.1.32) as

$$|||\nabla(u-v)||| \leq \mathrm{C}\|j'(v) - f + \operatorname{div} y\| + |||A\nabla v - y|||_*. \tag{8.1.33}$$

In the next section, we show that the method applied can be extended to a considerably wider class of problems.

8.2 General elliptic problem. Variational method.

A wide class of variational problems related to various physical models can be given in the following abstract form:

$$\inf_{v \in u_0 + V_0} J(v, \Lambda v), \quad J(v) := G(\Lambda v) + F(v), \qquad \forall v \in V, \tag{8.2.1}$$

where G and F are convex continuous functionals, V is a reflexive Banach space, and Λ is a linear continuous operator that maps V to another reflexive Banach space Y. In particular, if

$$G(y) = \frac{1}{2}(\mathcal{A}y, y), \qquad F(v) = \langle \ell, v\rangle, \qquad \ell \in V_0^*,$$

then (8.2.1) coincides with (7.1.1). Also, we assume that

$$c\|\Lambda w\|_Y \geq \|w\|_V, \qquad \forall v \in V_0, \tag{8.2.2}$$

where c is a positive constant, and J is coercive on $u_0 + V_0$, i.e., for $v \in u_0 + V_0$

$$J(v, \Lambda v) \to +\infty \quad \text{as} \quad \|v\|_V \to +\infty. \tag{8.2.3}$$

In this case, the problem (8.2.1) has a solution u (e.g., see [121]).

In ([276, 277, 282]), a posteriori error estimates for this class of problems were derived by the variational method. In order to discuss them, we need to introduce additional notation.

As before, the product of $v \in V$ and v^* in the topologically dual space V^* is denoted by $\langle v^*, v\rangle$. We denote the space topologically dual to Y by Y^* and the corresponding pairing by $\langle\!\langle y^*, y\rangle\!\rangle$. We note that the spaces Y and Y^* are essentially different (unlike the case considered in Chapter 7). For this reason, throughout this chapter we mark functions from Y^* by stars. The operator

$$\Lambda^* : V^* \to Y^*$$

satisfying the relation

$$\langle\!\langle y^*, \Lambda w\rangle\!\rangle = \left\langle \Lambda^* y, w\right\rangle, \qquad \forall w \in V, \tag{8.2.4}$$

is *conjugate* to Λ.

By G^* we denote the Fenchel conjugate of G (cf. (1.4.38)), which is defined by the relation

$$G^*(y^*) = \sup_{y\in Y}\Big(\langle\!\langle y^*, y\rangle\!\rangle - G(y)\Big).$$

Definition 8.4. We say that G and G^* are uniformly convex in the balls

$$B_\delta := B(0,\delta) \in Y \quad \text{and} \quad B^*_{\delta*} := B(0,\delta^*) \in Y^*,$$

respectively, if they satisfy the relations

$$G\left(\frac{y_1 + y_2}{2}\right) + \Phi_\delta\left(\frac{y_1 - y_2}{2}\right) \le \frac{1}{2}\big(G(y_1) + G(y_2)\big), \tag{8.2.5}$$

$$G^*\left(\frac{y_1^* + y_2^*}{2}\right) + \Phi^*_{\delta*}\left(\frac{y_1^* - y_2^*}{2}\right) \le \frac{1}{2}\left(G^*(y_1) + G^*(y_2)\right), \tag{8.2.6}$$

where $\Phi_\delta : Y \to \mathbb{R}_+$ and $\Phi^*_{\delta*} : Y^* \to \mathbb{R}_+$ are certain nonnegative functionals vanishing at the zero elements of Y and Y^*, respectively.

It follows directly from (8.2.5) that any uniformly convex functional G is convex. Moreover, the functional Φ_δ reinforces the usual convexity inequality. For this reason, sometimes it is called the *forcing* functional (e.g., see R. Glowinski [153]).

Let $v \in u_0 + V$ be an approximation of u. The variational method of deriving computable upper bounds for the quantity $\|\Lambda(v - u)\|_Y$ is based on the inequality

$$\Phi\left(\frac{\Lambda(v-u)}{2}\right) \le \frac{1}{2}\big(J(v,\Lambda v) - J(u,\Lambda u)\big), \tag{8.2.7}$$

which can be viewed as a generalization of (2.3.1) for problems with uniformly convex functionals. If G is uniformly convex, then it is easy to prove that (8.2.7) holds with $\Phi = \Phi_\delta$, provided that Λu and Λv belong to $B(0,\delta)$. Indeed,

$$\Phi_\delta\left(\frac{\Lambda(v-u)}{2}\right) \le \frac{1}{2}\left(G(\Lambda v) + G(\Lambda u)\right) - G\left(\frac{\Lambda(u+v)}{2}\right),$$

$$0 \le \frac{1}{2}\left(F(v) + F(u)\right) - F\left(\frac{u+v}{2}\right).$$

Since

$$G\left(\frac{\Lambda(u+v)}{2}\right)+F\left(\frac{u+v}{2}\right)\geq G(\Lambda u)+F(u),$$

we obtain (8.2.7).

We note that for problems with superquadratic growth, (8.2.7) holds for the whole space, so that Φ does not depend on δ.

Another key relation used in the derivation of a posteriori estimates is

$$\inf\mathcal{P}:=J(u,\Lambda u)=I^*(p^*,\Lambda^*p^*)=\sup\mathcal{P}^*,$$

where I^* is the functional of the so-called *dual variational problem* $\mathcal{P}^*$ and p^* is the corresponding solution that maximizes I^* on a set of admissible functions (cf. (7.1.9)). Typically, this set consists of the functions that satisfy certain differential relations. We will not give here a detailed exposition of the variational method (which can be found in [282] and [244]) and pass to a discussion of the error majorant derived with the help of it. This majorant provides an upper bound of the differences $v-u$ (for the primal problem) and y^*-p^* (for the dual one) evaluated in terms of the functionals Φ and Φ^*.

Theorem 8.5 ([276, 282]). *Let u be a minimizer of the problem, the functionals F and G satisfy the above conditions, and $\Lambda u\in B_\delta$ and $p^*\in B^*_{\delta*}$. Then for any $v\in u_0+V_0$ and $y^*\in Y^*$ such that $\Lambda v\in B_\delta$ and $y^*\in B^*_{\delta*}$, the estimate*

$$\begin{aligned}\Phi_\delta\left(\frac{\Lambda(v-u)}{2}\right)+\Phi^*_{\delta*}\left(\frac{y^*-p^*}{2}\right)&\leq\overline{\mathfrak{M}}_{\mathrm{I}}(v,y^*)\\&=\frac{1}{2}(D_G(\Lambda v,y^*)+D_F(v,\Lambda^*y^*))\end{aligned}\tag{8.2.8}$$

holds, where

$$\begin{aligned}D_F(v,\Lambda^*y^*)&:=F(v)+F^*(-\Lambda^*y^*)+\langle\Lambda^*y^*,v\rangle,\\D_G(\Lambda v,y^*)&:=G(\Lambda v)+G^*(y^*)-\langle\!\langle y^*,y\rangle\!\rangle.\end{aligned}$$

The functionals D_F and D_G are nonnegative (this fact follows from the definition of a polar functional; cf. 1.4). They play an important role in a posteriori analysis of various variational problems (e.g., see [57, 61, 139, 281, 282, 289, 293, 319, 320, 321]).

By Proposition 1.2, we know that the relation

$$F(v)+F^*(-\Lambda^*y^*)+\langle\Lambda^*y^*,v\rangle=0\tag{8.2.9}$$

is equivalent to

$$-\Lambda^*y^*\in\partial F(v)\tag{8.2.10}$$

and the relation

$$G(\Lambda v) + G^*(y^*) - \langle\!\langle y^*, \Lambda v \rangle\!\rangle = 0 \tag{8.2.11}$$

is equivalent to

$$y^* \in \partial G(\Lambda v). \tag{8.2.12}$$

Note that (8.2.10) and (8.2.12) are *duality relations*, which hold if and only if v and y^* coincide with u and p^*.

Remark 8.6. We observe that for nonlinear problems, it is natural to perform error control with the help of special functionals Φ. For linear problems, these functionals coincide with the standard energy norms.

Example. As an example, we apply (8.2.8) to the reaction-diffusion equation with mixed Dirichlet–Robin boundary conditions. It can be represented as the variational problem

$$\inf_{w \in V_0} J(v, \nabla v)$$

for the functional

$$J(v, \nabla v) = \int_\Omega \Big(\frac{1}{2}|\nabla v|^2 + \frac{\delta}{2}|v|^2\Big)\, dx + \int_{\Gamma_2} \Big(\frac{\alpha}{2}|v|^2 - g v\Big)\, ds,$$

where δ and α are positive real numbers and V_0 is a subspace of $H^1(\Omega)$, which contains functions vanishing at Γ_1.

It is not difficult to show that the minimizer u of this variational problem satisfies the relations

$$\begin{aligned} -\Delta u + \delta u &= 0 \qquad \text{in } \Omega, \\ u &= 0 \qquad \text{on } \Gamma_1, \\ \frac{\partial u}{\partial n} + \alpha u - g &= 0 \qquad \text{on } \Gamma_2. \end{aligned}$$

On Γ_2 the solution satisfies the so-called *Robin* boundary condition. An a posteriori estimate for this problem follows from (8.2.8) if Λ is associated with the operator ∇v and the functionals G and F are defined by the relations

$$\begin{aligned} G(\Lambda w) &= \int_\Omega \frac{1}{2}|\nabla v|^2\, dx, \\ F(v) &= \int_\Omega \frac{\delta}{2}|v|^2\, dx + \int_{\Gamma_2} (\frac{\alpha}{2}|v|^2 - g v)\, ds. \end{aligned}$$

In the case considered, Y and Y^* are identified with $L^2(\Omega, \mathbb{R}^d)$,

$$\langle y^*, y\rangle := \int_\Omega y^* \cdot y \, dx,$$

$$G^*(-y^*) = \sup_y \int_\Omega \Big(-y^* \cdot y - \frac{1}{2}|y|^2\Big)\, dx = \int_\Omega \frac{1}{2}|y^*|^2 \, dx.$$

By the integration-by-parts formula

$$\int_\Omega y^* \cdot \nabla v \, dx = \int_\Omega -\operatorname{div} y^* v \, dx + \int_{\Gamma_2} (y^* n) v \, ds, \qquad \forall v \in V_0,$$

we conclude that $\Lambda^* y^*$ should be understood as $\{-\operatorname{div} y^* \,|_\Omega,\ y^* \cdot n \,|_{\Gamma_2}\}$ and

$$\langle \Lambda^* y^*, v\rangle = \int_\Omega -\operatorname{div} y^* v \, dx + \int_{\Gamma_2} (y^* \cdot n) v \, ds,$$

provided that y^* possesses necessary regularity. By a direct substitution, we find that

$$G(\Lambda v) + G^*(-y^*) + \langle y^*, \Lambda v\rangle = \int_\Omega \Big(\frac{1}{2}|\nabla v|^2 + \frac{1}{2}|y^*|^2 + \nabla v \cdot y^*\Big)\, dx$$

and

$$\begin{aligned}
F^*(\Lambda^* y^*) &= \sup_{v \in V_0} \Big\{ \int_\Omega (-\operatorname{div} y^*) v \, dx + \int_{\Gamma_2} (y^* \cdot n) v \, ds - F(v) \Big\} \\
&= \sup_{v \in V_0} \Big\{ \int_\Omega (-\operatorname{div} y^*) v \, dx + \int_{\Gamma_2} (y^* \cdot n) v \, ds - \int_\Omega \frac{\delta}{2}|v|^2 \, dx \\
&\qquad - \int_{\Gamma_2} \Big(\frac{\alpha}{2}|v|^2 - g v\Big)\, ds \Big\} \\
&\le \sup_{v \in L^2(\Omega)} \int_\Omega \Big((-\operatorname{div} y^*) v - \frac{\delta}{2}|v|^2\Big)\, dx \\
&\qquad + \sup_{\eta \in L^2(\Gamma_2)} \int_{\Gamma_2} \Big((y^* \cdot n)\eta - \frac{\alpha}{2}|\eta|^2 + g\eta\Big)\, ds \\
&= \int_\Omega \frac{1}{2\delta} |\operatorname{div} y^*|^2 \, dx + \int_{\Gamma_2} \frac{1}{2\alpha} |y^* \cdot n + g|^2 ds.
\end{aligned}$$

From the above relations, we conclude that

$$\begin{aligned}
&F(v) + F^*(\Lambda^* y^*) - \langle \Lambda^* y^*, v\rangle \\
&\qquad \le \int_\Omega \frac{1}{2\delta} (\operatorname{div} y^* + \delta v)^2 \, dx + \int_{\Gamma_2} \frac{1}{2\alpha} |y^* \cdot n + g - \alpha v|^2 \, ds, \qquad (8.2.13)
\end{aligned}$$

and

$$G(\Lambda v) + G^*(-y^*) + \langle y^*, \Lambda v \rangle = \int_\Omega \frac{1}{2} |\nabla v + y^*|^2 \, dx, \tag{8.2.14}$$

which yields both terms of the error majorant. It is easy to see that they vanishes if and only if

$$\operatorname{div} y^* + \delta v = 0 \qquad \text{in } \Omega, \tag{8.2.15}$$

$$y^* \cdot n + g - \alpha v = 0 \qquad \text{on } \Gamma_2, \tag{8.2.16}$$

$$y^* = -\nabla v \qquad \text{in } \Omega. \tag{8.2.17}$$

Since $v = 0$ on Γ_N, the relations (8.2.15)–(8.2.17) mean that v coincides with the exact solution u and y^* coincides with p^*.

8.3 General elliptic problem. Nonvariational method

In this section, we derive an upper bound of the error from the variational inequality associated with the problem (8.2.1). The method applied is a generalization of that was used in Section 8.1.2 for variational inequalities with nondifferentiable terms. For the sake of simplicity, we assume that G is Gâteaux differentiable. Then, the variational inequality follows from the variational statement with the help of well-known arguments. Indeed, set $w = u + \lambda(v - u)$ where $v \in u_0 + V_0$ and λ is a positive number. Then $w \in u_0 + V_0$ and we have

$$J(w, \Lambda w) - J(u, \Lambda u) = G(\Lambda w) + F(w) - G(\Lambda u) - F(u) \geq 0. \tag{8.3.1}$$

Since

$$F(w) - F(u) = F(u + \lambda(v - u)) - F(u) \leq \lambda(F(v) - F(u)),$$

we rewrite (8.3.1) in the form

$$\frac{1}{\lambda}\left(G\left(\Lambda(u + \lambda(v - u))\right) - G(\Lambda u)\right) + F(v) - F(u) \geq 0, \qquad \forall v \in u_0 + V_0.$$

Let $\lambda \to 0$, then we arrive at the inequality

$$\langle\!\langle G'(\Lambda u), \Lambda(v - u) \rangle\!\rangle + F(v) - F(u) \geq 0. \tag{8.3.2}$$

From (8.3.2), it follows that

$$\langle\!\langle G'(\Lambda u) - G'(\Lambda v), \Lambda(v - u) \rangle\!\rangle + F(v) - F(u) \geq \langle\!\langle G'(\Lambda v), \Lambda(u - v) \rangle\!\rangle, \tag{8.3.3}$$

which is equivalent to

$$\begin{aligned}\Upsilon(\Lambda(v-u)) &:= \langle\langle G'(\Lambda v) - G'(\Lambda u), \Lambda(v-u)\rangle\rangle \\ &\le F(v) - F(u) + \langle\langle G'(\Lambda v), \Lambda(v-u)\rangle\rangle \\ &\le \langle v^*, v-u\rangle + \langle\langle G'(\Lambda v), \Lambda(v-u)\rangle\rangle \\ &\le \langle\langle G'(\Lambda v) - y^*, \Lambda(v-u)\rangle\rangle + \langle \Lambda^* y^* + v^*, v-u\rangle, \end{aligned} \tag{8.3.4}$$

where v^* is an element of the set $\partial F(v)$ and y^* is a function in Y^*.

Since G is a convex functional, its derivative is a monotone operator. Therefore, the quantity $\Upsilon(\Lambda(v-u)) := \langle\langle G'(\Lambda v) - G'(\Lambda u), \Lambda(v-u)\rangle\rangle$ is nonnegative and provides a certain measure of the error.

We can deduce computable bounds of errors by one of the methods discussed below.

The first method. Note that

$$\langle\langle G'(\Lambda v) - y^*, \Lambda(v-u)\rangle\rangle \le \|G'(\Lambda v) - y^*\|_{Y^*} \|\Lambda(v-u)\|_Y \tag{8.3.5}$$

and

$$\langle \Lambda^* y^* + v^*, v-u\rangle \le \|\Lambda^* y^* + v^*\|_{V^*} \|v-u\|_V. \tag{8.3.6}$$

In view of (8.2.2),

$$\|u-v\|_V \le c\|\Lambda(v-u)\|_Y, \tag{8.3.7}$$

and we obtain the estimate

$$\gamma(\Lambda(v-u)) \le \|G'(\Lambda v) - y^*\|_{Y^*} + c\|\Lambda^* y^* + v^*\|_{V^*}, \tag{8.3.8}$$

where

$$\gamma(\Lambda(v-u)) := \frac{\Upsilon(\Lambda(v-u))}{\|\Lambda(v-u)\|_Y}$$

is a nonnegative error functional. The second norm on the right-hand side of (8.3.8) is a norm of the space topologically dual to V, which may be incomputable. Therefore, it is desirable to represent the estimate in a somewhat different form. One can make this if the second product in (8.3.4) is estimated by the inequality

$$\langle \Lambda^* y^* + v^*, v-u\rangle \le \|\Lambda^* y^* + v^*\|_{U^*} \|v-u\|_U, \tag{8.3.9}$$

where U and U^* is a pair of dual spaces the norms of which are defined by explicitly computable integrals (e.g., L^p and $L^{p'}$ with indices satisfying the condition $1/p + 1/p' = 1$), V is continuously embedded in U, and

$$\|w\|_U \le C_{UV}\|w\|_V, \qquad \forall w \in V, \tag{8.3.10}$$

where C_{UV} does not depend on w. Then, there exists a constant C (which is less than $C_{UV}c$) such that

$$\|w\|_U \leq \mathrm{C}\|\Lambda w\|_Y, \qquad \forall w \in V_0, \tag{8.3.11}$$

and (8.3.8) is replaced by

$$\gamma(\Lambda(v-u)) \leq \mathfrak{M}_{\mathrm{II}}(v, y^*) := \|G'(\Lambda v) - y^*\|_{Y^*} + \mathrm{C}\|\Lambda^* y^* + v^*\|_{U^*}. \tag{8.3.12}$$

It is easy to note that the right-hand side of (8.3.12) vanishes if and only if

$$y^* = G'(\Lambda v) \quad \text{and} \quad -\Lambda^* y^* \in \partial F(v),$$

which is equivalent to (8.2.10) and (8.2.12).

Assume that G is a differentiable functional. Then

$$G\Big(y_1 + \frac{y_2 - y_1}{2}\Big) \geq G(y_1) + \frac{1}{2}\langle\langle G'(y_1), y_2 - y_1\rangle\rangle,$$
$$\tfrac{1}{2}\langle\langle G'(y_2), y_2 - y_1\rangle\rangle \geq \tfrac{1}{2}G(y_2) - \frac{1}{2}G(y_1),$$

and we find that

$$\tfrac{1}{2}\langle\langle G'(y_2) - G'(y_1), y_2 - y_1\rangle\rangle \geq \tfrac{1}{2}\, G(y_2) + \tfrac{1}{2}\, G(y_1) - G\Big(\frac{y_1 + y_2}{2}\Big). \tag{8.3.13}$$

In view of (8.2.5), we have

$$\langle\langle G'(y_2) - G'(y_1), y_2 - y_1\rangle\rangle \geq 2\Phi_\delta\left(\frac{y_2 - y_1}{2}\right) \tag{8.3.14}$$

for $y_1,\ y_2 \in B_\delta$. By (8.3.14), we conclude that

$$\gamma(\Lambda(v-u)) \geq \frac{2\Phi_\delta\big(\frac{\Lambda(v-u)}{2}\big)}{\|\Lambda(v-u)\|_Y}. \tag{8.3.15}$$

Thus, properties of the error functional in (8.3.12) are determined by properties of the forcing functional.

Remark 8.7. If G does not have the Gâteaux derivative at a certain point, then the above relation holds, provided that it is replaced by an element of the corresponding subdifferential set $\partial G(\Lambda v)$.

Remark 8.8. In addition to Φ_δ and Υ, errors can be estimated in terms of the quantity

$$D_G(\Lambda v, p^*) = G(\Lambda v) + G^*(p^*) - \langle\langle p^*, \Lambda v\rangle\rangle,$$

which is a certain measure of that how accurately Λv reproduces $p^* = \Lambda u$. Relations between the error measures $\Phi_\delta(\Lambda(v-u))$, $\Upsilon(\Lambda(v-u))$, and $D_G(\Lambda v, p^*)$ were studied in [244, 286, 298] (also, see comments and references in Section 8.4.1 related to properties of $D_G(\Lambda v, p^*)$). In particular, it is easy to see that

$$\begin{aligned} D_G(\Lambda v, p^*) &= G(\Lambda v) + G^*(p^*) - \langle\!\langle p^*, \Lambda v \rangle\!\rangle \\ &\leq \langle\!\langle G'(\Lambda v), \Lambda(v-u) \rangle\!\rangle + G(\Lambda u) + G^*(p^*) - \langle\!\langle G'(\Lambda u), \Lambda v \rangle\!\rangle \\ &= \langle\!\langle G'(\Lambda v), \Lambda(v-u) \rangle\!\rangle + \langle\!\langle G'(\Lambda u), \Lambda u \rangle\!\rangle - \langle\!\langle G'(\Lambda u), \Lambda v \rangle\!\rangle \\ &= \Upsilon(\Lambda(v-u)). \end{aligned}$$

The second method. Let $\varphi : \mathbb{R} \to \mathbb{R}$ be a nonnegative function and φ^* be the Fenchel conjugate to φ. We derive another estimate from (8.3.4), using the following arguments. Take a number $\alpha > 0$ and apply Young's inequality to the first term on the right-hand side of (8.3.4). We have

$$\begin{aligned} &\langle\!\langle G'(\Lambda v) - y^*, \Lambda(v-u) \rangle\!\rangle \\ &\quad \leq \alpha \, \|\Lambda(v-u)\|_Y \left\| \frac{G'(\Lambda v) - y^*}{\alpha} \right\|_{Y*} \\ &\quad \leq \alpha\varphi\left(\|\Lambda(v-u)\|_Y\right) + \alpha\phi^*\left(\left\| \frac{G'(\Lambda v) - y^*}{\alpha} \right\|_{Y*} \right). \end{aligned} \tag{8.3.16}$$

Analogously, let ψ and ψ^* be another pair of nonnegative functions, where ψ^* is conjugate to ψ. Take $\beta > 0$ and apply Young's inequality to the second term in (8.3.4). With the help of (8.3.11) we find that

$$\begin{aligned} \langle \Lambda^* y^* + v^*, v - u \rangle &\leq \beta \left\| \frac{\Lambda^* y^* + v^*}{\beta} \right\|_{V*} c\|\Lambda(v-u)\|_Y \\ &\leq \beta c \psi\left(\|\Lambda(v-u)\|_Y\right) + \beta c \psi^* \left(\left\| \frac{\Lambda^* y^* + v^*}{\beta} \right\|_{V*} \right). \end{aligned} \tag{8.3.17}$$

By (8.3.4), (8.3.16), and (8.3.17), we deduce the estimate

$$\begin{aligned} &\Upsilon_{\alpha\beta}(\Lambda(v-u)) \\ &\quad \leq \alpha\varphi^* \left(\left\| \frac{G'(\Lambda v) - y^*}{\alpha} \right\|_{Y*} \right) + \beta c \, \psi^* \left(\left\| \frac{\Lambda^* y^* + v^*}{\beta} \right\|_{V*} \right), \end{aligned} \tag{8.3.18}$$

where

$$\Upsilon_{\alpha\beta}(\Lambda(v-u)) := \Upsilon(\Lambda(v-u)) - \alpha\varphi\left(\|\Lambda(v-u)\|_Y\right) - \beta c \psi\left(\|\Lambda(v-u)\|_Y\right),$$

and it is assumed that α, β, ϕ, and ψ are selected in such a way that $\Upsilon_{\alpha\beta}$ is a nonnegative functional.

If (8.3.9) holds, then we obtain

$$\Upsilon_{\alpha\beta}(\Lambda(v-u)) \leq \alpha\varphi^*\left(\left\|\frac{G'(\Lambda v)-y^*}{\alpha}\right\|_{Y*}\right) + \beta C\psi^*\left(\left\|\frac{\Lambda^* y^* + v^*}{\beta}\right\|_{U*}\right)$$
$$=: \overline{\mathfrak{M}}_{\mathrm{III}}(v, y^*; \alpha, \beta). \tag{8.3.19}$$

Let the functional Υ satisfy the relation

$$\Upsilon(\Lambda(v-u)) \geq 2\phi\left(\|\Lambda(v-u)\|_Y\right), \tag{8.3.20}$$

where ϕ is a nonnegative increasing function of the energy norm. Then,

$$\Upsilon_{\alpha\beta}(\|\Lambda(v-u)\|_Y) \geq (2-\alpha-\beta)\phi(\|\Lambda(v-u)\|_Y),$$

and it is natural to set $\varphi = \psi = \phi$. In this case, (8.3.19) has the form

$$(2-\alpha-\beta)\phi(\|\Lambda(v-u)\|_Y)$$
$$\leq \alpha\phi^*\left(\left\|\frac{G'(\Lambda v)-y^*}{\alpha}\right\|_{Y*}\right) + \beta C\phi^*\left(\left\|\frac{\Lambda^* y^* + v^*}{\beta}\right\|_{U*}\right). \tag{8.3.21}$$

Particular case. We show that for the linear diffusion problem, the general estimates given above lead to well-known estimates established in Chapter 4 for the problem: Find $u \in V_0 = \mathring{H}^1(\Omega)$ such that

$$a(u, w) = (f, w)_U, \qquad \forall w \in V_0.$$

In this case,

$$\Lambda v = \nabla v, \quad G(v) = \frac{1}{2}a(v,v), \quad \text{and} \quad F(v) = \int_\Omega f v\, dx,$$

where $f \in U = U^* = L^2(\Omega)$. Then

$$G'(v) = A\nabla v, \qquad \Upsilon(\Lambda(v-u)) = |||\nabla(u-v)|||^2,$$
$$\Lambda^* y^* = \operatorname{div} y^*, \qquad \Lambda^* y^* + v^* = \operatorname{div} y^* + f,$$

and (8.3.8) gives the estimate

$$|||\nabla(u-v)||| \leq |||y^* - A\nabla v|||_* + C\|\operatorname{div} y^* + f\|. \tag{8.3.22}$$

Now, we derive this estimate with the help of (8.3.20). Note that (cf. (8.3.20))

$$\Upsilon(\nabla(v-u)) = 2\phi(\|\nabla(v-u)\|),$$

where $\phi(t) = \frac{1}{2}t^2$. By (8.3.21), we have

$$\alpha\phi^*\left(\left\|\frac{G'(\Lambda v) - y^*}{\alpha}\right\|_{Y^*}\right) = \frac{1}{2\alpha}|||y^* - A\nabla v|||_*^2,$$

$$\beta C\phi^*\left(\left\|\frac{\Lambda^* y^* + v^*}{\beta}\right\|_{U}\right) = C\frac{1}{2\beta}\left\|\operatorname{div} y^* + f\right\|^2,$$

and (8.3.21) implies the estimate

$$(2-\alpha-\beta)|||\nabla(v-u)|||^2 \leq \frac{1}{\alpha}|||y^* - A\nabla v|||_*^2 + \frac{C}{\beta}\left\|\operatorname{div} y^* + f\right\|^2. \quad (8.3.23)$$

Set $\alpha = \frac{1}{1+\gamma}$ and $\beta = \frac{\gamma}{1+\gamma}$, where $\gamma > 0$. Then $\alpha + \beta = 1$, and we arrive at the estimate

$$|||\nabla(v-u)|||^2 \leq (1+\gamma)|||y^* - A\nabla v|||_*^2 + C\frac{1+\gamma}{\gamma}\left\|\operatorname{div} y^* + f\right\|^2, \quad (8.3.24)$$

which is a special form of (4.1.14).

8.4 A posteriori estimates for special classes of nonlinear elliptic problems

8.4.1 α-Laplacian

Let α be a real number greater than 1, and let $\alpha^* = \frac{\alpha}{\alpha-1}$ be the corresponding conjugate number. Consider the problem: Find $u \in V := W^{1,\alpha}(\Omega)$ such that $u = u_0$ on Γ and

$$\operatorname{div}\left(|\nabla u|^{\alpha-2}\nabla u\right) + f = 0 \quad \text{in } \Omega. \quad (8.4.1)$$

The weak statement of the problem is given by the integral identity

$$\int_\Omega \left(|\nabla u|^{\alpha-2}\nabla u \cdot \nabla w - fw\right) dx = 0, \qquad \forall\, w \in V_0, \quad (8.4.2)$$

where f is assumed to be of class $L^{\alpha^*}(\Omega)$, V_0 is the subspace of V formed by the functions vanishing at the boundary.

Variational method. By the variational method, a posteriori estimates were derived in [278, 61]. In this analysis, we rest upon the minimization problem P

$$\inf_{w \in u_0 + V_0} J_\alpha(w), \quad J_\alpha(w) := \int_\Omega \left(\frac{1}{\alpha}|\nabla w|^\alpha - fw\right) dx. \quad (8.4.3)$$

As usual, by $u_0 + V_0$ we denote the subspace of V containing functions $w = u_0 + w_0$, where $w_0 \in V_0$ and u_0 is a given function in V. The existence and uniqueness of a minimizer follow from the strict convexity of J_α. In the variational method, we need to consider the dual variational problem. For this purpose, we introduce the Lagrangian

$$L(v, y^*) := \int_\Omega \left(\nabla v \cdot y^* - \frac{1}{\alpha^*} |y^*|^{\alpha^*} - fv \right) dx$$

and note that

$$J_\alpha(v) = \sup_{y^* \in Y^*} L(v, y).$$

The dual variational functional is defined by the relation

$$I^*_{\alpha^*}(y^*) := \inf_{v \in u_0 + V_0} L(v, y^*)$$

and has the form

$$I^*_{\alpha^*}(y^*) := \begin{cases} \int_\Omega \left(\nabla u_0 \cdot y^* - \frac{1}{\alpha^*} |y^*|^{\alpha^*} - fu_0 \right) dx & \text{if } \ y^* \in Q^*_f, \\ -\infty & \text{if } \ y^* \notin Q^*_f, \end{cases}$$

where

$$Q^*_f := \left\{ y^* \in Y^* \mid \int_\Omega y^* \cdot \nabla w \, dx = \int_\Omega fw \, dx, \quad \forall w \in V_0 \right\}$$

and $Y^* := L^{\alpha^*}(\Omega; \mathbb{R}^d)$. The dual problem $\mathcal{P}^*$ associated with the primal problem (8.4.3) is as follows: Find $p^* \in Y^*$ such that P^*

$$I^*_{\alpha^*}(p^*) = \sup_{y^* \in Y^*} I^*_{\alpha^*}(y^*). \tag{8.4.4}$$

Theorem 8.9. *The problems $\mathcal{P}$ and $\mathcal{P}^*$ have unique solutions u and p^*, respectively;*

$$J_\alpha(u) = I_{\alpha^*}(p^*), \tag{8.4.5}$$

$$p^* = |\nabla u|^{\alpha-2} \nabla u \qquad \text{a.e. in } \Omega, \tag{8.4.6}$$

$$\nabla u = |p^*|^{\alpha^*-2} p^* \qquad \text{a.e. in } \Omega. \tag{8.4.7}$$

The proof of Theorem 8.9 follows from well-known results of convex analysis (e.g., see I. Ekeland and R. Themam [121]). If $\alpha = 2$, then we arrive at the quadratic functional, which generates a linear elliptic equation. Properties of problems related to the superquadratic ($\alpha > 2$) and subquadratic ($\alpha \in (1, 2)$) cases are rather different. For this reason, we discuss them separately.

Estimates for problems with superquadratic growth. Variational problems with superquadratic growth were among the first studied in the context of the variational approach to a posteriori error estimation (see [278]). For $\alpha \in (1, +\infty)$, computable upper bounds of the error were derived in M. Bildhauer and S. Repin [61]. These results are based on the uniform convexity of J_α (see S. Sobolev [336] and P. Mosolov and V. Myasnikov [239]). For $\alpha \geq 2$, the uniform convexity follows from the first Clarkson's inequality

$$\int_\Omega \left(\left| \frac{y_1 + y_2}{2} \right|^\alpha + \left| \frac{y_1 - y_2}{2} \right|^\alpha \right) dx \leq \frac{1}{2} \|y_1\|^\alpha_{\alpha,\Omega} + \frac{1}{2} \|y_2\|^\alpha_{\alpha,\Omega}, \quad \forall y_1, y_2 \in Y. \tag{8.4.8}$$

Theorem 8.10. *For any $v \in u_0 + V_0$, the following estimate holds:*

$$\|\nabla(v - u)\|^\alpha_{\alpha,\Omega} \leq \alpha 2^{\alpha-1} \left(I_\alpha(v) - I^*_\alpha(q^*) \right), \qquad \forall q^* \in Q^*_f. \tag{8.4.9}$$

Proof. We use (8.4.8). Setting $y_1 = \nabla u$ and $y_2 = \nabla v$, we obtain

$$\begin{aligned}
\int_\Omega |\nabla(v - u)|^\alpha dx &\leq 2^{\alpha-1} \left(\int_\Omega |\nabla v|^\alpha dx + \int_\Omega |\nabla u|^\alpha dx - 2 \int_\Omega \left| \frac{\nabla(u + v)}{2} \right|^\alpha dx \right) \\
&= \alpha 2^{\alpha-1} \left(I_\alpha(v) + I_\alpha(u) - 2 I_\alpha \left(\frac{u + v}{2} \right) \right) \\
&\leq \alpha 2^{\alpha-1} \left(I_\alpha(v) - I_\alpha(u) \right),
\end{aligned}$$

By (8.4.5), we conclude that

$$I_\alpha(u) \geq I^*_\alpha(q^*), \qquad \forall q^* \in Q^*_f,$$

which leads to (8.4.9). □

Remark 8.11. We note that (8.4.9) can be viewed as a generalized form of (2.3.1) related to variational problems with power growth.

In [61], the difference $I_\alpha(v) - I^*_{\alpha^*}(q^*)$ was analyzed and rearranged into a computable form.

Theorem 8.12. *Let $\alpha \geq 2$ and $v \in u_0 + V_0$. For any function $y^* \in Y^*$ that has divergence summable with power α^* and for any real number $\beta > 0$,*

$$\frac{1}{\alpha 2^{\alpha-1}} \|\nabla(v - u)\|^\alpha_{\alpha,\Omega} \leq \overline{\mathfrak{M}}_\alpha(v, y^*, \beta) := M_1(v, y^*, \beta) + M_2(y^*, \beta), \tag{8.4.10}$$

where

$$M_1(\nabla v, y^*, \beta) = D_\alpha(\nabla v, y^*) + \frac{\beta^\alpha}{\alpha} \| |y^*|^{\alpha^*-2} y^* - \nabla v \|^\alpha_{\alpha,\Omega},$$
$$M_2(y^*, \beta) = C^{\alpha^*}_{\alpha F} \Big(\frac{1}{\alpha^* \beta^{\alpha^*}} + 2^{2-\alpha^*}(3-\alpha^*) \Big) \| f + \operatorname{div} y^* \|^{\alpha^*}_{\alpha^*,\Omega},$$

$C_{\alpha F}$ *is the constant in the Friedrichs type inequality*

$$\|w\|_{\alpha,\Omega} \le C_{\alpha F} \|\nabla w\|_{\alpha,\Omega},$$

and the functional $D_\alpha \colon Y \times Y^* \to \mathbb{R}^+_0$ *is defined by the relation*

$$D_\alpha(y, y^*) := \int_\Omega \Big(\frac{1}{\alpha}|y|^\alpha + \frac{1}{\alpha^*}|y^*|^{\alpha^*} - y \cdot y^* \Big)\, dx.$$

It is easy to see that the right-hand side of (8.4.10) vanishes if and only if the relations

$$|y^*|^{\alpha^*-2} y^* = \nabla v, \tag{8.4.11}$$
$$\operatorname{div} y^* + f = 0 \tag{8.4.12}$$

hold almost everywhere in Ω. Since the solution of the problem (8.4.3) is unique, the relations (8.4.11) and (8.4.12) are equivalent to the fact that $v = u$ and $y^* = p^*$. The functional D_α can be viewed as a certain measure of the error in the duality relations (8.4.6) and (8.4.7). Indeed, this functional is nonnegative and vanishes if and only if (cf. Proposition 1.2)

$$y = |y^*|^{\alpha^*-2} y^* \quad \text{and} \quad y^* = |y|^{\alpha-2} y.$$

The second term of M_1 possesses the same properties. Therefore, M_1 is *a measure of the error in the duality relations*. The term $M_2(y^*, \beta)$ penalizes *the violation of the relation* $\operatorname{div} y^* + f = 0$.

Comments. It should be noted that we cannot prove that there always exist y^* and β such that the right-hand side of (8.4.10) coincides with the left-hand one. However, such a property may hold for some special quantities introduced to characterize the accuracy of an approximate solution (this question is discussed in [61, 244, 282, 286, 298] and some other publications). For example, instead of the L^α-norm of $\nabla(v - u)$ we can take a special measure of the error defined by the compound functional $D_G(\nabla v, p^*)$, which in our case has the form

$$\begin{aligned} D_\alpha(\nabla v, p^*) &= \int_\Omega \Big(\frac{1}{\alpha}|\nabla v|^\alpha + \frac{1}{\alpha^*}|p^*|^{\alpha^*} - \nabla v \cdot p^* \Big)\, dx = J_\alpha(v) - I_{\alpha^*}(p^*) \\ &= J_\alpha(v) - J_\alpha(u). \end{aligned}$$

The majorant $\overline{\mathfrak{M}}_\alpha$ is deduced as an upper bound of $J_\alpha(v) - I_{\alpha^*}(p^*)$ and, therefore, provides an upper estimate for $D_\alpha(\nabla v, p^*)$. It is easy to see that for $y^* = p^*$, the term $M_2(y^*, \beta)$ vanishes and $M_1(v, y^*, \beta)$ coincides with $D_\alpha(\nabla v, p^*)$. The latter quantity is a nonnegative functional, which can be regarded as a certain measure of the error. In terms of such a measure, the majorant $\overline{\mathfrak{M}}_\alpha$ is sharp. Similar results hold for other nonlinear problems (see above-cited publications).

If $\alpha = 2$, then the quantity

$$d_\alpha(v) := D_2(\nabla v, p^*)$$

coincides with $\frac{1}{2}\|\nabla(v-u)\|^2_{2,\Omega}$ and gives a natural energy norm of the error. If $\alpha \neq 2$, then $d_\alpha(v)$ is not a norm. Nevertheless,

$$d_\alpha(v) := D_\alpha(\nabla v, p^*) \geq 0,$$

and $d_\alpha(v) = 0$ if and only if

$$\nabla v = |p^*|^{\alpha^*-2} p^* = \nabla u \qquad \text{a.e. in } \Omega.$$

In some cases, $d_\alpha(v)$ can be regarded as a certain weighted norm of the deviation from the exact solution. For example, for $\alpha = 3$, we have

$$d_3(v) \geq \frac{1}{3} \int_\Omega (|\nabla v| + 2|\nabla u|)|\nabla(v-u)|^2 \, dx.$$

However, in some cases it is important to get realistic error bounds in terms of the energy norm. In general, this is a more complicated task. We shortly discuss it in Section 8.5.

Estimates for problems with subquadratic growth. For $\alpha < 2$, it is more convenient to derive a posteriori estimates within the framework of Problem $(\mathcal{P})^*$. In this case, $\alpha^* > 2$ and the dual functional

$$(-I^*_{\alpha^*})(y^*) = \int_\Omega \Big(- \nabla u_0 \cdot y^* + \frac{1}{\alpha^*}|y^*|^{\alpha^*} + f u_0 \Big) \, dx$$

is uniformly convex on Y^*. Therefore, for any q_1^* and $q_2^* \in Q_f^*$ we apply Clarkson's inequality and obtain

$$(-I^*_{\alpha^*})\left(\frac{q_1^* + q_2^*}{2}\right) + \int_\Omega \left(\frac{q_1^* - q_2^*}{2}\right)^{\alpha^*} dx \leq \frac{1}{2}\Big((-I^*_{\alpha^*})(q_1^*) + (-I^*_{\alpha^*})(q_2^*)\Big),$$

which allows us to derive an upper bound of $\|p^* - y^*\|_{\alpha^*,\Omega}$, where y^* is an approximate one. The assertion below is an analog of Theorem 8.12 for the case $\alpha^* > 2$.

Theorem 8.13. *For any function* $v \in u_0 + V_0$*, any vector-valued function* $y^* \in Y^*$ *such that* $\operatorname{div} y^*$ *is summable with power* α^**, and any* $\beta > 0$*, the following estimate holds:*

$$\|p^* - y^*\|_{\alpha^*,\Omega}^{\alpha^*} \leq \overline{\mathfrak{M}}_{\alpha^*}(v, y^*, \beta) := M_1(v, y^*, \beta) + M_2(y^*, \beta). \tag{8.4.13}$$

Here $M_1(v, y^*, \beta) = 2\alpha^* \mu^2 \Big(D_\alpha(\nabla v, y^*) + \frac{\beta}{2} \| |y^*|^{\alpha^*-2} y^* - \nabla v \|_{\alpha,\Omega}^2 \Big)$,

$$\begin{aligned} M_2(y^*, \beta) = 2\alpha^* \mu^2 \Big(&\frac{1}{2\beta} (\mathrm{r}(y^*))^2 \\ &+ (\alpha^* - 1)\kappa\mu \Big(\mu \|y^*\|_{\alpha^*,\Omega}^{\alpha^*-2} + (\mathrm{r}(y^*))^{\alpha^*-2} \Big) (\mathrm{r}(y^*))^2 \Big) + 2^{-}(\mathrm{r}(y^*))^{\mathrm{ff}^*}, \end{aligned}$$

$\mu = 2^{\alpha^*-2}$*, and* $\mathrm{r}(y^*)$ *is defined by the relation* $\mathrm{r}(y^*) := C_{\alpha F} \|f + \operatorname{div} y^*\|_{\alpha^*,\Omega}$.

The reader interested in a more detailed discussion of functional a posteriori estimates for variational problems with power growth functionals is referred to [61, 57, 139, 244, 278].

8.4.2 Problems with nonlinear boundary conditions

Boundary conditions in general form. In many cases, the commonly used Dirichlet or Neumann boundary conditions cannot properly describe the behavior of a model and should be replaced by more sophisticated conditions that reflect real physical situations. Typical examples are presented by problems with unilateral boundary conditions and friction (e.g., see G. Duvaut and G.-L. Lions [120] and P. Panagiotopoulos [259]). The corresponding boundary value problems are formulated as variational inequalities and can be solved by known numerical methods (e.g., see R. Glowinski [153] and R. Glowinski, J.-L. Lions, and R. Trémolierès [155]).

We study the case of nonlinear boundary conditions with the paradigm of the problem

$$\operatorname{div} A\nabla u + f = 0 \qquad \text{in } \Omega, \tag{8.4.14}$$

where A satisfies (4.1.4) and

$$u(x) = u_0(x), \qquad x \in \Gamma_1, \tag{8.4.15}$$

$$-A\nabla u \cdot n(x) \in \partial j(u(x)), \qquad x \in \Gamma_2. \tag{8.4.16}$$

Henceforth, we assume that $f \in L_2(\Omega)$, $u_0 \in H^1(\Omega)$, and the boundary Γ consists of two disjoint measurable parts Γ_1 and Γ_2. Also, we assume that it is piecewise smooth, so that one can uniquely define the unit outward normal at almost all points of Γ. By $u_{,n}$ we denote the normal derivative of u and $j : \mathbb{R}^d \to \mathbb{R}$ is a convex lower semicontinuous functional, which determines the so-called *boundary dissipative potential* (e.g., see [259]). The relation (8.4.16) is the general form of a wide spectrum of boundary conditions. Similar relations are often used in continuum mechanics if it is necessary to model unilateral boundary contact or contact with friction. In this case, boundary conditions are represented in the form

$$-\sigma_n(x) \in \partial j(u(x)), \qquad x \in \Gamma_2, \tag{8.4.17}$$

where σ is the stress tensor and u is the displacement. Problem (8.4.12)–(8.4.15) is a simplified version of the elasticity model, in which u is a scalar-valued function and (8.4.17) is replaced by (8.4.14). However, from the mathematical point of view these two problems are quite similar.

We recall that any $v \in V = H^1(\Omega, \mathbb{R}^d)$ has a trace on Γ denoted by γv (cf. (1.4.18) and (1.4.19)), where $\gamma \in \mathcal{L}\big(H^1(\Omega), H^{1/2}(\Gamma)\big)$ is the trace operator. In the preceding sections, the operator γ was omitted in the formulas associated with boundary relations. In this section, we keep it in explicit form. Let V stand for $H^1(\Omega)$ and

$$V_0 := \{v \in V \mid \gamma v = 0 \text{ a.e. on } \Gamma_1\},$$

which is a subspace of V. The set $\gamma(V_0)$ is denoted by $\mathfrak{T}$ (it is a subspace of $H^{1/2}(\Gamma)$). The corresponding dual space $\mathfrak{T}^*$ contains traces (on Γ_2) of the functions from the space $H(\Omega, \mathrm{div})$. Indeed, for any smooth y^* and any $v \in V_0$, we have the relation

$$\int_{\Gamma_2} (y_n^*)\, v\, ds = \int_\Omega (y^* \cdot \nabla v + (\mathrm{div} y^*) v)\, dx, \tag{8.4.18}$$

where $y_n^* := y^* \cdot n$. For any $y^* \in H(\Omega, \mathrm{div})$, the right-hand side of this identity is a linear continuous functional $\ell_{y^*} : V_0 \to \mathbb{R}$ satisfying the relations

$$\ell_{y^*}(v) = 0, \qquad \forall v \in \mathring{H}^1(\Omega), \tag{8.4.19}$$

$$\left|\ell_{y^*}(v)\right| \le c_\mu \left\|y^*\right\|_{\mathrm{div}} \|\gamma v\|_{H^{1/2},\Gamma}\,. \tag{8.4.20}$$

In fact, ℓ_{y^*} is a linear continuous mapping defined on a factor space of V_0 two elements of which are considered as different only if they have different traces on Γ. Indeed, $\ell_{y^*}(v_1) = \ell_{y^*}(v_2)$ if $v_1, v_2 \in V_0$ and $\gamma v_1 = \gamma v_2$ on Γ. For this reason, ℓ_{y^*} can be identified with a certain element in $\mathfrak{T}^*$, which we denote by $\gamma^* y_n^*$ and call the *normal trace of* y^* *on* Γ_2. The value of the functional $\xi^* \in \mathfrak{T}^*$ on $\xi \in \mathfrak{T}$ is denoted by $\langle \xi^*, \xi\rangle_{\Gamma_2}$. Then, (8.4.18) has the form

$$\langle \gamma^* y_n^*, \gamma v\rangle_{\Gamma_2} = \int_\Omega (y^* \cdot \nabla v + \mathrm{div}\, y^* \cdot v)\, dx. \tag{8.4.21}$$

The norm of $\gamma^* y_n^*$ is defined by the standard relation

$$\left\|\gamma^* y_n^*\right\|_{\mathfrak{T}^*} := \sup_{v \in V_0} \frac{\langle \gamma^* y_n^*, \gamma v\rangle_{\Gamma_2}}{\|\gamma v\|_{\mathfrak{T}}} = \sup_{v \in V_0} \frac{\int_\Omega (y^* \cdot \nabla v + \mathrm{div}\, y^* \cdot v)\, dx}{\|\gamma v\|_{\mathfrak{T}}}. \tag{8.4.22}$$

In view of (8.4.22),

$$\left\|\gamma^* y_n^*\right\|_{\mathfrak{T}^*} \le c_\mu \left\|y^*\right\|_{\mathrm{div}}\,. \tag{8.4.23}$$

Conjugate functionals defined on spaces of traces. For any $\xi \in \mathfrak{T}$ we define the functional

$$\mathfrak{J}(\xi) := \int_{\Gamma_2} j(\xi)\, ds.$$

We assume that the integrand $j : \mathbb{R}^d \to \mathbb{R}^d$ is a nonnegative, continuous, and convex functional such that $j(0) = 0$ and $\operatorname{dom} j := \{p \in \mathbb{R}^d \mid j(p) < +\infty\} \neq \emptyset$, so that j belongs to the class of so-called *proper* convex functionals.

In this case, the functional $\mathfrak{J}(\xi)$ is also nonnegative, convex, and lower semicontinuous on $\mathfrak{T}$. Since γ is a bounded linear operator, the functional $\mathfrak{J}(\gamma v)$ also possesses the above properties as a functional on V_0.

By definition, $\partial\mathfrak{J}(v)$ contains elements $\xi^* \in \mathfrak{T}$ such that

$$\mathfrak{J}(v + w) - \mathfrak{J}(v) \leq \langle \xi^*, w\rangle_{\Gamma_2}, \qquad \forall w \in V_0, \tag{8.4.24}$$

and

$$\mathfrak{J}^*(\xi^*) := \sup_{\xi \in \mathfrak{T}} \left\{\langle \xi^*, \xi\rangle_{\Gamma_2} - \mathfrak{J}(\xi)\right\}$$

is the functional *conjugate* to the functional $\mathfrak{J}$.

By recalling (8.4.21), we find that

$$\mathfrak{J}^*(\gamma^* y_n^*) = \sup_{w \in V_0} \left\{\int_\Omega \left(y^* \cdot \nabla w + \operatorname{div} y^* \cdot w\right) dx - \mathfrak{J}(\gamma w)\right\}$$

and define the *compound functional*

$$D_{\Gamma_2}(\gamma v, \gamma^* y_n^*) := \mathfrak{J}(\gamma v) + \mathfrak{J}^*(\gamma y_n^*) - \langle \gamma v, \gamma y_n^*\rangle_{\Gamma_2}$$

generated by traces on Γ_2. It is easy to see that

$$\begin{aligned} D_{\Gamma_2}(\gamma v, \gamma^* y_n^*) := \sup_{w \in V_0} \Bigg(\int_\Omega & \left(y^* \cdot \nabla(w - v) + \operatorname{div} y^* \cdot (w - v)\right) dx \\ & + \int_{\Gamma_2} (j(\gamma v) - j(\gamma w))\, ds \Bigg) \geq 0. \end{aligned}$$

Moreover, if $D_{\Gamma_2}(\gamma v, \gamma^* y_n^*) = 0$, then (see Proposition 1.2) $\gamma^* y_n^* \in \partial\mathfrak{J}(\gamma v)$. If y_n^* is sufficiently regular (e.g., $\gamma^* y_n^* \in L_2(\Gamma_2, \mathbb{R}^d)$), then $\mathfrak{J}^*$ has an explicit form

$$\mathfrak{J}^*(\gamma^* y_n^*) = \int_{\Gamma_2} j^*(\gamma^* y_n^*)\, ds,$$

where $j^*(\xi^*) = \sup_{\xi \in \mathbb{R}^d} \{\xi^* \cdot \xi - j(\xi)\}$.

Variational inequality. Now we can state the problem as a variational inequality: Find $u \in u_0 + V_0$ such that

$$a(u, w-u) + \mathfrak{J}(w) - \mathfrak{J}(u) \geq \int_\Omega f(w-u)\,dx \tag{8.4.25}$$

holds for all $w \in u_0 + V_0$. This problem is equivalent to the variational problem: Find $u \in u_0 + V_0$ such that

$$J(u) = \inf_{w \in u_0 + V_0} J(w), \quad J(w) = \frac{1}{2} a(w, w) + \mathfrak{J}(w) - \int_\Omega f w\,dx. \tag{8.4.26}$$

The functional J is strictly convex, continuous, and coercive on V, and the set $u_0 + V_0$ is a convex closed subset of V. Therefore, the variational problem (8.4.25) is uniquely solvable.

Assume that the function u is sufficiently regular (e.g., has first and second derivatives in the classical sense). Then, from (8.4.24) it follows that

$$\int_\Omega (\operatorname{div} A\nabla u + f)(w-u)\,dx \leq \int_{\Gamma_2} (j(w) - j(u))\,ds + \int_{\Gamma_2} A\nabla u \cdot n(w-u)\,ds$$

for any $w \in u_0 + V_0$. Set $w = u$ on Γ. Then, this relation implies (8.4.14). Hence, we find that

$$\int_{\Gamma_2} (j(w) - j(u) + A\nabla u \cdot n)(w-u)\,ds \geq 0, \qquad \forall w \in u_0 + V_0,$$

and we arrive at (8.4.16).

Estimates of the difference between the exact and approximate solutions to (8.4.25) can be obtained by any of the two methods (variational and nonvariational).

A posteriori estimates. The nonvariational method. Let $v \in u_0 + V_0$ be an approximation of u. We substitute it into (8.4.25) and rewrite in the form

$$a(u-v, v-u) + \mathfrak{J}(v) - \mathfrak{J}(u) \geq \int_\Omega f(v-u)\,dx - a(v, v-u). \tag{8.4.27}$$

By (8.4.24), we find that

$$a(u-v, u-v) \leq \langle v^*, v-u \rangle_{\Gamma_2} + \int_\Omega f(u-v)\,dx + a(v, v-u), \tag{8.4.28}$$

where v^* is an element of the set $\partial\mathfrak{J}(v)$. In view of (8.4.21),

$$\langle \gamma^* y_n^*, \gamma(v-u) \rangle_{\Gamma_2} = \int_\Omega (y^* \cdot \nabla(v-u) + \operatorname{div} y^* \cdot (v-u))\,dx,$$

and we rearrange the right-hand side of (8.4.28) as follows:

$$a(u-v,u-v) \le \langle v^* + \gamma^* y_n^*, \gamma(v-u)\rangle_{\Gamma_2} + \int_\Omega (\operatorname{div} y^* + f)(u-v)\,dx + \int_\Omega (A\nabla v - y^*)\cdot\nabla(u-v)\,dx. \tag{8.4.29}$$

Now we recall that (cf. (8.4.22) and (4.1.9))

$$\begin{aligned}\langle v^* + \gamma^* y_n^*, \gamma(v-u)\rangle_{\Gamma_2} &\le \left\| v^* + \gamma^* y_n^* \right\|_{\mathfrak{T}^*} \|\gamma(v-u)\|_{\mathfrak{T}} \\ &\le C_{T\Gamma_2} \left\| v^* + \gamma^* y_n^* \right\|_{\mathfrak{T}^*} \|\nabla(u-v)\|.\end{aligned}$$

By this relation, we obtain

$$|||\nabla(u-v)||| \le |||A\nabla v - y|||_* + \frac{1}{c_1}\left(C_{F\Omega}\|\operatorname{div} y^* + f\| + \left\| v^* + \gamma^* y_n^* \right\|_{\mathfrak{T}^*}\right). \tag{8.4.30}$$

This estimate yields the general form of the upper bound.

Assume that the boundary potential is sufficiently regular, so that we can write

$$\mathfrak{J}(v) - \mathfrak{J}(u) = \int_{\Gamma_2} (j(v) - j(u))\,ds \le \int_{\Gamma_2} \gamma(v-u)\,v^*\,ds, \tag{8.4.31}$$

where v^* (which is determined by $\partial j(v)$) can be identified with a square integrable function defined on Γ_2 (we assume that such an element exists). In addition, assume that $\gamma^* y_n^*$ can be identified with a square integrable function defined on Γ_2. Then, (8.4.30) implies the estimate

$$|||\nabla(u-v)||| \le |||A\nabla v - y|||_* + \frac{1}{c_1}\left(C_{F\Gamma_1}\|\operatorname{div} y^* + f\| + C_{T\Gamma_2}\|v^* + \gamma^* y_n^*\|_{\Gamma_2}\right), \tag{8.4.32}$$

where $C_{F\Gamma_1}$ and $C_{T\Gamma_2}$ are defined by (4.1.8) and (4.1.9). This estimate is a generalization of (4.1.12). It shows that the error bound is represented as the sum of three terms that penalize the relations

$$\operatorname{div} y^* + f = 0 \quad \text{in } \Omega, \tag{8.4.33}$$

$$y^* = A\nabla v \quad \text{in } \Omega, \tag{8.4.34}$$

$$-\gamma^* y_n^* \subset \partial j(v) \quad \text{on } \Gamma_2. \tag{8.4.35}$$

A posteriori estimates. The variational method. Since the minimizer u to the problem $\mathcal{P}$ satisfies (8.4.25), we find that

$$\begin{aligned} J(v) - J(u) &= \frac{1}{2}a(v-u,v-u) + a(u,v-u) - \langle f, v-u\rangle + \mathfrak{J}(v) - \mathfrak{J}(u) \\ &\ge \frac{1}{2}a(v-u,v-u), \qquad \forall v \in u_0 + V_0, \end{aligned} \tag{8.4.36}$$

which implies the basic "deviation" estimate

$$\frac{1}{2}|||\nabla(u-v)|||^2 \leq J(v) - \inf \mathcal{P}, \qquad \forall v \in u_0 + V_0. \tag{8.4.37}$$

The right-hand side of (8.4.37) can be estimated from above by a method similar to that discussed in Section 3.1.

Let v be an arbitrary approximation in $u_0 + V_0$. Then, the first upper bound of the error for problems with nonlinear boundary conditions is given by the estimate

$$\frac{1}{2}|||\nabla(u-v)|||^2 \leq (1+\beta)M_1(v,y^*) + M_2(\gamma v, \xi^*) + \frac{1+\beta}{\beta}M_3(y^*,\xi^*), \tag{8.4.38}$$

where y^*, ξ^*, and β are arbitrary elements of the sets $L^2(\Omega, \mathbb{M}^{d\times d})$, $\mathfrak{T}^*$, and $\mathbb{R}_+$, respectively. The functionals M_1, M_2, and M_3 are defined by the relations

$$\begin{aligned} M_1(v,y^*) &= \frac{1}{2}\int_\Omega (A\nabla v\cdot\nabla v + A^{-1}y^*\cdot y^* - 2\nabla v\cdot y^*)\,dx,\\ M_2(\gamma v,\xi^*) &= \mathfrak{J}(\gamma v) + \mathfrak{J}^*(\xi^*) - \langle \xi^*, \gamma v\rangle_{\Gamma_2},\\ M_3(y^*,\xi^*) &= \frac{1}{2}\inf_{\eta^*\in Q^*_{\ell_{\xi^*}}} |||\eta^* - y^*|||_*^2, \end{aligned}$$

where

$$Q^*_{\ell_{\xi^*}} := \left\{ y^* \in L^2(\Omega, \mathbb{M}^{d\times d}) \mid \int_\Omega y^*\cdot\nabla w\,dx = \ell_{\xi^*}(w), \qquad \forall w \in V_0 \right\},$$

and

$$\ell_{\xi^*}(w) := \int_\Omega f\cdot w\,dx - \langle \xi^*, \gamma w\rangle_{\Gamma_2}.$$

The proof of (8.4.35) can be found in S. Repin and J. Valdman [319].

It is clear that the quantities M_1, M_2, and M_3 are nonnegative. The quantity $M_1(v,y^*)$ vanishes if and only if v and y^* satisfy the relation (8.4.34) and $M_2(\gamma v,\xi^*) = 0$ if and only if $\xi^* \in \partial\mathfrak{J}(\gamma v)$ on Γ_2. Thus, M_2 is a measure of the error in the boundary condition (8.4.14) computed on Γ_2 for the function $-\xi^* \in \mathfrak{T}^*$ (which can be thought of as an image of the normal component of the flux) and the trace of v . The quantity $M_3(y^*)$ vanishes if and only if $y^* \in Q^*_{\ell_{\xi^*}}$, i.e., if

$$\int_\Omega y^*\cdot\nabla w\,dx = \int_\Omega f\cdot w\,dx - \langle \xi^*, \gamma w\rangle_{\Gamma_2}, \qquad \forall w \in V_0.$$

Since

$$\int_\Omega y^*\cdot\nabla w\,dx = \langle \gamma^* y^*_n, \gamma w\rangle_{\Gamma_2} - \int_\Omega \operatorname{div} y^*\cdot w\,dx,$$

we conclude that $M_3(y^*)$ vanishes if and only if the equilibrium equation (8.4.33) and the relation $\gamma^* y_n^* = -\xi^*$ hold (in a generalized sense).

Estimate (8.4.35) can be represented in a form, which is more convenient from the practical point of view. Assume that

$$\xi^* \in L_2(\Gamma_2), \quad y^* \in H(\Omega, \mathrm{div}), \quad \text{and} \quad \gamma^* y_n^* \in L_2(\Gamma_2, \mathbb{R}^d). \tag{8.4.39}$$

Then, the term M_3 can be estimated from above by a directly computable quantity (using the same arguments as in Lemmas 3.2 and 7.3) and we obtain

$$\tfrac{1}{2}|||\nabla(u-v)|||^2 \le (1+\beta)M_1(v,y^*) + M_2(\gamma v, \xi^*) + \frac{1+\beta}{2c_1^2\beta}\Big(C_{F\Gamma_1}\|\mathrm{r}_\Omega(y^*)\| \\ + C_{T\Gamma_2}\int_{\Gamma_2}(\xi^* + \gamma^* y_n^*)^2 ds\Big)^2. \tag{8.4.40}$$

Here $\beta > 0$ and $\mathrm{r}_\Omega(y^*) := \mathrm{div}\, y^* + f$. We see that (8.4.40) has the same principal structure as (8.4.38). The difference is that the functions y^* and ξ^* are integrable functions. As before, the majorant vanishes if and only if the relations (8.4.33)–(8.4.35) hold.

Particular forms of (8.4.40) Estimate (8.4.40) has particular forms, which deserve special comments.

1. The first form arises if we set

$$\xi^* = -\gamma^* y_n^*, \tag{8.4.41}$$

i.e., if we define ξ^* (which is an image of the true boundary flux), using a known approximation of the dual variable. In this case, the last integral on the right-hand side of (8.4.40) vanishes, and we arrive at the estimate

$$\frac{1}{2}|||\nabla(u-v)|||^2 \le (1+\beta)M_1(v,y^*) + M_2(\gamma v, \gamma^* y_n^*) \\ + \frac{1+\beta}{2c_1^2\beta}C_{F\Gamma_1}^2\|\mathrm{r}_\Omega(y^*)\|^2. \tag{8.4.42}$$

2. Another option is to take ξ^* in accordance with the relation

$$\xi^* \in \partial j(v), \tag{8.4.43}$$

i.e., define it using the approximation v. In this case, the second integral on the right-hand side of (8.4.40) vanishes and we obtain

$$\frac{1}{2}|||\nabla(u-v)|||^2 \le (1+\beta)M_1(v,y^*) + \frac{1+\beta}{2c_1^2\beta}\Big(C_{F\Gamma_1}^2\|\mathrm{r}_\Omega(y^*)\|^2 \\ + C_{T\Gamma_2}\int_{\Gamma_2}(v^* + \gamma^* y_n^*)^2 ds\Big)^2. \tag{8.4.44}$$

Note that $M_1(v, y^*) = \frac{1}{2}|||A\nabla v - y^*|||_*^2$. Therefore, (8.4.44) is the squared form of (8.4.32).

3. We can try to select ξ^* in the best possible way. For this purpose, we estimate the last term of (8.4.40) by means of Young's inequality. Then, we get the following inequality, which involves a new positive constant α:

$$\begin{aligned}\tfrac{1}{2}|||\nabla(u-v)|||^2 \leq (1+\beta)M_1(v, y^*) + M_2(\gamma v, \xi^*)\\ + \tfrac{1}{2}\left(1+\tfrac{1}{\beta}\right)(1+\alpha)C^2_{F\Gamma_1}\|\mathrm{r}_\Omega(y^*)\|^2\\ + \tfrac{1}{2}\left(1+\tfrac{1}{\beta}\right)\left(1+\tfrac{1}{\alpha}\right)C^2_{T\Gamma_2}\int_{\Gamma_2}(\xi^* + \gamma^* y_n^*)^2\,ds.\end{aligned} \tag{8.4.45}$$

We gather the terms related to Γ_2 and denote them by

$$I_{\Gamma_2}(\gamma v, \gamma^* y_n^*, \xi^*) = \int_{\Gamma_2}\left(j(\gamma v) + j^*(\xi^*) - \xi^* \cdot \gamma v + \tfrac{\theta}{2}\left|\gamma^* y_n^* + \xi^*\right|^2\right)ds, \tag{8.4.46}$$

where $\theta = \left(1+\frac{1}{\beta}\right)\left(1+\frac{1}{\alpha}\right)C^2_{T\Gamma_2}$.

To minimize the right-hand side of (8.4.46), we need to minimize I_{Γ_2} with respect to ξ^*, i.e., to solve the problem

$$\inf_{\xi^* \in L^2(\Gamma_2)}\left(j^*(\xi^*) + \frac{\theta}{2}|\xi^*|^2 - \xi^*\gamma v + \xi^*\gamma^* y_n^*\right).$$

Under the assumptions made, the corresponding minimum has the form

$$M_{\Gamma_2}(\gamma v, \gamma^* y_n^*, \theta) := \int_{\Gamma_2}\left(j(\gamma v) + \frac{\theta}{2}\left|\gamma^* y_n^*\right|^2 - \rho(\gamma v - \theta\gamma^* y_n^*)\right)ds,$$

where $\rho : \mathbb{R}^d \to \mathbb{R}$ is the function conjugate to $j^*(\xi^*) + \frac{\theta}{2}\xi^{*2}$. Now, we find that

$$\begin{aligned}\tfrac{1}{2}|||\nabla(u-v)|||^2 \leq (1+\beta)M_1(v, y^*) + M_{\Gamma_2}(\gamma v, \gamma^* y_n^*, \theta)\\ + \tfrac{1}{2}\left(1+\tfrac{1}{\beta}\right)(1+\alpha)C^2_{F\Gamma_1}\|\mathrm{r}_\Omega(y^*)\|^2.\end{aligned} \tag{8.4.47}$$

Neumann type boundary condition. The boundary condition of this type corresponds to the case

$$\mathfrak{J}(\xi) := \langle \eta^*, \xi \rangle_{\Gamma_2}, \tag{8.4.48}$$

where $\eta^* \in \mathfrak{T}^*$. In particular, if η^* is associated with a function $-F \in L^2(\Gamma_2)$, then

$$j(\gamma u) = -F \cdot \gamma u, \qquad \partial j(\gamma u) = -F,$$

and (8.4.48) is the Neumann boundary condition

$$\gamma(n \cdot p^*) = F \qquad \text{a.e. on } \Gamma_2. \tag{8.4.49}$$

Since

$$\mathfrak{J}(\xi) = -\int_{\Gamma_2} F \cdot \xi \, ds$$

and

$$\mathfrak{J}^*(\xi^*) = \begin{cases} 0 & \text{if } \xi^* = -F \text{ a.e. on } \Gamma_2, \\ +\infty & \text{otherwise,} \end{cases}$$

we find that

$$\begin{aligned} I_{\Gamma_2}(\gamma v, \gamma^* y_n^*, \xi^*) &= \int_{\Gamma_2} \left(-F \cdot \gamma v + 0 + F \cdot \gamma v + \tfrac{\theta}{2} \left| \gamma^* y_n^* - F \right|^2 \right) ds \\ &= \frac{\theta}{2} \int_{\Gamma_2} \left| \gamma^* y_n^* - F \right|^2 ds. \end{aligned}$$

Hence, (8.4.47) has the form

$$\begin{aligned} ||| \nabla (u - v) |||^2 \leq (1+\beta) 2 M_1(v, y^*) + \left(1 + \tfrac{1}{\beta}\right) (1+\alpha) C_{F\Gamma_1}^2 \mathrm{r}_\Omega{}^2(y^*) \\ + \left(1 + \tfrac{1}{\beta}\right) \left(1 + \tfrac{1}{\alpha}\right) C_{T\Gamma_2}^2 \int_{\Gamma_2} \left| \gamma^* y_n^* - F \right|^2 ds. \end{aligned} \tag{8.4.50}$$

The minimization of the right-hand side with respect to α leads to the estimate

$$\begin{aligned} ||| \nabla (u - v) |||^2 \leq (1+\beta) ||| A \nabla v - y^* |||_*^2 \\ + \left(1 + \tfrac{1}{\beta}\right) \left(C_{F\Gamma_1} \| \mathrm{r}_\Omega(y^*) \| + C_{T\Gamma_2} \| \gamma^* y_n^* - F \|_{\Gamma_2} \right)^2, \end{aligned} \tag{8.4.51}$$

which is the squared form of (4.1.12).

Friction type boundary condition. In this case,

$$j(\gamma v) = \mu |\gamma v|, \qquad \mathfrak{J}(\xi) = -\int_{\Gamma_2} \mu |\xi| \, ds, \qquad \mu > 0, \tag{8.4.52}$$

and

$$j^*(\xi^*) = \begin{cases} 0 & \text{if } |\xi^*| \leq \mu, \\ +\infty & \text{otherwise.} \end{cases} \tag{8.4.53}$$

It is easy to see that

$$\partial j(\xi) = \begin{cases} +\mu & \text{if } \xi > 0, \\ t \in [-\mu, +\mu] & \text{if } \xi = 0, \\ -\mu & \text{if } \xi < 0. \end{cases}$$

Therefore, if $|p^* \cdot n| < \mu$, then (cf. (8.4.16)) implies $\gamma v = 0$; γv can take nonzero values are only if $|p^* \cdot n|$ attains limit values stated by the constant μ.s

Set $\xi^* = -\gamma^* y_n^*$ and impose the condition $|\gamma^* y_n^*| \leq \mu$. Then,

$$M_2(\gamma v, -\gamma^* y_n^*) = \int_{\Gamma_2} \big(\mu|\gamma v| + (\gamma^* y_n^*)(\gamma v)\big)\, ds, \tag{8.4.54}$$

provided that $\gamma^* y_n^*$ is a square integrable function on Γ_2. Now the estimate (8.4.45) takes the form

$$\tfrac{1}{2}|||\nabla(u-v)|||^2 \leq (1+\beta)M_1(v,y^*) + \int_{\Gamma_2} \big(\mu|\gamma v| + (\gamma^* y_n^*)(\gamma v)\big)\, ds$$
$$+ \tfrac{1}{2}\left(1+\tfrac{1}{\beta}\right)(1+\alpha)C^2_{F\Gamma_1}\mathrm{r}_\Omega{}^2(y^*), \tag{8.4.55}$$

where $|\gamma^* y_n^*| \leq \mu$.

Assume that the right-hand side of (8.4.55) is zero. Then

$$y^* = A\nabla v \qquad \text{in } \Omega, \tag{8.4.56}$$

$$\operatorname{div} y^* + f = 0 \qquad \text{in } \Omega, \tag{8.4.57}$$

$$\mu|\gamma v| + (\gamma^* y_n^*)(\gamma v) = 0 \qquad \text{on } \Gamma_2. \tag{8.4.58}$$

We note that the last relation models boundary conditions of the friction type. Indeed, if $|\gamma^* y_n^*| < \mu$, then (8.4.58) means that $\gamma v = 0$. If $\gamma^* y_n^* = \mu$, then from (8.4.58) it follows that $\gamma v < 0$. If $\gamma^* y_n^* = -\mu$ then (8.4.58) implies $\gamma v > 0$.

Winkler type boundary condition. Set

$$j(v) = \frac{1}{2}\kappa\,|v|^2, \tag{8.4.59}$$

where κ and is a positive constant. This relation can be viewed as a simplified variant of the Winkler's boundary condition widely used in solid mechanics. In this condition, on Γ_2 a body is connected with an elastic foundation, which provides a certain response to the boundary deflections (such a condition can be modeled by a large amount of springs connected with Γ_2). In view of (8.4.59), we have

$$-\gamma^* p_n^* = \kappa(\gamma u) \qquad \text{on } \Gamma_2 \tag{8.4.60}$$

and

$$j^*(\xi^*) = \sup_{\xi\in\mathbb{R}^d}\left\{\xi^*\cdot\xi - \frac{1}{2}\kappa\,|\xi|^2\right\} = \frac{1}{2\kappa}\,|\xi^*|^2. \tag{8.4.61}$$

Consider the quantity

$$I_{\Gamma_2} = \frac{1}{2}\int_{\Gamma_2} \left(\kappa\,|\gamma v|^2 + \frac{1}{\kappa}\,|\xi^*|^2 - 2\xi^*\cdot\gamma v + \theta\,|\gamma^* y_n^* + \xi^*|^2\right) ds.$$

The minimization of this quantity over ξ^* leads to the condition

$$\left(\frac{1}{\kappa}+\theta\right)\xi^* = \gamma v - \theta\gamma^* y_n^* \;\Rightarrow\; \xi^* = \frac{\kappa(\gamma v - \theta\gamma^* y_n^*)}{1+\kappa\theta}.$$

This gives a simple expression for I_{Γ_2}:

$$I_{\Gamma_2} = \frac{1}{2}\int_{\Gamma_2}\frac{\theta}{1+\kappa\theta}(\kappa(\gamma v) + \gamma^* y_n^*)^2\,ds.$$

By (8.4.36), we obtain the estimate

$$\begin{aligned}|||\nabla(u-v)|||^2 \le (1+\beta)2M_1(v,y^*) + \int_{\Gamma_2}\frac{\theta}{1+\kappa\theta}(\kappa(\gamma v)+\gamma^* y_n^*)^2\,ds \\ + \left(1+\tfrac{1}{\beta}\right)(1+\alpha)C_1^2\mathrm{r}_\Omega{}^2(y^*).\end{aligned} \tag{8.4.62}$$

If $\kappa = 0$ (i.e., if we consider homogeneous Neumann boundary condition), then the second term on the right-hand side of (8.4.62) has the form

$$\left(1+\tfrac{1}{\beta}\right)\left(1+\tfrac{1}{\alpha}\right)C_{T\Gamma_2}^2\int_{\Gamma_2}(\gamma^* y_n^*)^2\,ds$$

and (8.4.62) gives the same estimate as (8.4.50) (with $F = 0$).

8.4.3 Generalized Newtonian fluids

Now, we briefly discuss error estimates for some nonlinear models of viscous fluids. In these models, the basic relations are as follows:

$$u_t - \operatorname{Div}\sigma + \operatorname{Div}(u\otimes u) = f - \nabla\mathrm{p} \qquad \text{in } \Omega, \tag{8.4.63}$$

$$\sigma \in \partial\pi(\varepsilon(u)), \tag{8.4.64}$$

where $\operatorname{div} u = 0$, $\varepsilon(u)$ is the symmetric part of ∇u, and π is the so-called *dissipative potential.*

Many physically motivated dissipative potentials have the form

$$\partial\pi(\varepsilon) = \mathcal{H}'(|\varepsilon|)\,\varepsilon, \tag{8.4.65}$$

where $\mathcal{H}(0) = 0$ and the prime denotes the derivative of $\mathcal{H}$ with respect to the argument. In particular, the classical Newtonian fluid relates to the case

$$\mathcal{H}(\zeta) = \frac{1}{2}\nu\zeta^2, \qquad \pi(\varepsilon) = \frac{1}{2}\nu\,|\varepsilon|^2. \tag{8.4.66}$$

Another well-known example is the Bingham fluid, where

$$\pi(\varepsilon) = \frac{1}{2}\nu\,|\varepsilon|^2 + k_*|\varepsilon|, \qquad k_* > 0. \tag{8.4.67}$$

For slow stationary flows, the classical statement of this type problem is as follows: Find a vector-valued function u (velocity), a scalar-valued function p (pressure) and a tensor-valued function σ (stress deviator) such that

$$-\operatorname{Div}\sigma = f - \nabla p \qquad \text{in } \Omega, \tag{8.4.68}$$

$$\operatorname{div} u = 0 \qquad \text{in } \Omega, \tag{8.4.69}$$

$$\sigma \in \partial\pi(\varepsilon(u)) \qquad \text{in } \Omega, \tag{8.4.70}$$

$$u = u_0 \qquad \text{on } \Gamma, \tag{8.4.71}$$

where f and u_0 are given functions satisfying the same conditions as in Chapter 6, and π is the dissipative potential that defines physical properties of a fluid. Consider the class of potentials

$$\pi(\varepsilon) = \frac{\nu}{2}\,|\varepsilon|^2 + \psi(\varepsilon),$$

where $\psi : \mathbb{M}^{d\times d} \to \mathbb{R}_+$ is a convex nonnegative function such that

$$\psi(0) = 0, \quad \psi(\varepsilon) \le c_1\,|\varepsilon|^2 + c_2, \qquad c_1 > 0. \tag{8.4.72}$$

Obviously, the cases $\psi \equiv 0$ and $\psi(\varepsilon) = k_*\,|\varepsilon|$ correspond to Newtonian and Bingham models, respectively. Investigation of mathematical properties of such type models was started in the 60s (e.g., see O. A. Ladyzhenskaya [212]). Variational methods for this and other classes of nonlinear problems are studied in the book by M. Fuchs and G. Seregin [140], where the reader will find a consequent exposition of the regularity theory and many references related to the subject.

For dissipative potentials of such a type, energy estimates of the distance between $v \in u_0 + \overset{\circ}{\mathrm{S}}{}^1(\Omega)$ and the exact solution u were derived in the author's papers [289, 292, 293], in M. Fuchs and S. Repin [139], and in M. Bildhauer, M. Fuchs, and S. Repin [57]. In those papers, the derivation of a posteriori estimates is based on the variational method.

In general, estimates of such a type have the following form:

$$\frac{\nu}{2}\,\|\varepsilon(v-u)\|^2 \le (1+\beta)D_1(\varepsilon(v),\tau_1) + D_2(\varepsilon(v),\tau_2) + \left(1+\tfrac{1}{\beta}\right)\tfrac{1}{2\nu}\widehat{C}^2_{F\Omega}\,\|\operatorname{div}(\tau_1+\tau_2) + f - \nabla \mathrm{q}\|^2, \tag{8.4.73}$$

where

$$D_1(\varepsilon(v),\tau_1) := \int_\Omega \left(\tfrac{\nu}{2}\,|\varepsilon(v)|^2 + \tfrac{1}{2\nu}\,|\tau_1|^2 - \varepsilon(v) : \tau_1\right)dx = \tfrac{1}{2\nu}\,\|\nu\varepsilon(v) - \tau_1\|^2,$$

$$D_2(\varepsilon(v),\tau_2) := \int_\Omega \left(\psi(\varepsilon(v)) + \psi^*(\tau_2) - \varepsilon(v) : \tau_2\right)dx,$$

ψ^* is the functional conjugate to ψ, $\tau_1 + \tau_2 \in H(\Omega, \operatorname{Div})$ holds, q is a function in $H^1(\Omega)\cap\widetilde{L}^2(\Omega)$, $\beta > 0$, and $\widehat{C}_{F\Omega}$ is a constant in the inequality $\|w\| \le \widehat{C}_{F\Omega}\|\varepsilon(w)\|$.

Assume that the right-hand side of (8.4.73) is equal to zero. Then

$$-\text{div}\,(\tau_1 + \tau_2) = f - \nabla \mathrm{q}$$

and, in addition,

$$\begin{aligned} \tau_1 &= \nu\varepsilon(v) && \text{a.e. in } \Omega, \\ \tau_2 &\in \partial\psi(\varepsilon(v)) && \text{a.e. in } \Omega. \end{aligned}$$

Since $v \in u_0 + \overset{\circ}{\mathrm{S}}{}^1(\Omega)$, we conclude that in such a case, v coincides with the exact solution u, $\tau_1 = \sigma_1$, and $\tau_2 = \sigma_2$.

Also, computable error bounds can be derived directly from the respective integral identity, which has the form

$$\begin{aligned} \int_\Omega \nu\varepsilon(u) : \varepsilon(w)\,dx &+ \int_\Omega \psi'(|\varepsilon(u)|)\frac{\varepsilon(u) : \varepsilon(w)}{|\varepsilon(u)|}\,dx \\ &= \int_\Omega f \cdot w\,dx, \qquad \forall w \in u_0 + \overset{\circ}{\mathrm{S}}{}^1(\Omega). \end{aligned} \tag{8.4.74}$$

Let $v \in u_0 + \overset{\circ}{\mathrm{S}}{}^1(\Omega)$ be an approximation of u. We transform (8.4.74) as follows:

$$\begin{aligned} \int_\Omega \nu\varepsilon(u - v) : \varepsilon(w)\,dx &+ \int_\Omega \left(\psi'(|\varepsilon(u)|)\frac{\varepsilon(u)}{|\varepsilon(u)|} - \psi'(|\varepsilon(v)|)\frac{\varepsilon(v)}{|\varepsilon(v)|}\right) : \varepsilon(w)\,dx \\ &= \int_\Omega (f \cdot w - \nu\varepsilon(v) : \varepsilon(w))\,dx - \int_\Omega \psi'(|\varepsilon(v)|)\frac{\varepsilon(v) : \varepsilon(w)}{|\varepsilon(v)|}\,dx. \end{aligned} \tag{8.4.75}$$

We reform the right-hand side of (8.4.75), introducing two symmetric tensor-valued functions τ_1 and τ_2 in $H(\Omega, \text{Div})$ and using the identities

$$\int_\Omega (w \cdot \text{Div}\,\tau_i + \varepsilon(w) : \tau_i)\,dx = 0, \qquad \forall w \in V_0, \ \ i = 1, 2.$$

We obtain

$$\begin{aligned} \int_\Omega \nu\varepsilon(u - v) : \varepsilon(w)\,dx &+ \int_\Omega \left(\psi'(|\varepsilon(u)|)\frac{\varepsilon(u)}{|\varepsilon(u)|} - \psi'(|\varepsilon(v)|)\frac{\varepsilon(v)}{|\varepsilon(v)|}\right) : \varepsilon(w)\,dx \\ = \int_\Omega (f + \text{Div}\,(\tau_1 + \tau_2)) \cdot w\,dx &+ \int_\Omega (\tau_1 + \mathrm{q}\mathbb{I} - \nu\varepsilon(v)) : \varepsilon(w)\,dx \\ &+ \int_\Omega \left(\tau_2 - \psi'(|\varepsilon(v)|)\frac{\varepsilon(v)}{|\varepsilon(v)|}\right) : \varepsilon(w)\,dx, \end{aligned} \tag{8.4.76}$$

where q is a square summable function. Set $w = u - v$ and note that

$$\int_\Omega (f + \text{Div}\,(\tau_1 + \tau_2)) \cdot (u - v)\,dx \le \widehat{C}_{F\Omega}\|f + \text{Div}\,(\tau_1 + \tau_2)\|\,\|\varepsilon(u - v)\|.$$

With the help of Young's inequality, we deduce the estimate

$$\begin{aligned}
(1-\alpha_1-\alpha_2-\alpha_3)&\int_\Omega \nu|\varepsilon(u-v)|^2\,dx \\
&+\int_\Omega\left(\psi'(|\varepsilon(u)|)\frac{\varepsilon(u)}{|\varepsilon(u)|}-\psi'(|\varepsilon(v)|)\frac{\varepsilon(v)}{|\varepsilon(v)|}\right):\varepsilon(w)\,dx \\
&=\frac{1}{4\alpha_1}\widehat{C}^2_{F\Omega}\|f+\operatorname{Div}(\tau_1+\tau_2)\|^2+\frac{1}{4\alpha_2}\|\tau_1+q\mathbb{I}-\nu\varepsilon(v)\|^2 \\
&+\frac{1}{4\alpha_3}\left\|\tau_2-\psi'(|\varepsilon(v)|)\frac{\varepsilon(v)}{|\varepsilon(v)|}\right\|,
\end{aligned}\tag{8.4.77}$$

where α_1, α_2, and α_3 are positive numbers such that $\alpha_1+\alpha_2+\alpha_3<1$.

Estimate (8.4.77) has the same principal structure as (8.4.73): its right-hand side consists of penalties for the violation of the constitutive law relations and the equilibrium equation formed by the total stress $\tau_1+\tau_2$. If the right-hand side of (8.4.77) vanishes, then

$$\begin{aligned}
f&=-\operatorname{Div}(\tau_1+\tau_2),\\
\tau_1&=-q\mathbb{I}+\nu\varepsilon(v),\\
\tau_2&=\psi'(|\varepsilon(v)|)\frac{\varepsilon(v)}{|\varepsilon(v)|}.
\end{aligned}$$

Since v is a solenoidal field satisfying the prescribed boundary condition, we conclude that v coincides with u.

Remark 8.14. The second term on the left-hand side of (8.4.77) is nonnegative and, therefore, can be removed. However, for certain ψ it may be evaluated in terms of $\varepsilon(u-v)$, which would make the overall estimate stronger.

8.5 Notes for the chapter

1. Estimates discussed in this chapter are valid for a wide class of nonlinear variational problems in continuum mechanics associated with the functional

$$J(v)=\int_\Omega g(\varepsilon(v))\,dx-\int_\Omega fv\,dx-\int_{\Gamma_2}Fv\,ds,\tag{8.5.1}$$

where g (internal energy function) has the form

$$g(\varepsilon)=\mathbb{L}\varepsilon:\varepsilon+\pi(\varepsilon).\tag{8.5.2}$$

In (8.5.1), $\mathbb{L}$ satisfies (5.1.5) and π is a nonnegative convex function. A minimizer of the functional (8.5.1) satisfies the relations (5.1.2)–(5.1.4), and (5.1.1) is replaced by a nonlinear constitutive relation. If π is a differentiable functional, then the latter relation has the form

$$\sigma = \mathbb{L}\varepsilon(u) + \pi'(\varepsilon(v)). \tag{8.5.3}$$

One example of such relations is offered by deformation plasticity theory (e.g., see A. Iljushin [183] or R. Temam [349]), which is based on the constitutive relation

$$\sigma = K_0 \mathrm{tr}(\varepsilon)\,\mathbb{I} + \gamma(|\varepsilon^D|)\,\varepsilon^D, \tag{8.5.4}$$

where

$$\gamma(t) = \begin{cases} 2\mu & \text{if } t \leq t_0 = k_*/\sqrt{2}\mu, \\ (2\mu - \delta)t_0 t^{-1} + \delta & \text{if } t > t_0, \end{cases}$$

K_0 and μ are positive (elasticity) constants, $k_* > 0$ is a plasticity module, and $\delta > 0$ is a hardening module. In this case, the integrand g has the form

$$g(\varepsilon) = \frac{1}{2}a(\varepsilon, \varepsilon) + \mu_1 \phi(|\varepsilon^D|),$$

where $\mu_1 = 1 - \delta/2\mu$,

$$a(\varepsilon_1, \varepsilon_2) = K_0 \mathrm{tr}(\varepsilon_1)\,\mathrm{tr}(\varepsilon_2) \; + \; \delta\varepsilon_1^D : \varepsilon_2^D, \quad \forall \varepsilon_1, \varepsilon_2 \in \mathbb{M}^{d\times d},$$

and

$$\phi(t) = \begin{cases} \mu\, t^2 & \text{if } |\, t \,| \leq t_0 = k_*/\sqrt{2}\mu, \\ k_*\left(\sqrt{2}\, t - k_*/2\mu\right) & \text{if } |\, t \,| > t_0. \end{cases}$$

Functional type a posteriori error estimates for variational problems with functionals of the type (8.5.1)–(8.5.2) were derived in S. Repin and L. Xanthis [320, 321].

A posteriori estimates for lower semicontinuous relaxations of some nonconvex variational problems (which are related to simple phase transitions models in the theory of solids) were obtained in the author's paper [281]). Estimates for the Ramberg–Osgood model and for elasto-plastic torsion problem have been recently derived in M. Bildhauer, M. Fuchs, and S. Repin [58] and [59], respectively. Estimates for an incremental evolutionary plasticity model were obtained in S. Repin and J. Valdman [318].

2. As we have seen, functional a posteriori estimates derived for different problems have certain common features. We summarize this experience and put forward the following conjecture:

> A majorant (guaranteed upper bound) of the deviation $v - u$, where v is an arbitrary function from the energy space (i.e., from the functional class that contains the generalized solution u) consists of terms, which can be thought of as *penalties for unconformity* in all basic relations. Relevant multipliers are defined by *constants in the embedding inequalities* for spaces pertaining to the mathematical statement of the problem.

Certainly, this conjecture is yet to be justified for many classes of boundary value problems (e.g., for those that are related to differential equations of a nondivergent type).

3. All the majorants that we discussed satisfy the following conditions:

$$\text{(a)} \quad |||\Lambda(u - v)||| \leq \overline{\mathfrak{M}}\,(v, y^*, \mathcal{D}) \text{ for any } y^* \in Y^* \text{ and } v \in u_0 + V_0,$$

$$\text{(b)} \quad \inf_{y^* \in Y^*} \overline{\mathfrak{M}}\,(u, y^*, \mathcal{D}) = 0, \tag{8.5.5}$$

$$\text{(c)} \quad \overline{\mathfrak{M}}\,(v_k, y_k^*, \mathcal{D}) \to \overline{\mathfrak{M}}\,(v, y^*, \mathcal{D}),$$

where $\{v_k\}$ and $\{y_k^*\}$ are arbitrary sequences in V and Y^*, respectively, such that $v_k \to v$ in V and $y_k^* \to y^*$ in Y^*. We note that the above-stated requirements are quite natural. Indeed, (a) means that the upper bound is guaranteed, (b) says that if v coincides with the exact solution u, then Y^* contains a counterpart function y^* (which is p^*) such that the majorant vanishes, and (c) is the continuity property. Obviously, practically valuable error majorants must satisfy (a)–(c). Quite similar conditions should be imposed on error minorants.

However, the conditions (a)–(c) do not guarantee that the majorant $\overline{\mathfrak{M}}\,(v, y^*, \mathcal{D})$ is equal to $|||\Lambda(u - v)|||$ for some y^* and other parameters involved in the majorant. In other words, we cannot guarantee that there is no irremovable gap between the left-hand and right-hand sides of (8.5.5 a).

Nevertheless, we can show that if $\overline{\mathfrak{M}}$ satisfies (a)–(c), then another majorant possesses such a property. For this purpose, we introduce an arbitrary function $w \in V_0$ and define a new majorant

$$\overline{M}(v, w, y^*, \mathcal{D}) := |||\Lambda w||| + \overline{\mathfrak{M}}\,(v + w, y^*, \mathcal{D}).$$

$\overline{M}\,(v, w, y^*, \mathcal{D})$ also contains only known functions and, therefore, is explicitly computable. This new majorant satisfies the properties (a)–(c). Indeed, for any element $v \in u_0 + V_0$,

$$|||\Lambda(u - v)||| \leq |||\Lambda w||| + |||\Lambda(u - v - w)||| \leq \overline{M}(v, w, y^*, \mathcal{D}).$$

Thus, $\overline{M}(v, w, y^*, \mathcal{D})$ yields a guaranteed upper bound of the error in terms of the energy norm. It is easy to see that $\overline{M}(v, 0, y^*, \mathcal{D}) = \overline{\mathfrak{M}}(v, y^*, \mathcal{D})$, so that

$$\inf_{\substack{y^* \in Y^* \\ w \in V_0}} \overline{M}(u, w, y^*, \mathcal{D}) = 0.$$

Moreover, the functional $\overline{M}(v, w, y^*, \mathcal{D})$ is continuous (i.e., the condition (c) holds). However, $\overline{M}(v, w, y^*, \mathcal{D})$ possesses one more property. Since

$$\begin{aligned} \inf_{\substack{y^* \in Y^* \\ w \in V_0}} \overline{M}(v, w, y^*, \mathcal{D}) &\leq \inf_{y^* \in Y^*} \overline{M}(v, u - v, y^*, \mathcal{D}) \\ &= |||\Lambda(u - v)||| + \inf_{y^* \in Y^*} \overline{\mathfrak{M}}(u, y^*, \mathcal{D}) \\ &= |||\Lambda(u - v)|||, \end{aligned} \tag{8.5.6}$$

we find that, in principle, the majorant $\overline{M}(v, w, y^*, \mathcal{D})$ can give an upper bound of the error with any desired accuracy.

Similar arguments lead to sharp lower bounds of the error: for any $w \in V_0$ we have

$$|||\Lambda w||| \leq |||\Lambda(u - v)||| + |||\Lambda(w - u + v)||| \leq |||\Lambda(u - v)||| + \overline{\mathfrak{M}}(v + w, y^*, \mathcal{D}).$$

Define the minorant

$$\underline{M}(v, w, y^*, \mathcal{D}) := |||\Lambda w||| - \overline{\mathfrak{M}}(v + w, y^*, \mathcal{D}).$$

We have

$$\begin{aligned} \sup_{\substack{y^* \in Y^* \\ w \in V_0}} \underline{M}(v, w, y^*, \mathcal{D}) &\geq \sup_{y^* \in Y^*} \underline{M}(v, u - v, y^*, \mathcal{D}) \\ &= |||\Lambda(u - v)||| - \inf_{y^* \in Y^*} \overline{\mathfrak{M}}(u, y^*, \mathcal{D}) \\ &= |||\Lambda(u - v)||| \end{aligned}$$

and, consequently, the minorant also has no "gap".

Using continuity property (c), it is not difficult to prove that the above-defined two-sided bounds of the error (which converge to the exact error) can be constructed by solving only finite-dimensional problems. Thus, we arrive at the following conclusion: if for a boundary value problem a majorant with properties (a)–(c) has been constructed, then (in principle) errors in the energy norm can be evaluated with any desirable accuracy. Certainly, this theoretical conclusion may have different value for different problems. It is very probable that for strongly nonlinear problem the practical computation of sharp estimates (in terms of the global error norm) may lead to high computational costs.

9 A posteriori estimates for other problems

9.1 Differential equations of higher order

Fourth order elliptic equation. A posteriori error estimation methods discussed in previous chapters can be applied to boundary value problems associated with higher order differential equations. As an example, we consider the 4th order elliptic problem

$$\operatorname{div}\operatorname{Div}(B\nabla\nabla u) = f \qquad \text{in } \Omega, \tag{9.1.1}$$

$$u = \frac{\partial u}{\partial n} = 0 \qquad \text{on } \Gamma. \tag{9.1.2}$$

Here

$$f \in L^2(\Omega), \quad B = \{b_{ijkl}\}, \quad b_{ijkl} = b_{jikl} = b_{klij} \in L^\infty(\Omega),$$

where the indices change from 1 to d, and

$$c_1^2|\eta|^2 \le B\eta : \eta \le c_2^2|\eta|^2, \qquad \forall \eta \in \mathbb{M}_s^{d\times d}. \tag{9.1.3}$$

Then, the inverse tensor B^{-1} exists and for any tensor-valued function τ with square summable components we define the norms

$$|||\tau|||^2 := \int_\Omega B\tau : \tau\, dx \quad \text{and} \quad |||\tau|||_*^2 := \int_\Omega B^{-1}\tau : \tau\, dx.$$

A posteriori estimates for this problem can be obtained from the general estimates considered in Chapter 7 if we define Λ as the Hessian operator and set

$$U = L^2(\Omega, \mathbb{M}_s^{d\times d}), \qquad V = H^2(\Omega),$$

$$V_0 = \left\{ w \in V \mid w = \frac{\partial w}{\partial n} = 0 \text{ on } \Gamma \right\}.$$

However, to make the exposition more transparent and to obtain a posteriori estimates of a different form, we derive them below directly from the integral identity

$$\int_\Omega B\nabla\nabla u : \nabla\nabla w\, dx = \int_\Omega f w\, dx, \qquad \forall w \in V_0, \tag{9.1.4}$$

that defines a generalized solution $u \in V_0$.

Let $v \in V_0$ be an approximation of u. By the identity

$$\int_\Omega (\tau : \nabla\nabla w - w \operatorname{div}\operatorname{Div}\tau)\, dx = 0,$$

where $w \in V_0$ and τ is an arbitrary tensor-valued function from the space

$$H(\operatorname{div}\operatorname{Div},\Omega) := \left\{\eta \in U \mid \operatorname{div}\operatorname{Div}\eta \in L^2(\Omega)\right\},$$

we transform (9.1.4) as follows:

$$\begin{aligned}\int_\Omega & B\nabla\nabla(u-v) : \nabla\nabla w\,dx \\ &= \int_\Omega (f - \operatorname{div}\operatorname{Div}\tau)w\,dx + \int_\Omega (\tau - B\nabla\nabla v) : \nabla\nabla w\,dx. \quad (9.1.5)\end{aligned}$$

Let $C_{1\Omega}$ denote the constant in the inequality

$$\|w\|_\Omega \le C_{1\Omega} |||\nabla\nabla w|||, \qquad \forall w \in V_0. \quad (9.1.6)$$

Set $w = u - v$. From (9.1.5) and (9.1.6), it follows that

$$\begin{aligned}|||\nabla\nabla(v-u)|||^2 &\le (1+\beta)|||B\nabla\nabla v - \tau|||_*^2 \\ &+ \left(1 + \frac{1}{\beta}\right) C_{1\Omega}^2 \|\operatorname{div}\operatorname{Div}\tau - f\|_\Omega^2, \quad (9.1.7)\end{aligned}$$

where β is a positive real number. This estimate is quite analogous to (3.2.8) and can be obtained as a particular case of (7.1.20). It corresponds to the following decomposition of (9.1.1):

$$\operatorname{div}\operatorname{Div}\sigma = f,$$
$$\sigma = B\nabla\nabla u.$$

However, the condition

$$\operatorname{div}\operatorname{Div}\tau \in L^2(\Omega)$$

is rather demanding (for example, if τ is constructed with the help of piecewise affine continuous approximations, then it does not satisfy this condition). To avoid arising technical difficulties, we introduce a new vector-valued function $y \in H(\Omega, \operatorname{div})$ and put (9.1.5) in the form

$$\begin{aligned}\int_\Omega & B\nabla\nabla(u-v) : \nabla\nabla w\,dx \\ &= \int_\Omega (f - \operatorname{div}\operatorname{Div}\tau)w\,dx + \int_\Omega (\tau - B\nabla\nabla v) : \nabla\nabla w\,dx \\ &\quad - \int_\Omega (w\operatorname{div} y + y \cdot \nabla w)\,dx \\ &= \int_\Omega (f - \operatorname{div} y)w\,dx + \int_\Omega (\operatorname{Div}\tau - y)\cdot\nabla w\,dx \\ &\quad + \int_\Omega (\tau - B\nabla\nabla v) : \nabla\nabla w\,dx. \quad (9.1.8)\end{aligned}$$

Let $C_{2\Omega}$ denote the constant in the inequality

$$\|\nabla w\|_\Omega \le C_{2\Omega} |||B\nabla\nabla w|||, \qquad \forall w \in V_0. \tag{9.1.9}$$

Set $w = u - v$ and estimate the right-hand side of (9.1.8), using (9.1.6) and (9.1.9). We obtain

$$\begin{aligned} |||\nabla\nabla(v-u)||| \le |||B\nabla\nabla v - \tau|||_* + C_{1\Omega}\|\operatorname{div} y - f\|_\Omega \\ + C_{2\Omega}\|\operatorname{Div}\tau - y\|_\Omega. \end{aligned} \tag{9.1.10}$$

Square both parts of the above estimate and apply Young's inequality. Then (9.1.10) implies another estimate:

$$\begin{aligned} |||\nabla\nabla(v-u)|||^2 &\le (1+\beta)|||B\nabla\nabla v - \tau|||_*^2 \\ &+ \frac{1+\beta}{\beta}\left(C_{1\Omega}\|\operatorname{div} y - f\|_\Omega + C_{2\Omega}\|\operatorname{Div}\tau - y\|_\Omega\right)^2. \end{aligned} \tag{9.1.11}$$

Estimates (9.1.10) and (9.1.11) have two "free" functions $y \in H(\operatorname{div},\Omega)$ and $\tau \in H(\operatorname{Div},\Omega)$, which can be viewed as images of the gradient and double gradient, respectively. They reflect the representation of (9.1.1) in the form

$$\begin{aligned} \operatorname{div} y &= f, \\ \operatorname{Div}\sigma &= y, \\ \sigma &= B\nabla\nabla u. \end{aligned}$$

Remark 9.1. Note that $C_{1\Omega}$ and $C_{2\Omega}$ are estimated by $c_1^{-1}C_{1\Omega_\Box}$ and $c_1^{-1}C_{2\Omega_\Box}$, respectively, where $C_{1\Omega_\Box}$ and $C_{2\Omega_\Box}$ are constants in the inequalities

$$\|w\|_{\Omega_\Box} \le C_{1\Omega_\Box}\|\nabla\nabla w\|_{\Omega_\Box}, \qquad \forall w \in V_0, \tag{9.1.12}$$

$$\|\nabla w\|_{\Omega_\Box} \le C_{2\Omega_\Box}\|\nabla\nabla w\|_{\Omega_\Box} \tag{9.1.13}$$

and $\Omega_\Box$ is a rectangular domain containing Ω.

Remark 9.2. A posteriori estimates for the equation (9.1.1) with other boundary conditions can be derived by the same arguments as for the second order problems we discussed in Chapters 4 and 5. Instead of $C_{1\Omega}$ and $C_{2\Omega}$ they involve constants in the inequalities analogous to (9.1.6) and (9.1.9) in which V_0 contains functions vanishing on the Dirichlet part of the boundary. Also, such estimates involve constants in the trace inequalities on other parts of the boundary (e.g., on the part related to the Neumann boundary condition). We leave this task to the reader as an exercise.

Example. Consider the application of the above a posteriori error estimates to a plate bending problem. In this case, $\Omega \subset \mathbb{R}^2$ is associated with the middle surface of a plate. Let the deformation of a plate be described by the Kirchhoff–Love model. Then $u = u(x_1, x_2)$ is the bending function, $\kappa(u) = \nabla\nabla u$ is the curvature tensor, which is connected with the bending moments τ by a linear constitutive law

$$\tau = B\kappa. \tag{9.1.14}$$

If the plate is made of an isotropic elastic material, then (9.1.14) has a simple form

$$\begin{aligned}\tau_{11} &= \mathsf{H}(\kappa_{11} + \bar{\nu}\kappa_{22}),\\ \tau_{22} &= \mathsf{H}(\kappa_{22} + \bar{\nu}\kappa_{11}),\\ \tau_{12} &= \mathsf{H}(1-\bar{\nu})\kappa_{12},\end{aligned}$$

where $\mathsf{H} = \frac{Eh^3}{12(1-\bar{\nu}^2)}$, $E > 0$, and $\bar{\nu} \in (0,1)$. Here E an $\bar{\nu}$ are elasticity constants and $h = h(x_1, x_2)$ is the thickness parameter. We see that B has the following nonzero components:

$$\begin{aligned}&b_{1111} = \mathsf{H}, \quad b_{2222} = \mathsf{H}, \quad b_{1122} = \bar{\nu}\mathsf{H},\\ &b_{2211} = \bar{\nu}\mathsf{H}, \quad b_{1212} = b_{2121} = (1-\bar{\nu})\mathsf{H}.\end{aligned}$$

The natural condition

$$h_1 \le h(x_1, x_2) \le h_2, \qquad (x_1, x_2) \in \Omega, \tag{9.1.15}$$

guarantees that the estimate (9.1.3) holds with

$$c_1^2 = \frac{Eh_1^3}{12(1+\bar{\nu})} \quad \text{and} \quad c_2^2 = \frac{Eh_2^3}{12(1-\bar{\nu})}.$$

The nonzero components the tensor $C = B^{-1}$ are as follows:

$$\begin{aligned}&c_{1111} = \frac{1}{\mathsf{H}(1-\bar{\nu}^2)}, \quad c_{2222} = \frac{1}{\mathsf{H}(1-\bar{\nu}^2)}, \quad c_{1122} = -\frac{\bar{\nu}}{\mathsf{H}(1-\bar{\nu}^2)},\\ &c_{2211} = -\frac{\bar{\nu}}{\mathsf{H}(1-\bar{\nu}^2)}, \quad c_{1212} = c_{2121} = \frac{1}{\mathsf{H}(1-\bar{\nu})}.\end{aligned}$$

Hence,

$$|\!|\!| B\kappa - \tau |\!|\!|_*^2 = \int_\Omega (B\kappa : \kappa + C\tau : \tau - 2\kappa : \tau)\, dx,$$

and the other parts of the error majorants (9.1.10)–(9.1.11) are directly computable.

Biharmonic equation. A somewhat different a posteriori estimate can be derived for the biharmonic problem

$$\Delta\Delta u = f \quad \text{in } \Omega$$

with the boundary conditions (9.1.2). In this case, the integral identity has the form

$$\int_\Omega \Delta u \Delta w \, dx = \int_\Omega f w \, dx, \qquad \forall w \in V_0.$$

Introduce a function $\xi \in H^2(\Omega)$. We have

$$\begin{aligned}\int_\Omega \Delta(u - v)\Delta w \, dx &= \int_\Omega (f w - \Delta v \Delta w) \, dx \\ &= \int_\Omega ((f - \Delta\xi)w + (\xi - \Delta v)\Delta w \, dx.\end{aligned}$$

From here, we find that

$$\|\Delta(u - v)\| \leq C_{3\Omega}\|f - \Delta\xi\| + \|\xi - \Delta v\|, \tag{9.1.16}$$

where $C_{3\Omega}$ is the constant in the inequality

$$\|w\| \leq C_{3\Omega}\|\Delta w\|, \qquad \forall w \in V_0.$$

Variational inequalities. Consider the problem (9.1.1)–(9.1.2) with the condition

$$u \in K := \{w \in H^2(\Omega) \mid w \geq \phi(x) \quad \text{a.e. in } \Omega\},$$

which arises if the solution must lie above the obstacle $\phi(x)$. For the sake of simplicity, we consider the case with one obstacle and assume that ϕ is sufficiently regular (e.g., continuous and piecewise smooth). Then, the solution satisfies the variational inequality

$$\int_\Omega B\nabla\nabla u : \nabla\nabla(w - u) \, dx \geq \int_\Omega f(w - u) \, dx, \qquad \forall w \in K. \tag{9.1.17}$$

Assume that $v \in K$ is a function, which is an approximation of u. Then,

$$|||u - v|||^2 \leq \int_\Omega (f(u - v) - B\nabla\nabla v : \nabla\nabla(u - v)) \, dx. \tag{9.1.18}$$

As in the linear case, we introduce two auxiliary functions $y \in H(\Omega, \mathrm{div})$ and $\tau \in H(\Omega, \mathrm{Div})$. We have

$$\begin{aligned}
|||u - v|||^2 &\leq \int_\Omega (f - \operatorname{div} y)(u - v)\, dx + \int_\Omega (\operatorname{Div} \tau - y) \cdot \nabla(u - v)\, dx \\
&\quad + \int_\Omega (\tau - B\nabla\nabla v) : \nabla\nabla(u - v)\, dx \\
&= \int_{\Omega_\phi^v} (f - \operatorname{div} y)(u - v)\, dx + \int_{\Omega_0^v} (f - \operatorname{div} y)(u - v)\, dx \\
&\quad + \int_\Omega (\operatorname{Div} \tau - y) \cdot \nabla(u - v)\, dx \\
&\quad + \int_\Omega (\tau - B\nabla\nabla v) : \nabla\nabla w\, dx, \qquad (9.1.19)
\end{aligned}$$

where

$$\Omega_\phi^v := \{x \in \Omega \mid v(x) = \phi(x)\} \quad \text{and} \quad \Omega_0^v := \{x \in \Omega \mid v > \phi(x)\}.$$

Since

$$\int_{\Omega_\phi^v} (f - \operatorname{div} y)(u - v)\, dx \leq \int_{\Omega_\phi^v} (f - \operatorname{div} y)_+ (u - v)\, dx,$$

we estimate the right-hand side of (9.1.19) in the same way as (9.1.8) and deduce the estimate

$$|||\nabla\nabla(v - u)||| \leq |||B\nabla\nabla v - \tau|||_* + C_{1\Omega}\|\mathrm{r}(v, y)\|_\Omega + C_{2\Omega}\|\operatorname{Div} \tau - y\|_\Omega, \quad (9.1.20)$$

where

$$\mathrm{r}(v, y) := \begin{cases} (f - \operatorname{div} y)_+ & \text{in } \Omega_\phi^v, \\ f - \operatorname{div} y & \text{in } \Omega_0^v. \end{cases}$$

Remark 9.3. If u is subject to two obstacles (i.e., $u \leq \psi$ as in Section 8.1), then the estimate is obtained quite analogously. In this case, we append the third branch and set $\mathrm{r}(v, y) = (f - \operatorname{div} y)_-$ on Ω_ψ^v.

Comments. Estimates (9.1.10) and (9.1.11) were derived in P. Neittaanmaäki and S. Repin [243] by variational techniqus. In [300], it was shown that the estimates follow from the corresponding integral identity. Numerical testing of these estimates was performed in the PhD thesis of M. Frolov [134], in which the estimate (9.1.16) was derived (see also [135]). In the context of the above-discussed a posteriori estimates, the classical Kirchhoff–Love plate model was considered in P. Neittaanmaäki and S. Repin [243] and in the author's paper [285]. Estimates of modeling errors arising if the

Kirchhoff–Love model is used instead of the 3D elasticity model has recently been derived in S. Repin and S. Sauter [306]. Estimates for the Reissner–Mindlin model were obtained by the variational and nonvariational methods in the papers by M. Frolov, P. Neittaanmäki, and S. Repin [138, 304]. A posteriori estimates for the fourth order elliptic equations with obstacles were obtained in M. Bildhauer, M. Fuchs, and S. Repin [60] with the help of the variational method. In M. Bildhauer and M. Fuchs [56] these results were extended to a wider class of nonlinear functionals.

9.2 Equations with the operator curl

Basic problem. The simplest version of the Maxwell's problem is given by the equation

$$\operatorname{curl} \mu^{-1} \operatorname{curl} u + \kappa^2 u = j \qquad \text{in } \Omega, \tag{9.2.1}$$

where Ω is a bounded domain in $\mathbb{R}^d$, j is a given current density, and μ is the permeability of a medium (may be a positive constant or a positive bounded function). The case $\kappa = 0$ corresponds to magnetostatics. Equation (9.2.1) with positive κ arises in semidiscrete approximations of the evolutionary Maxwell's problem.

On Γ the condition

$$n \times u = 0 \tag{9.2.2}$$

is stated. In this section, $V(\Omega)$ is the space $H(\Omega, \operatorname{curl})$,

$$V_0(\Omega) := \{w \in H(\Omega, \operatorname{curl}) \mid | n \times w = 0 \quad \text{on } \Gamma\}$$

and

$$\ell(w) := \int_\Omega j \cdot w \, dx.$$

First, we consider the case $\kappa > 0$. Multiply (9.2.1) by a smooth vector-valued function w that satisfies (9.2.2) and integrate over Ω. We have

$$\int_\Omega (\operatorname{curl} \mu^{-1} \operatorname{curl} u \cdot w + \kappa^2 u \cdot w) \, dx = \ell(w).$$

By the relation

$$\int_\Omega (\operatorname{curl} v) \cdot w \, dx = \int_\Omega v \cdot (\operatorname{curl} w) \, dx - \int_\Gamma (v \times n) \cdot w \, ds, \tag{9.2.3}$$

we express the integral identity as follows

$$\int_\Omega (\mu^{-1} \operatorname{curl} u \cdot \operatorname{curl} w + \kappa^2 u \cdot w) \, dx - \int_\Gamma (\mu^{-1} \operatorname{curl} u \times n) \cdot w \, ds = \ell(w).$$

Note that

$$w \cdot (\operatorname{curl} u \times n) = -\operatorname{curl} u \cdot (w \times n).$$

Therefore, if $w \in V_0$, then the boundary term vanishes. Since smooth functions are dense in V, we conclude that the generalized solution can be defined as a vector-valued function $u \in V_0$ such that integral identity

$$\int_\Omega \left(\mu^{-1} \operatorname{curl} u \cdot \operatorname{curl} w + \kappa^2 u \cdot w\right) dx = \ell(w) \tag{9.2.4}$$

holds for $w \in V_0$.

If $\kappa = 0$, then the additional compatibility condition

$$\operatorname{div} j = 0 \quad \text{in } \Omega, \qquad j \cdot n = 0 \quad \text{on } \Gamma \tag{9.2.5}$$

is necessary to have a well-posed problem. In this case, a respective generalized solution is defined by the integral identity

$$\int_\Omega \mu^{-1} \operatorname{curl} u \cdot \operatorname{curl} w \, dx = \ell(w), \qquad \forall w \in V_0. \tag{9.2.6}$$

Since $\operatorname{curl} \nabla\psi = 0$ (see Section 1.4), a solution should be understood as an element of the factor space, in which the functions are equivalent if their difference is a gradient field. To ensure the uniqueness of a solution, the Coulomb gauge condition

$$u \in V_{00} := \left\{ v \in V_0 \mid \int_\Omega v \cdot \nabla\phi \, dx = 0, \qquad \forall \phi \in \mathring{H}^1(\Omega) \right\}, \tag{9.2.7}$$

is usually attracted.

Let $v \in V_0(\Omega)$ be an approximation of u. Rewrite (9.2.4) in the form

$$\begin{aligned} &\int_\Omega \left(\mu^{-1} \operatorname{curl}(u - v) \cdot (\operatorname{curl} w) + \kappa^2 (u - v) \cdot w\right) dx \\ &\quad = \int_\Omega \left(j \cdot w - \mu^{-1} (\operatorname{curl} v) \cdot (\operatorname{curl} w) - \kappa^2 v \cdot w\right) dx, \qquad \forall w \in V_0(\Omega). \end{aligned} \tag{9.2.8}$$

By (9.2.3), we obtain

$$\begin{aligned} &\int_\Omega \left(\mu^{-1} \operatorname{curl}(u - v) \cdot (\operatorname{curl} w) + \kappa^2 (u - v) \cdot w\right) dx \\ &\quad = \int_\Omega \left((j - \kappa^2 v - \operatorname{curl} y) \cdot w + (y - \mu^{-1} \operatorname{curl} v) \cdot \operatorname{curl} w\right) dx, \end{aligned} \tag{9.2.9}$$

where $y \in V(\Omega)$. Introduce the norm

$$|||w|||^2 := \int_\Omega \left(\mu^{-1} |\operatorname{curl} w|^2 + \kappa^2 |w|^2\right) dx.$$

Setting $w = u - v$ in (9.2.9), we obtain the relation

$$|||u - v|||^2 = \int_\Omega (j - \kappa^2 v - \operatorname{curl} y) \cdot (u - v)\, dx + \int_\Omega (y - \mu^{-1} \operatorname{curl} v) \cdot \operatorname{curl}(u - v)\, dx,$$

which leads to the estimate

$$|||u - v|||^2 \le \left\| \frac{1}{\kappa} (j - \kappa^2 v - \operatorname{curl} y) \right\| \|\kappa(u - v)\| + \|\mu^{1/2}(y - \mu^{-1}\operatorname{curl} v)\| \|\mu^{-1/2}\operatorname{curl}(u - v)\|.$$

Hence, we find that

$$|||u - v|||^2 \le \overline{\mathfrak{M}}^2_{\mathrm{MAX}}(v, y), \tag{9.2.10}$$

where

$$\overline{\mathfrak{M}}^2_{\mathrm{MAX}}(v, y) := \left\| \frac{1}{\kappa} (j - \kappa^2 v - \operatorname{curl} y) \right\|^2 + \|\mu^{1/2}(y - \mu^{-1}\operatorname{curl} v)\|^2$$

is the majorant for this type of Maxwell's problem. It is easy to see that

$$\inf_{\substack{v \in V_0, \\ y \in H(\Omega, \mathrm{curl})}} \overline{\mathfrak{M}}_{\mathrm{MAX}}(v, y) = 0,$$

and the exact lower bound is attained if and only if

$$\operatorname{curl} y + \kappa^2 v = j \qquad \text{a.e. in } \Omega, \tag{9.2.11}$$

$$y = \mu^{-1}\operatorname{curl} v \qquad \text{a.e. in } \Omega. \tag{9.2.12}$$

Since $v \times n = 0$ on Γ, (9.2.11) and (9.2.12) mean that v coincides with the exact solution u and y coincides with $\mu^{-1}\operatorname{curl} u$.

For any $y \in V_0$ the quantity $\overline{\mathfrak{M}}^2_{\mathrm{MAX}}(v, y)$ gives an upper bound of the error. It is easy to observe that

$$\begin{aligned}\inf_{y \in V_0} \overline{\mathfrak{M}}^2_{\mathrm{MAX}}(v, y) &\le \overline{\mathfrak{M}}^2_{\mathrm{MAX}}(v, \mu^{-1}\operatorname{curl} u) \\ &= \left\| \frac{1}{\kappa} (j - \kappa^2 v - \operatorname{curl} \mu^{-1}\operatorname{curl} u) \right\|^2 + \|\mu^{-1/2}\operatorname{curl}(u - v)\|^2 \\ &= \|\kappa(u - v)\|^2 + \|\mu^{-1/2}\operatorname{curl}(u - v)\|^2 = |||u - v|||^2.\end{aligned}$$

Therefore, the estimate (9.2.10) has no gap between the left- and right-hand sides. A practically computable estimate can be determined if $\overline{\mathfrak{M}}^2_{\mathrm{MAX}}(v, y)$ is minimized over a finite-dimensional subspace $V_m \subset V(\Omega)$.

A lower bound of $|||u - v|||$ follows from the relation

$$
\begin{aligned}
&\sup_{w\in V_0} \int_\Omega \Big(\mu^{-1}\mathrm{curl}\,(u-v)\cdot \mathrm{curl}\,w + \kappa^2 w\cdot(u-v) \\
&\qquad\qquad - \frac{1}{2}(\mu^{-1}\mathrm{curl}\,w\cdot\mathrm{curl}\,w + \kappa^2 w\cdot w\Big)\,dx \\
&\le \sup_{\substack{\tau\in L^2(\Omega,\mathbb{R}^d)\\ \eta\in L^2(\Omega,\mathbb{R}^d)}} \int_\Omega \Big(\mu^{-1}\mathrm{curl}\,(u-v)\cdot\tau - \frac{1}{2}\mu^{-1}\tau\cdot\tau + \kappa^2\eta\cdot(u-v) - \frac{1}{2}\kappa^2\eta\cdot\eta\Big)\,dx \\
&= \frac{1}{2}|||u-v|||^2 \\
&= \int_\Omega \Big(\mu^{-1}\mathrm{curl}\,(u-v)\cdot\mathrm{curl}\,(u-v) + \kappa^2(u-v)\cdot(u-v) \\
&\qquad\qquad - \frac{1}{2}(\mu^{-1}|\mathrm{curl}\,(u-v)|^2 + \kappa^2|u-v|^2\Big)\,dx \\
&\le \sup_{w\in V_0} \int_\Omega \Big(\mu^{-1}\mathrm{curl}\,(u-v)\cdot\mathrm{curl}\,w + \kappa^2 w\cdot(u-v) \\
&\qquad\qquad - \frac{1}{2}(\mu^{-1}\mathrm{curl}\,w\cdot\mathrm{curl}\,w + \kappa^2 w\cdot w\Big)\,dx.
\end{aligned}
$$

Thus, we conclude that

$$
\begin{aligned}
\frac{1}{2}|||u-v|||^2 = \sup_{w\in V_0} &\int_\Omega \Big(\mu^{-1}\mathrm{curl}\,(u-v)\cdot\mathrm{curl}\,w + \kappa^2 w\cdot(u-v) \\
&\qquad - \frac{1}{2}(\mu^{-1}\mathrm{curl}\,w\cdot\mathrm{curl}\,w + \kappa^2 w\cdot w\Big)\,dx.
\end{aligned}
$$

By (9.2.4), we obtain

$$
\begin{aligned}
|||u-v|||^2 &\ge M^2_\ominus(v,w) \\
&:= \int_\Omega \Big(2j\cdot w - \mu^{-1}|\mathrm{curl}\,w|^2 - \kappa^2|w|^2 \\
&\qquad\qquad - 2\mu^{-1}\mathrm{curl}\,v\cdot\mathrm{curl}\,w - 2\kappa^2 v\cdot w\Big)\,dx. \qquad (9.2.13)
\end{aligned}
$$

For any $w \in V_0$ the quantity $M^2_\ominus(v,w)$ provides a lower bound of the error. Certainly, the sharpest bound is given by

$$\sup_{w\in V_0} M^2_\ominus(v,w).$$

It is not difficult to prove that this quantity coincides with the squared error (to prove that it suffices to set $w = u - v$). A practically computable lower bound can be determined if maximization is performed over a finite-dimensional subspace $V_{0m} \subset V_0$,

$\dim V_{0m} = m$. Then, finding the quantity

$$\sup_{w \in V_{0m}} M_{\ominus}^2(v, w)$$

requires solving a quadratic type maximization problem.

Now, we consider the problem (9.2.6). Since u is defined up to a gradient field, we consider only the divergent part of v, which leads to the assumption $v \in V_{00}$. Our aim is to estimate the quantity

$$|[u - v]| := \|\operatorname{curl}(u - v)\|.$$

First, we recall a result in the theory of functions in $H(\Omega, \operatorname{curl})$ (e.g., see [152, 236, 327]).

Lemma 9.4. *Let Ω be a Lipschitz simply connected bounded domain. There exists a constant C_Ω such that*

$$\|w\| \leq C_\Omega \|\operatorname{curl} w\|, \qquad \forall\, w \in V_0, \tag{9.2.14}$$

provided that $\operatorname{div} w = 0$ *in* Ω.

By (9.2.6) we observe that

$$\int_\Omega \operatorname{curl}(u - v) \cdot \operatorname{curl} w \, dx = \int_\Omega (j \cdot w - \operatorname{curl} v \cdot \operatorname{curl} w)\, dx, \qquad \forall w \in V_0. \tag{9.2.15}$$

Since

$$\int_\Omega (\operatorname{curl} y \cdot w - y \cdot \operatorname{curl} w)\, dx = 0, \qquad \forall w \in V_0,$$

we rearrange (9.2.15) as follows:

$$\int_\Omega \operatorname{curl}(u - v) \cdot \operatorname{curl} w \, dx = \int_\Omega \Big((j - \operatorname{curl} y) \cdot w - (\operatorname{curl} v - y) \cdot \operatorname{curl} w \Big)\, dx.$$

Set $w = u - v \in V_0$. We obtain

$$|[u - v]|^2 \leq \|j - \operatorname{curl} y\| \, \|u - v\| + \|\operatorname{curl} v - y\| \, \|\operatorname{curl}(u - v)\|.$$

In view of (9.2.14), this estimate leads to the following upper bound of the error:

$$|[u - v]| \leq C_\Omega \|j - \operatorname{curl} y\| + \|\operatorname{curl} v - y\|. \tag{9.2.16}$$

Note that in (9.2.16) v occurs only as the argument of the operator curl. Therefore, it should be understood as an estimate of the factor-norm for functions in the factor space equivalent up to a gradient field.

Comments. Approximation methods for the Maxwell's equation were investigated by many authors (e.g., see P. Monk [236] and R. Hiptmair [174]). Parallel multigrid solvers were studied in G. Haase, M. Kuhn, and U. Langer [164]). A posteriori estimates were derived in R. Beck, R. Hiptmair, R. Hoppe, and B. Wohlmuth [45] in the context of the residual approach and in D. Braess and J. Schöberl [69] with the help of the equilibrated approach. A posteriori estimates for nonconforming approximations of $H(\text{curl})$ elliptic partial differential equations were studied in P. Houston, I. Perugia, and D. Schotzau [181].

Estimates (9.2.10) and (9.2.16) were obtained in the author's papers [296, 297] (estimate (9.2.10) was independently derived by A. Hannukainen [168]). We note that (9.2.10) can also be derived from the general a posteriori estimate given in [280] (see also [244]) for convex variational problems related to the functional $G(\Lambda v) + F(v)$ (see Chapter 7). The problem is encompassed in the general framework if we set $\Lambda = \text{curl}$,

$$G(y) = \int_\Omega \frac{1}{2\mu}|y|^2\,dx, \quad \text{and} \quad F(v) = \int_\Omega \left(\frac{\kappa^2}{2}|v|^2 - j\cdot v\right)dx.$$

Advanced forms of a posteriori estimates for the Maxwell's problem has been recently obtained in P. Neittaanmäki and S. Repin [245].

9.3 Evolutionary problems

A posteriori error control for evolutionary problems is a substantial topic that requires a special consideration. The goal of this section is to give only an idea of how to extend methods considered in previous chapters to evolutionary problems. Below we shortly discuss a posteriori estimates that follow from the evolutionary integral identity. As before, we do this with the paradigm of the linear diffusion problem.

9.3.1 The linear evolutionary problem

Consider the classical initial-boundary value problem for the heat equation: Find $u(x,t)$ such that

$$u_t - \Delta u = f \qquad \text{in } Q_T, \tag{9.3.1}$$

$$u(x,0) = \phi, \qquad x \in \Omega, \tag{9.3.2}$$

$$u(x,t) = 0, \qquad (x,t) \in S_T, \tag{9.3.3}$$

where $Q_T := \Omega \times (0,T)$ is a space-time cylinder and $S_T := \Gamma \times [0,T]$. This problem is one of the simplest evolutionary problems, which is often used to model various diffusion type processes (e.g., heat transfer). With the paradigm of this problem, we explain how to derive computable estimates (in terms of a space-time norm) of the difference between u and any admissible approximation v. Our analysis follow the lines of [287]. In A. Gaevskaya and S. Repin [144], the method was applied to a more general class of linear parabolic problems (this paper also contains results of numerical tests).

We begin by introducing some spaces of functions defined on Q_T. Spaces of functions that map $(0,T)$ into a Banach space X are called Bochner spaces. For example, if $p \in [1,+\infty)$ and $\|\cdot\|_X$ denotes the norm of X, then the Bochner space $L^p((0,T),X)$ is the Banach space of mappings g such that

$$\|g\|_{L^p((0,T),X)} := \left(\int_0^T \|g(\cdot,t)\|_X^p \, dt\right)^{1/p} < \infty.$$

In particular, $L^2(\,(0,T);\ H^1(\Omega))$ consists of H^1-functions (with respect to spatial variables) the norms of which are L^2-functions with respect to $t \in (0,T))$. Let $H_0^1(Q_T)$ be the subspace of $H^1(Q_T)$ that contain functions with zero traces on S_T. By $\mathcal{V}(Q_T)$, we define the Banach space of functions from $L^2(\,(0,T);\ H^1(\Omega))$ having finite norm

$$\|w\|_{\mathcal{V}}^2 := \operatorname{vrai}\max_{t\in(0,T)} \|w(\cdot,t)\|_\Omega^2 + \|\nabla w\|_{2,Q_T}^2.$$

The space $\mathcal{V}^{1,0}(Q_T) = C([0,T];L^2(\Omega)) \cap L^2(\,(0,T);\ H^1(\Omega))$ is a subspace of $\mathcal{V}(Q_T)$. For all $t \in [0,T]$, elements of this space have traces (which are square summable functions on cross-sections of Q_T) that continuously change with respect to $t \in [0,T]$. By $\overset{\circ}{V}{}^{1,0}(Q_T)$, we denote another subspace of $\mathcal{V}(Q_T)$, which is the intersection of $\mathcal{V}^{1,0}(Q_T)$ and $L^2((0,T);\overset{\circ}{H}{}^1(\Omega))$. For elements of the space $\mathcal{V}^{1,0}(Q_T)$, we define the quantity

$$|\!|\!|w|\!|\!|_{(\kappa,\epsilon)}^2 := \kappa\|w(\cdot,T)\|_\Omega^2 + \epsilon\|\nabla w\|_{Q_T}^2, \qquad \mu,\epsilon > 0,$$

which will be used as a measure of the difference between the exact solutions and approximations.

Also, we use the space $H_0^{1,\Delta}(Q_T)$ that consists of functions $w \in L^2(Q_T)$ having finite norm

$$\|w\|_{2,0}^{1,\Delta} := \int_{Q_T} (w^2 + w_t^2 + |\nabla w|^2 + (\Delta w)^2)\, dx\, dt$$

and vanishing on S_T.

In the context of the well-known theory for parabolic type problems, a function $u \in \overset{\circ}{V}{}^{1,0}(Q_T)$ is called a (generalized) solution of (9.3.1)–(9.3.3) if it satisfies the following integral identity:

$$\int_{Q_T} \nabla u \cdot \nabla w \, dx\, dt - \int_{Q_T} u w_t \, dx\, dt + \int_\Omega (u(x,T)w(x,T) - u(x,0)w(x,0))\, dx$$
$$= \int_{Q_T} f w \, dx\, dt, \qquad \forall w \in H_0^1(Q_T). \tag{9.3.4}$$

We recall the classical solvability results (e.g., see Ladyzhenskaya [217]) for this problem.

Theorem 9.5. *Let Ω be a bounded connected domain with Lipschitz continuous boundary Γ.*

(i) Let $f \in L^2(Q_T)$ and $\phi(x) \in \mathring{H}^1(\Omega)$. Then the problem (9.3.1)–(9.3.3) is uniquely solvable in the space $H_0^{1,\Delta}(Q_T)$.

(ii) If $f \in L^1((0,T), L^2(\Omega))$ and $\phi \in L^2(\Omega)$, then u belongs to the class $\mathring{V}^{1,0}(Q_T)$.

9.3.2 First form of the error majorant

Assume that

$$f \in L^2(Q_T), \quad \phi \in H^1(\Omega). \tag{9.3.5}$$

In this case, Theorem 9.5 guarantees the existence of a solution u that satisfies (9.3.4). Let $v \in H_0^1(Q_T)$ be a given function. In particular, v may be an approximation of u obtained by a semidiscrete approximation of (9.3.1)–(9.3.3). We are interested in deriving an upper bound of the deviation $u - v$ evaluated in the norm $\|\!|\cdot|\!\|$ or in terms of the quantity $[\cdot]_{(\gamma,\delta)}$. From (9.3.4) we obtain

$$\begin{aligned}\int_{Q_T} \nabla(u-v)\cdot\nabla w\,dx\,dt - \int_{Q_T}(u-v)w_t\,dx\,dt&\\
+\int_\Omega ((u(x,T)-v(x,T))w(x,T)-(u(x,0)-v(x,0))w(x,0))\,dx&\\
=\int_{Q_T}(fw-\nabla v\cdot\nabla w - v_t w)\,dx\,dt, \qquad \forall w \in H_0^1(Q_T).&\end{aligned}$$

Set $w = u - v$ and note that

$$\begin{aligned}\int_\Omega |\, w(x,T)\,|^2\,dx - \int_\Omega |\, w(x,0)\,|^2\,dx&\\
=\int_{Q_T} w\,w_t\,dx\,dt + \tfrac12\|w(\cdot,T)\|_\Omega^2 - \tfrac12\|w(\cdot,0)\|_\Omega^2.&\end{aligned}$$

This yields the integral relation

$$\begin{aligned}\|\nabla(u-v)\|_{2,Q_T}^2 + \tfrac12\|u(\cdot,T)-v(\cdot,T)\|_\Omega^2&\\
=\int_{Q_T}(f(u-v)-\nabla v\cdot\nabla(u-v)-v_t(u-v))\,dxdt&\\
+\tfrac12\|u(\cdot,0)-v(\cdot,0)\|_\Omega^2,&\end{aligned} \tag{9.3.6}$$

which presents the energy balance in terms of deviations from the exact solution u. The relation (9.3.6) can be regarded as a generalization of the well-known *energy-balance equation* for the heat equation (see, e.g. [217]). It is easy to see that the classical energy balance equation follows from (9.3.6) if we set $v \equiv 0$. Subsequently, we use (9.3.6) as a starting point of the analysis. Introduce a new vector-valued function $y(x,t) \in Y(Q_T)$, where

$$Y(Q_T) := \{y(x,t) = \{y_i(x,t)\} \mid y_i \in L^2(Q_T),\ 1 \le i \le d\},$$

and rearrange (9.3.6) as follows:

$$\begin{aligned}\|\nabla(u-v)\|^2_{Q_T} + \tfrac{1}{2}\|u(\cdot,T) - v(\cdot,T)\|^2_\Omega - \tfrac{1}{2}\|u(\cdot,0) - v(\cdot,0)\|^2_\Omega \\ = \int_{Q_T} (f(u-v) - v_t(u-v) - y\cdot\nabla(u-v))\,dx\,dt \\ + \int_{Q_T} (y - \nabla v)\cdot\nabla(u-v)\,dx\,dt. \qquad (9.3.7)\end{aligned}$$

For almost all $t \in [0,T]$, we can define a linear functional $F_t : \overset{\circ}{H}{}^1(\Omega) \to \mathbb{R}$ by the relation

$$F_t(\widetilde{w}; v, y) := \int_\Omega (f\widetilde{w} - v_t\widetilde{w} - y\cdot\nabla\widetilde{w})\,dx.$$

The quantity

$$| F_t(v,y) | = \sup_{\substack{\widetilde{w}\in\overset{\circ}{H}{}^1(\Omega)\\ \widetilde{w}\neq 0}} \frac{\int_\Omega (f\widetilde{w} - v_t\widetilde{w} - y\cdot\nabla\widetilde{w})\,dx}{\|\nabla\widetilde{w}\|_\Omega}$$

is finite (it is bounded by $C_{F\Omega}(\|f(\cdot,t)\|_\Omega + \|v_t(\cdot,t)\|_\Omega) + \|y(\cdot,t)\|_\Omega$) and can be viewed as a norm of this functional (note that $| F_t(v,y) |$ is square integrable on $(0,T)$).

Now, we put (9.3.7) in the form

$$\begin{aligned}\|\nabla(u-v)\|^2_{Q_T} + \tfrac{1}{2}\|(u-v)(\cdot,T)\|^2_\Omega - \tfrac{1}{2}\|(u-v)(\cdot,0)\|^2_\Omega \\ = \int_0^T F_t((u-v); v, y)\,dt + \int_{Q_T} (y - \nabla v)\cdot\nabla(u-v)\,dx\,dt. \qquad (9.3.8)\end{aligned}$$

Let δ and μ be two given constants such that

$$0 < \delta \le 2, \quad 0 < \mu < 1. \qquad (9.3.9)$$

Define the set

$$L^\infty_\mu(0,T) := \{\beta(t) \in L^\infty(0,T) \mid \beta(t) \ge \mu \quad \text{for almost all } t \in (0,T)\}.$$

Take two scalar-valued functions $\alpha_1(t)$ and $\alpha_2(t)$ such that

$$\alpha_1(t) = \frac{1}{\delta}\left(1 + \frac{1}{\beta(t)}\right) \quad \text{and} \quad \alpha_2(t) = \frac{1}{\delta}\left(1 + \beta(t)\right). \tag{9.3.10}$$

In view of the Young–Fenchel inequality, we have

$$\int_0^T F_t((u-v), v, y)\, dt \le \int_0^T \left(\tfrac{\alpha_1(t)}{2} \,|\, F_t(v,y) \,|^2 + \tfrac{1}{2\alpha_1(t)} \|\nabla(u-v)\|_\Omega^2\right) dt,$$

$$\int_{Q_T} (\nabla v - y)\cdot \nabla(u-v)\, dx\, dt \le \int_0^T \left(\tfrac{\alpha_2(t)}{2} \|\nabla v - y\|_\Omega^2 + \tfrac{1}{2\alpha_2(t)} \|\nabla(u-v)\|_\Omega^2\right) dt.$$

Note that $\alpha_1(t)$ and $\alpha_2(t)$ satisfy the relation

$$\frac{1}{\alpha_1(t)} + \frac{1}{\alpha_2(t)} = \delta.$$

Now, by (9.3.9) and the inequalities we deduce the estimate

$$\begin{aligned}
(2-\delta)\|\nabla(u-v)\|_{Q_T}^2 &+ \|(u-v)(\cdot,T)\|_\Omega^2 \\
&\le \|v(\cdot,0) - \phi\|_\Omega^2 + \frac{1}{\delta}\int_0^T (1+\beta(t))\|y - \nabla v\|_\Omega^2\, dt \\
&\quad + \frac{1}{\delta}\int_0^T \left(1 + \tfrac{1}{\beta(t)}\right) |\, F_t(v,y) \,|_\Omega^2\, dt.
\end{aligned} \tag{9.3.11}$$

This estimate is valid for any $\beta(t) \in L_\mu^\infty(0,T)$ and $\delta \in (0,2]$. It is not difficult to observe that the right-hand side of (9.3.11) vanishes if and only if

$$\begin{aligned}
F_t(w; v, y) &= 0 && \text{for all } w \in \overset{\circ}{H}{}^1(\Omega) \text{ and almost all } t \in (0,T), \\
y &= \nabla v && \text{a.e. in } Q_T, \\
v(0,x) &= \phi(x) && \text{for a.e. } x \in \Omega.
\end{aligned}$$

These relations mean that $v \in H^1(Q_T)$ satisfies the initial and boundary conditions and for almost all $t \in (0,T)$ satisfies the relation

$$\int_\Omega (fw - v_t w - \nabla v \cdot \nabla w)\, dx = 0, \qquad \forall w \in \overset{\circ}{H}{}^1(\Omega),$$

which shows that v is a solution of the problem.

To make the estimates computable, we should replace the norm of F_t by an explicitly computable quantity. For this purpose, we take y in a certain subspace of $Y(Q_T)$. Namely, suppose

$$y \in Y_{\text{div}}(Q_T) := \{y \in Y(Q_T) \mid \operatorname{div} y \in L^2(\Omega) \text{ for a.e. } t \in (0,T)\}. \tag{9.3.12}$$

Then, for almost all $t \in (0, T)$,

$$\int_\Omega w(x,t) \operatorname{div} y(x,t)\, dx = -\int_\Omega y(x,t) \cdot \nabla w(x,t)\, dx,$$

and we have

$$| F_t(v, y) | \leq C_{F\Omega} \| f - v_t + \operatorname{div} y \|_\Omega. \tag{9.3.13}$$

Then, by (9.3.11), we arrive at the following result.

Theorem 9.6. *Let the conditions (9.3.5) and (9.3.9) be satisfied. Then*

$$|\!|\!| u - v |\!|\!|^2_{(1,2-\delta)} \leq \overline{\mathfrak{M}}^2_{\mathrm{EV1}}(\beta, \delta, v,\ y) \tag{9.3.14}$$

where

$$\begin{aligned} \overline{\mathfrak{M}}^2_{\mathrm{EV1}}(\beta, \delta, v,\ y) &:= \int_\Omega |v(x,0) - \phi(x)|^2\, dx \\ &+ \frac{1}{\delta} \int_{Q_T} \left((1+\beta)|y - \nabla v|^2 + C^2_{F\Omega} \left(1 + \tfrac{1}{\beta}\right) |f - v_t + \operatorname{div} y|^2 \right) dx\, dt \end{aligned}$$

and $y \in Y_{\mathrm{div}}(Q_T)$ and $\beta \in L^\infty_\mu(0, T)$.

Consider two particular forms of (9.3.14), which deserve a special discussion. Assume that v satisfies the initial condition and set $\delta = 1$. Then we obtain

$$\begin{aligned} \|\nabla(u - v)\|^2_{Q_T} &+ \|(u - v)(\cdot, T)\|^2_\Omega \\ &\leq \overline{\mathfrak{M}}^2_{\mathrm{EV1}}(\beta, 1, v,\ y) \\ &= \int_0^T (1 + \beta(t)) \|y - \nabla v\|^2_\Omega\, dt \\ &\qquad + C_{F\Omega} \int_0^T \left(1 + \tfrac{1}{\beta(t)}\right) \|f - v_t + \operatorname{div} y\|^2_\Omega\, dt. \end{aligned} \tag{9.3.15}$$

This estimate can be viewed as an analog of (3.2.8) derived for the stationary diffusion problem. Note that in (9.3.15), $\beta = \beta(t)$ is a positive function (in the stationary case, β is a positive constant).

If $\delta = 2$, then we arrive at another estimate:

$$\begin{aligned} \|(u - v)(\cdot, T)\|^2_\Omega &\leq \overline{\mathfrak{M}}^2_{\mathrm{EV1}}(\beta, 2, v,\ y) \\ &= \frac{1}{2} \int_0^T (1 + \beta(t)) \|y - \nabla v\|^2_\Omega\, dt \\ &\qquad + \frac{C_{F\Omega}}{2} \int_0^T \left(1 + \tfrac{1}{\beta(t)}\right) \|f - v_t + \operatorname{div} y\|^2_\Omega\, dt, \end{aligned} \tag{9.3.16}$$

which yields an upper bound of the error on the top of the space-time cylinder. Since the right-hand side of (9.3.16) is a monotone function with respect to T, the right-hand side also gives an upper bound of the quantity

$$\operatorname{vrai}\max_{t\in[0,T]} \|(u-v)(\cdot,t)\|^2_\Omega.$$

Proposition 9.7. *For any $\delta \in (0,2]$ and $\beta \in L^\infty_\mu(0,T)$, the variational problem*

$$\inf_{\substack{v\in H^1_0(Q_T)\\ y\in Y_{\mathrm{div}}(Q_T)}} \overline{\mathfrak{M}}^2_{\mathrm{EV1}}(\beta,\delta,v,\ y) \tag{9.3.17}$$

has a solution. The exact lower bound of this problem is equal to zero. It is attained if and only if $v = u$ and $y = \nabla u$.

Proof. The existence of a pair $(v,y) \in H^1_0(Q_T)\times Y_{\mathrm{div}}(Q_T)$ minimizing the functional $\overline{\mathfrak{M}}^2_{\mathrm{EV1}}(\beta,\delta,v,\ y)$ is proved straightforwardly. Indeed, set $v = u$ and $y = \nabla u$. Since $u \in H^{1,\Delta}_{2,1}$, we see that $\operatorname{div}\nabla u \in L^2(Q_T)$ and, therefore, $y \in Y_{\mathrm{div}}(Q_T)$. In this case, $\overline{\mathfrak{M}}_\delta(v,\ y,\beta) = 0$, so that the exact lower bound is attained.

Assume that $\overline{\mathfrak{M}}^2_{\mathrm{EV1}}(\beta,\delta,v,\ y) = 0$. Then, the function $v(x,t)$ satisfies the initial and boundary conditions. In addition, for almost all $(x,t) \in Q_T$ the relations

$$\nabla v = y \in Y_{\mathrm{div}}(Q_T) \tag{9.3.18}$$

and

$$\operatorname{div} y - v_t + f = 0 \tag{9.3.19}$$

hold. Hence, v is the exact solution. □

Corollary 9.8. *By (9.3.15) and (9.3.16), we find that*

$$\|u-v\|^2_{\mathcal{V}} \le \overline{\mathfrak{M}}^2_{\mathrm{EV1}}(\beta,1,v,\ y) + \overline{\mathfrak{M}}^2_{\mathrm{EV1}}(\beta,2,v,\ y). \tag{9.3.20}$$

Remark 9.9. The majorant $\overline{\mathfrak{M}}^2_{\mathrm{EV1}}(\beta,\delta,v,\ y)$ is well defined for $v \in H^1_0(Q_T)$, $f \in L^2(Q_T)$, $v(x,0) \in L^2(\Omega)$, and $\phi(x) \in L^2(\Omega)$. Using arguments close to those often used in the theory of PDE's (e.g., see [217] (Chapter 2, §2)) one can extend the estimates (9.3.15), (9.3.16), and (9.3.20) to wider sets of functions.

9.3.3 Second form of the error majorant

Now we reform the right-hand side of (9.3.7) by other means and deduce an advanced form of the error majorant. Present the right-hand side of this relation as sum of three terms, which are

$$I_1 = \int_{Q_T} (f(u-v) - v_t(u-v) - \vartheta_t(u-v) - y \cdot \nabla(u-v))\, dx\, dt,$$

$$I_2 = \int_{Q_T} (y - \nabla v + \nabla \vartheta) \cdot \nabla(u-v)\, dx\, dt,$$

$$I_3 = \int_{Q_T} (\vartheta_t(u-v) - \nabla \vartheta \cdot \nabla(u-v))\, dx\, dt.$$

Here, $y \in Y(Q_T)$ and $\vartheta \in H_0^1(Q_T)$ are some arbitrary functions (later we discuss how one can chose these functions in order to obtain optimal estimates).

For almost all $t \in (0, T)$, we define a linear functional

$$F_t(\,\cdot\,; v, \vartheta, y) : \overset{\circ}{H}{}^1(\Omega) \to \mathbb{R}$$

by the relation

$$F_t(\widetilde{w}; v, \vartheta, y) := \int_\Omega (f\widetilde{w} - v_t\widetilde{w} - \vartheta_t\widetilde{w} - y \cdot \nabla\widetilde{w})\, dx.$$

It is easy to observe that the quantity

$$| F_t(v, \vartheta, y) | := \sup_{\substack{\widetilde{w} \in \overset{\circ}{H}{}^1(\Omega),\\ \widetilde{w} \neq 0}} \frac{\int_\Omega (f\widetilde{w} - v_t\widetilde{w} - \vartheta_t\widetilde{w} - y \cdot \nabla\widetilde{w})\, dx}{\|\nabla\widetilde{w}\|_\Omega}$$

is finite. It defines a norm of this functional and generates the estimate

$$I_1 \le \int_0^T | F_t(v, \vartheta, y) |\, \|\nabla(u-v)\|_\Omega\, dt.$$

The term I_2 is estimated by

$$\int_0^T \|y - \nabla v - \nabla\vartheta\|_\Omega\, \|\nabla(u-v)\|_\Omega\, dt$$

and I_3 is represented in the form

$$\begin{aligned} I_3 &= \int_{Q_T} (\nabla\vartheta \cdot \nabla v - \vartheta_t\, v)\, dx\, dt - \int_{Q_T} (\nabla\vartheta \cdot \nabla u - \vartheta_t u)\, dx\, dt \\ &= \int_{Q_T} (\nabla\vartheta \cdot \nabla v - \vartheta_t\, v - f\vartheta)\, dx\, dt + \int_\Omega (u(x,T)\vartheta(x,T) - u(x,0)\vartheta(x,0))\, dx \\ &= \mathcal{F}(v, \vartheta) + \int_\Omega ((u-v)(x,T)\vartheta(x,T) - (u-v)(x,0)\vartheta(x,0))\, dx, \end{aligned}$$

where

$$\mathcal{F}(v,\vartheta) := \int_{Q_T} (\nabla v \cdot \nabla\vartheta + v_t\vartheta - f\vartheta)\,dx.$$

Since

$$\begin{aligned}&\int_\Omega ((u-v)(x,T)\vartheta(x,T) - (u-v)(x,0)\vartheta(x,0))\,dx\\ &\qquad \le \frac{1}{2\gamma}\|(u-v)(\cdot,T)\|^2_\Omega + \frac{\gamma}{2}\|\vartheta(\cdot,T)\|^2_\Omega - \int_\Omega (\phi(x) - v(x,0))\vartheta(x,0)\,dx,\end{aligned}$$

we deduce the estimate

$$\begin{aligned}&(2-\delta)\|\nabla(u-v)\|^2_{Q_T} + \left(1-\frac{1}{\gamma}\right)\|(u-v)(\cdot,T)\|^2_\Omega\\ &\quad \le \gamma\|\vartheta(\cdot,T)\|^2_\Omega + 2\mathcal{F}(v,\vartheta)\\ &\qquad + \frac{1}{\delta}\int_0^T \left((1+\beta)\|y - \nabla v + \nabla\vartheta\|^2_\Omega + \left(1+\tfrac{1}{\beta}\right) | F_t(v,\vartheta,y) |^2_\Omega\right) dt\\ &\qquad + 2\mathcal{F}(v,\vartheta) + \int_\Omega \left(|\phi(x) - v(x,0)|^2 - 2\vartheta(x,0)(\phi(x) - v(x,0))\right) dx.\end{aligned}$$

Here $\vartheta(x,t) \in H^1_0(Q_T)$, $y \in Y(Q_T)$, $\beta(t) \in L^\infty_\mu(0,T)$, $\gamma \ge 1$, and $\delta \in (0,2]$.

As in the previous section, we find a computable majorant of the error, provided that $y \in Y_{\rm div}(Q_T)$. In this case,

$$| F_t(v,\vartheta,y) | \le C_{F\Omega}\|f - v_t - \vartheta_t + \operatorname{div} y\|_\Omega,$$

and we arrive at the estimate

$$|||u-v|||^2_{(1-\frac{1}{\gamma},2-\delta)} \le \overline{\mathfrak{M}}^2_{\rm EV2}(\beta,\gamma,\delta,v,\vartheta,\ y), \tag{9.3.21}$$

where

$$\begin{aligned}\overline{\mathfrak{M}}^2_{\rm EV2}(\beta,\gamma,\delta,v,\vartheta,y) := \gamma\|\vartheta(\cdot,T)\|^2_\Omega &+ \int_\Omega |\phi(x) - v(x,0)|^2\,dx\\ &+ \frac{1}{\delta}\int_0^T \Big((1+\beta)\|y - \nabla v + \nabla\vartheta\|^2_\Omega\\ &+ \tfrac{C^2_\Omega(1+\beta)}{\beta}\|f - v_t - \vartheta_t + \operatorname{div} y\|^2_\Omega\Big)\,dt\\ &+ 2\mathcal{F}(v,\vartheta) - 2\int_\Omega \vartheta(x,0)(\phi(x) - v(x,0))\,dx.\end{aligned}$$

If $v(x,0) = \phi(x)$ then the majorant has a simplified form:

$$\overline{\mathfrak{M}}^{\,2}_{\mathrm{EV2}}(\beta,\gamma,\delta,v,\vartheta,y) := \gamma\|\vartheta(\cdot,T)\|^2_\Omega + 2\mathcal{F}(v,\vartheta) + \frac{1}{\delta}\int_0^T \Big((1+\beta)\|y-\nabla v+\nabla\vartheta\|^2_\Omega + \frac{C^2_\Omega(1+\beta)}{\beta}\|f-v_t-\vartheta_t+\operatorname{div} y\|^2_\Omega\Big)\,dt. \quad (9.3.22)$$

Remark 9.10. If $\vartheta \equiv 0$ and $\gamma \to +\infty$, then $\overline{\mathfrak{M}}^{\,2}_{\mathrm{EV2}}$ is reduced to $\overline{\mathfrak{M}}^{\,I}_{\mathrm{EV}}$. In practice, the estimate (9.3.21) (which contains a "correction function" ϑ) gives a sharper upper bound than (9.3.14). Using ϑ, one can reduce the residual term $f - v_t + \operatorname{div} y$, which may be difficult to make small by operating only with y (such a situation arises if v_t significantly differs from u_t).

Remark 9.11. By setting $\delta = 1$ and $\delta = 2$, we obtain the estimates

$$\|\nabla(u-v)\|^2_{Q_T} \le \overline{\mathfrak{M}}^{\,2}_{\mathrm{EV2}}(\beta,\gamma,1,v,\vartheta,\ y), \quad (9.3.23)$$

$$\operatorname{vrai}\max_{t\in[0,T]} \|(u-v)(\cdot,t)\|^2_\Omega \le \frac{\gamma}{\gamma-1}\overline{\mathfrak{M}}^{\,2}_{\mathrm{EV2}}(\beta,\gamma,2,v,\vartheta,\ y). \quad (9.3.24)$$

The proposition below shows that the majorant $\overline{\mathfrak{M}}^{\,2}_{\mathrm{EV2}}$ generates a variational problem the exact lower bound of which is attained on the solution of our problem.

Proposition 9.12. *For any $\delta \in (0,2]$, $\gamma \ge 1$, and $\beta \in L^\infty_\mu(0,T)$, the variational problem*

$$\inf_{\substack{v\in H^1_0(Q_T)\\ w\in H^1_0(Q_T),\\ y\in Y_{\mathrm{div}}(Q_T)}} \overline{\mathfrak{M}}^{\,2}_{\mathrm{EV2}}(\beta,\gamma,\delta,v,\vartheta,\ y), \quad (9.3.25)$$

has a solution. The exact lower bound of this problem is equal to zero and is attained if $v = u$, $w = 0$, and $y = \nabla u$.

Proof. Obviously, the infimum in (9.3.25) is majorated by $\overline{\mathfrak{M}}^{\,2}_{\mathrm{EV2}}(\beta,\gamma,\delta,v,0,\ y)$, which coincides with $\overline{\mathfrak{M}}^{\,I}_{\mathrm{EV}}(\beta,\delta,v,\ y)$. Therefore, the result follows from Proposition 9.7. □

9.3.4 Equivalence of the deviation and majorant

Now we focus on another property of the majorant $\overline{\mathfrak{M}}^{\,2}_{\mathrm{EV2}}$. Suppose $v(x,0) = \phi(x)$ and consider the quantity

$$M^2_\oplus(\gamma,\delta,v) := \inf_{\substack{\beta\in L^\infty_\mu(0,T)\\ \vartheta\in H^1_0(Q_T)\\ y\in Y_{\mathrm{div}}(Q_T)}} \overline{\mathfrak{M}}^{\,2}_{\mathrm{EV2}}(\beta,\gamma,\delta,v,\vartheta,\ y),$$

which is an upper bound of the error (see (9.3.21)). We are aimed at showing that this bound is realistic, i.e., it does not lead to a large overestimation of the actual value of the norm of the true error. For this purpose, we estimate $M_{\oplus}(\gamma,\delta,v)$ from above and show that this estimate is equivalent to the error.

Since $u \in H_0^{1,\Delta}(Q_T)$, one can put $y = \nabla u \in Y_{\rm div}(Q_T)$. Moreover, we set $\vartheta = u - v$. Then

$$f - v_t - \vartheta_t + \operatorname{div} y = f - u_t + \Delta u = 0,$$

and we find that

$$M^2_{\oplus}(\gamma,\delta,v) \le \gamma \|(u-v)(\cdot,T)\|^2_{\Omega} + 2\mathcal{F}(v,u-v) + \frac{4}{\delta}\int_0^T (1+\beta)\|\nabla(u-v)\|^2_{\Omega}\,dt.$$

Note that

$$\begin{aligned}\mathcal{F}(v,u-v) &= \int_{Q_T} (\nabla v \cdot \nabla(u-v) + v_t(u-v) - f(u-v))\,dx\,dt\\ &= \int_{Q_T} (\nabla u \cdot \nabla(u-v) + u_t(u-v) - f(u-v))\,dx\,dt\\ &\quad - \int_{Q_T} \big(|\nabla(u-v)|^2 + (u-v)_t(u-v)\big)\,dx\,dt.\end{aligned}$$

The first integral on the right-hand side is equal to zero. Hence,

$$\begin{aligned}M^2_{\oplus}(\gamma,\delta,v) &\le \int_0^T (4\frac{1+\beta}{\delta} - 2)\|\nabla(u-v)\|^2_{\Omega}\,dt\\ &\quad + \gamma\|(u-v)(\cdot,T)\|^2_{\Omega} - 2\int_{Q_T} (u-v)_t(u-v)\,dx\,dt\\ &\le \int_0^T (4\frac{1+\mu}{\delta} - 2)\|\nabla(u-v)\|^2_{\Omega}\,dt + (\gamma-1)\|(u-v)(\cdot,T)\|^2_{\Omega}.\end{aligned}$$

Set $\delta' = 2 - \delta$. Then, we obtain

$$M^2_{\oplus}(\gamma,\delta,v) \le \frac{2}{\delta}(\delta' + 2\mu)\|\nabla(u-v)\|^2_{Q_T} + (\gamma-1)\|(u-v)(\cdot,T)\|^2_{\Omega}.$$

Recall (9.3.21). We observe that for any $v \in H_0^1(Q_T)$

$$|||u-v|||^2_{(\gamma',\delta')} \le M^2_{\oplus}(\gamma,\delta,v) \le |||u-v|||^2_{(\gamma'',\delta'')} \le \kappa |||u-v|||^2_{(\gamma',\delta')},$$

where $\gamma' = \frac{\gamma-1}{\gamma}$, $\delta'' = \frac{2}{\delta}(\delta' + 2\mu)$, $\gamma'' = \gamma - 1$, and $\kappa = \max\left\{\gamma, \frac{2}{\delta}\left(1 + \frac{2\mu}{\delta'}\right)\right\}$. This relation means that the quantity $M_{\oplus}(\gamma,\delta,v)$ is equivalent to a certain measure of $u - v$.

9.3.5 Comments

Generalizations. The derivation method considered above is extendable to other parabolic equations of the form

$$u_t - \Lambda^* \mathcal{A} \Lambda u = f \qquad \text{in } Q_T, \tag{9.3.26}$$

$$u(x,0) = \phi, \qquad x \in \Omega, \tag{9.3.27}$$

$$u(x,t) = 0, \qquad (x,t) \in S_T, \tag{9.3.28}$$

which are generated by the elliptic operator $\Lambda^* \mathcal{A} \Lambda$ (cf. Section 7.1). Generalized solutions of such problems are defined by the integral identitiy

$$\int_0^T (\mathcal{A}\Lambda u, \Lambda w)\, dt - \int_{Q_T} u w_t \, dx\, dt + \int_\Omega (u(x,T)w(x,T) - u(x,0)w(x,0))\, dx$$
$$= \int_{Q_T} f w \, dx\, dt, \qquad \forall w \in V(Q_T), \tag{9.3.29}$$

where $(\cdot\,,\cdot)$ is the scalar product associated with the spatial part of the operator and $V(Q_T)$ is a suitable space of trial functions. Then, an upper bound of $u - v$ is derived by a procedure close to that used for the diffusion equation. Below we give a sketch of it. Let v be an approximation of u (which satisfies the boundary conditions and belongs to the corresponding energy space). We have

$$\int_0^T (\mathcal{A}\Lambda(u-v), \Lambda w)\, dt - \int_{Q_T} (u-v) w_t \, dx\, dt$$
$$+ \int_\Omega (u(x,T) - v(x,T)) w(x,T)\, dx - \int_\Omega (u(x,0) - v(x,0)) w(x,0)\, dx$$
$$= \int_0^T \int_\Omega \Big(f w - v_t w \Big)\, dx\, dt - \int_0^T (\mathcal{A}\Lambda v, \Lambda w)\, dt.$$

Set $w = u - v$ and note that

$$\int_{Q_T} w\, w_t \, dx\, dt = \frac{1}{2} \left(\|w(\cdot,T)\|^2_\Omega - \|w(\cdot,0)\|^2_\Omega \right).$$

We arrive at the relation

$$\int_0^T |||\Lambda(u-v)|||^2\, dt + \tfrac{1}{2} \|u(\cdot,T) - v(\cdot,T)\|^2_\Omega$$
$$= \int_0^T \int_\Omega \Big(f w - v_t w \Big)\, dx\, dt - \int_0^T (\mathcal{A}\Lambda v, \Lambda w)\, dt$$
$$+ \tfrac{1}{2} \|u(\cdot,0) - v(\cdot,0)\|^2_\Omega. \tag{9.3.30}$$

Further transformations are based on spatial properties of the operator Λ and its conjugate counterpart Λ^*. Suppose y is such that we can represent $\Lambda^* y$ as an integrable function and write

$$\int_0^T (y, \Lambda w)\, dt = \int_0^T \int_\Omega \Lambda^* y\, w\, dx\, dt,$$

where $w \in V$. Then, we have

$$\begin{aligned}
&\int_0^T |||\Lambda(u-v)|||^2\, dt + \tfrac{1}{2}\|u(\cdot,T) - v(\cdot,T)\|_\Omega^2 \\
&\quad = \int_0^T (y - \mathcal{A}\Lambda v, \Lambda(u-v))\, dt + \int_{Q_T} (f - \Lambda^* y + v_t)(u-v)\, dx\, dt \\
&\qquad + \tfrac{1}{2}\|\phi - v(\cdot,0)\|_\Omega^2,
\end{aligned} \tag{9.3.31}$$

where (as in Chapter 7) $|||\cdot|||$ denotes the spatial energy norm generated by $\mathcal{A}$. Note that

$$\int_{Q_T} (f - \Lambda^* y + v_t)(u-v)\, dx\, dt \le \mathrm{C} \int_0^T \|f - \Lambda^* y + v_t\|_\Omega |||\Lambda(u-v)|||\, dt,$$

where C is the constant in the inequality $\|w\|_\Omega \le \mathrm{C}|||\Lambda w|||$.

Define α_1 and α_2 by (9.3.10) and use the estimates

$$\begin{aligned}
&\int_{Q_T} (f - \Lambda^* y + v_t)(u-v)\, dx\, dt \\
&\quad \le \int_0^T \frac{\alpha_1(t)}{2} \mathrm{C}^2 \|f - \Lambda^* y + v_t\|_\Omega^2\, dt + \int_0^T \frac{1}{2\alpha_1(t)} |||\Lambda(u-v)|||^2\, dt
\end{aligned}$$

and

$$\begin{aligned}
&\int_0^T (y - \mathcal{A}\Lambda v, \Lambda(u-v)) \\
&\quad \le \int_0^T \frac{\alpha_2(t)}{2} |||y - \mathcal{A}\Lambda v|||_*^2\, dt + \int_0^T \frac{1}{2\alpha_2(t)} |||\Lambda(u-v)|||^2\, dt.
\end{aligned}$$

Then, we obtain an estimate analogous to (9.3.14) in which $\|y - \nabla v\|_\Omega^2$ is replaced by $|||y - \mathcal{A}\Lambda v|||_*$, $C_{F\Omega}$ by C, and div y by $\Lambda^* y$. We note that the above-discussed method can be applied to other evolutionary problems (see [162, 301]).

Practical applications. Finally, we briefly comment on possible applications of the estimates derived. As for elliptic problems, the simplest way for the practical exploitation of the above estimates consists of using post-processed fluxes of approximate solutions. Let v be an admissible approximation. Set $y = R\nabla v$, where

$R : Y(Q_T) \rightarrow Y_{\text{div}}(Q_T)$ is a post-processing operator (which performs necessary regularization of y). Then the quantity $\overline{\mathfrak{M}}^2_{\text{EV1}}(\beta, \delta, v, R\nabla v)$ yields a directly computable bound of the respective error norm, provided that δ and β lie in admissible sets.

Usually, the numerical analysis of evolutionary problems is based on using a sequence of consequently refining meshes. In this case, a directly computable error estimate can be obtained in the same way as for elliptic problems (cf. Section 3.6.2). Suppose v is an approximate solution of the heat equation computed on a coarse mesh with mesh-size τ for the time variable and h for spatial variables. Let v_{ref} be another approximate solution computed on a finer mesh $(\tau_{\text{ref}}, h_{\text{ref}})$ and R_{ref} denote a post-processing operator on the refined mesh. Then, the estimates (9.3.14) and (9.3.21) provide guaranteed and easily computable bounds of the approximation errors related to v:

$$|||u - v|||^2_{(1,\delta')} \leq \overline{\mathfrak{M}}^2_{\text{EV1}}(\beta, \delta, R_{\text{ref}}(\nabla u_{\text{ref}})) \tag{9.3.32}$$

and

$$|||u - v|||^2_{(\gamma',\delta')} \leq \overline{\mathfrak{M}}^2_{\text{EV2}}(\beta, \gamma, \delta, v, v_{\text{ref}} - v, R_{\text{ref}}(\nabla u_{\text{ref}})). \tag{9.3.33}$$

These estimates can be viewed as justified quantitative forms of the Runge's rule. Minimization with respect to $\beta(t) \in L^\infty_\mu(0, T)$ (which is not difficult to perform) will make the above estimates sharper.

If a more accurate error bound is required, then the majorant $\overline{\mathfrak{M}}^2_{\text{EV1}}(\beta, \delta, v, y)$ (or $\overline{\mathfrak{M}}^2_{\text{EV2}}(\beta, \gamma, \delta, v, \vartheta, y)$) should be minimized with respect to β, γ, δ, y (and ϑ). Certainly, such a procedure needs additional computational efforts. A rational way consists of using the majorants on each step of time integration (instead of the whole interval $(0, T]$). For example, if $\delta = 2$, then we rewrite (9.3.14) for the interval $(t_k, t_{k+1}]$ and obtain

$$e^2_{k+1} = e^2_k + \frac{1}{2}\int_{t_k}^{t_{k+1}} (1+\beta) \int_\Omega \left(|y - \nabla v|^2 + \frac{C^2_{F\Omega}}{\beta} |f - v_t + \operatorname{div} y|^2 \right) dx\, dt,$$

where e_k denotes the error at $t = t_k$. This formula gives a way for controling the accumulation of errors on time-steps. If the minimization with respect to y and β does not reduce the estimate below an acceptable level, then the corresponding interval (or even several neighboring intervals) should be diminished.

9.4 A posteriori estimates for optimal control problems

Functional a posteriori error estimates open new ways of error estimation for some classes of optimal control problems. Let $\psi \in L^\infty(\Omega)$, $\sigma^d \in L^2(\Omega, \mathbb{R}^d)$, and $f \in L^2(\Omega)$ be given functions, and

$$\mathsf{U} := \{\mathsf{v} \in L^2(\Omega) \,|\, \mathsf{v} \leq \psi \text{ a.e. in } \Omega\}.$$

Our goal is to find a control function $\mathsf{u} \in \mathsf{U}$ and a state function η_{u}[1] defined by the boundary value problem

$$-\Delta \eta_{\mathsf{v}} = \mathsf{v} + f \qquad \text{a.e. in } \Omega, \tag{9.4.1}$$

$$\mathsf{v} = 0 \qquad \text{on } \Gamma \tag{9.4.2}$$

such that the cost functional

$$J_1(\eta, \mathsf{v}) := \frac{1}{2}\|\nabla \eta - \sigma^d\|^2 + \frac{a}{2}\|\mathsf{v} - \mathsf{u}^d\|^2, \qquad a > 0,$$

attains its minimal value $J(\eta_{\mathsf{u}}, \mathsf{u})$. Another version of such a problem is generated by the functional

$$J_2(\eta, \mathsf{v}) := \frac{1}{2}\|\eta_{\mathsf{v}} - \eta^d\|^2 + \frac{a}{2}\|\mathsf{v} - \mathsf{u}^d\|^2,$$

where $\eta^d \in L^2(\Omega)$.

It is well known that under the above assumptions, the optimal control problem (with the functional J_1 or J_2) has a unique solution (e.g., see J.-L. Lions [221]). In this section, we show that functional a posteriori estimates allow us to obtain guaranteed and computable bounds for the cost functional and for errors of approximations of the state and control functions. First, we deduce an upper bound for the cost functional, which leads to an unconstrained minimization problem. In this problem, the differential equation (9.4.1) (whose presence is the major difficulty of the above optimal control problem) does not appear explicitly. This new minimization problem can be solved by well-known methods (e.g., by means of direct minimization). We prove that the sequence of the so-obtained upper bounds converges to the exact value of the cost functional and that the associated states and controls converge to the exact state and control, respectively. These results have been established in the papers A. Gaevskaya, R. Hoppe, and S. Repin [142, 143], where the reader will also find numerical results illustrating the reliability and efficiency of the approach.

9.4.1 Two-sided bounds for cost functionals

Upper bounds. Assume that $\mathsf{v} \in \mathsf{U}$ is an admissible control function computed by some numerical procedure and η_{v} is the corresponding state function. Since the latter is a solution of the state boundary value problem, we do not know it exactly and instead must operate with a certain approximation $\eta \in \overset{\circ}{H}{}^1(\Omega)$ (which may not satisfy the

[1] In the literature devoted to optimal control problems, the control function is traditionally denoted by u. However, in other parts of the book this letter was used to denote the exact solution of a boundary value problem. We try to follow the style accepted and, at the same time, to avoid a collision of notation. For this reason, in this section, we use special fonts (u, v) for the denotation of control functions.

differential equation). By the triangle inequality, we obtain the following upper bound for the cost functional:

$$J_1(\eta_{\mathsf{v}}, \mathsf{v}) \leq \frac{1}{2}\left(\|\nabla\eta - \sigma^d\| + \|\nabla(\eta_{\mathsf{v}} - \eta)\|\right)^2 + \frac{a}{2}\|\mathsf{v} - \mathsf{u}^d\|^2. \qquad (9.4.3)$$

Now, we estimate the term $\|\nabla(\eta_{\mathsf{v}} - \eta)\|$ by (4.1.13) and find that

$$\|\nabla(\eta_{\mathsf{v}} - \eta)\| \leq \|\tau - \nabla\eta\| + C_{F\Omega}\|\operatorname{div}\tau + \mathsf{v} + f\|,$$

where τ is an arbitrary function in $H(\Omega, \operatorname{div})$.

In view of (9.4.3) and (9.4.4), we have

$$J_1(\eta_{\mathsf{v}}, \mathsf{v}) \leq \overline{J}_1(\eta, \tau, \mathsf{v}) := \frac{a}{2}\|\mathsf{v} - \mathsf{u}^d\|^2 + \frac{1}{2}\Big(\|\nabla\eta - \sigma^d\| + \|\tau - \nabla\eta\| + C_{F\Omega}\|\operatorname{div}\tau + \mathsf{v} + f\|\Big)^2. \qquad (9.4.4)$$

The functional $\overline{J}_1(\eta, \tau, \mathsf{v})$ is directly computable and provides an upper bound of the cost functional for any $\eta \in V_0 := \overset{\circ}{H}{}^1(\Omega)$, $\tau \in H(\Omega, \operatorname{div})$, and $\mathsf{v} \in \mathsf{U}$. It is easy to see that (9.4.4) has no gap between its left- and right-hand sides. Indeed, set $\mathsf{v} = \mathsf{u}$, $\tau = \nabla\eta_{\mathsf{u}}$, and $\eta = \eta_{\mathsf{u}}$. Then,

$$\overline{J}_1(\eta, \tau, \mathsf{v}) = \frac{a}{2}\|\mathsf{u} - \mathsf{u}^d\|^2 + \frac{1}{2}\|\nabla\eta_{\mathsf{u}} - \sigma^d\|^2,$$

i.e., $\overline{J}_1(\eta, \tau, \mathsf{v})$ coincides with the value of the cost functional computed for the exact solution.

Using Young's inequality with positive parameters α and β, we can represent this bound in terms of a quadratic functional:

$$J_1(\eta_{\mathsf{u}}, \mathsf{u}) \leq J_1(\eta_{\mathsf{v}}, \mathsf{v}) \leq \overline{J}_1(\alpha, \beta; \eta, \tau, \mathsf{v}), \qquad (9.4.5)$$

where

$$\overline{J}_1(\alpha, \beta; \eta, \tau, \mathsf{v}) := \frac{1+\alpha}{2}\|\nabla\eta - \sigma^d\|^2 + \frac{(1+\alpha)(1+\beta)}{2\alpha}\|\tau - \nabla\eta\|^2 + \frac{(1+\alpha)(1+\beta)}{2\alpha\beta}C_{F\Omega}^2\|\operatorname{div}\tau + \mathsf{v} + f\|^2 + \frac{1}{2}\|\mathsf{v} - \mathsf{u}^d\|^2. \qquad (9.4.6)$$

Setting $\mathsf{v} = \mathsf{u}$, $\tau = \nabla\eta_{\mathsf{u}}$, and $\eta = \eta_{\mathsf{u}}$ and letting α go to zero, we find that $\overline{J}_1(\alpha, 0; \eta, \tau, \mathsf{v})$ also tends to $J_1(\eta_{\mathsf{u}}, \mathsf{u})$.

Hence, we conclude that

$$J_1(\eta_{\mathsf{u}}, \mathsf{u}) = \inf_{\substack{\eta \in V_0,\, \mathsf{v} \in \mathsf{U}, \\ \tau \in H(\Omega, \operatorname{div}),\, \alpha, \beta \in \mathbb{R}_+}} \overline{J}_1(\alpha, \beta; \eta, \tau, \mathsf{v}). \qquad (9.4.7)$$

The majorants $\overline{J}_1(\eta, \tau, \mathsf{v})$ and $\overline{J}_1(\alpha, \beta; \eta, \tau, \mathsf{v})$ can be used for finding *guaranteed upper bounds* for the cost functional. For example, we can take η and v as approximate solutions computed by a certain optimization procedure and additionally minimize the majorant with respect to τ (and to the parameters β and α). In the simplest case, we can take τ as a post-processed flux ∇v and perform a simple minimization with respect to α and β. The respective value $\overline{J}$ gives an upper bound of the cost functional. It should be outlined that the quantity $J_1(\mathsf{v}, \eta)$ may not provide such a guaranteed upper bound because η is not the exact solution of (9.4.1).

For the functional J_2, the majorant can be easily derived by applying the same method. We have

$$J_2(\eta, \mathsf{v}) = \frac{1}{2}\|\eta - \eta_d\|^2 + \frac{a}{2}\|\mathsf{v} - \mathsf{u}^d\|^2 \leq \overline{J}_2(\alpha, \beta; \eta, \tau, \mathsf{v}), \tag{9.4.8}$$

where

$$\overline{J}_2(\alpha, \beta; \eta, \tau, \mathsf{v}) := \frac{1+\alpha}{2}\|\eta - \eta^d\|^2 + \frac{(1+\alpha)(1+\beta)}{2\alpha} C_{F\Omega}^2 \|\tau - \nabla\eta\|^2 \\ + \frac{(1+\alpha)(1+\beta)}{2\alpha\beta} C_{F\Omega}^4 \|\operatorname{div}\tau + \mathsf{v} + f\|^2 + \frac{a}{2}\|\mathsf{v} - \mathsf{u}^d\|^2.$$

Finding the sharpest upper bound for cost functionals requires the minimization of $\overline{J}_1$ (or $\overline{J}_2$) over $\eta, \tau, \mathsf{v}, \alpha$, and β, where the variables are taken in the above-stated sets and are formally independent. Below, we show (with the paradigm of the majorant $\overline{J}_1$) that the amount of independent variables can be reduced.

It is easy to observe that the minimization of $\overline{J}_1$ with respect to v is equivalent to the problem

$$\inf_{\mathsf{v} \in \mathsf{U}} \mathcal{M}(\alpha, \beta; \tau, \mathsf{v}) = \widehat{\mathcal{M}}(\alpha, \beta; \tau),$$

where

$$\mathcal{M}(\alpha, \beta; \tau, \mathsf{v}) := \frac{C_{\alpha\beta}}{2}\|\operatorname{div}\tau + \mathsf{v} + f\|^2 + \frac{1}{2}\|\mathsf{v} - \mathsf{u}^d\|^2$$

and $C_{\alpha\beta} = C_{F\Omega}^2 \frac{(1+\alpha)(1+\beta)}{\alpha\beta}$. This problem is reduced to the minimization of the integrand of $\mathcal{M}$ at almost all $x \in \Omega$. If no constraints are imposed on the control function (i.e., $\mathsf{U} = L^2(\Omega)$), then the respective minimizer $\widehat{\mathsf{v}}$ is easy to find. It satisfies the relation

$$\widehat{\mathsf{v}}(x) = \frac{1}{C_{\alpha\beta} + a}\left(a\mathsf{u}^d(x) - C_{\alpha\beta}(\operatorname{div}\tau(x) + f(x))\right),$$

which implies

$$\widehat{\mathcal{M}}(\alpha, \beta; \tau) = \frac{C_{\alpha\beta}\, a}{2(C_{\alpha\beta} + a)}\|\operatorname{div}\tau + \mathsf{u}^d + f\|^2. \tag{9.4.9}$$

If U contains a finite constraint ψ, then

$$\widehat{\mathsf{v}}(x) = \begin{cases} \widehat{\mathsf{v}}(x) & \text{if } x \in \Omega_0, \\ \psi(x) & \text{if } x \in \Omega_\psi, \end{cases}$$

where $\Omega_\psi := \{x \in \Omega \mid \widehat{\mathsf{v}}(x) > \psi(x)\}$ and $\Omega_0 := \Omega \setminus \Omega_\psi$. In this case,

$$\widehat{\mathcal{M}}(\alpha, \beta; \tau, \mathsf{v}) = \frac{C_{\alpha\beta}\, a}{2(C_{\alpha\beta} + a)} \|\operatorname{div} \tau + \mathsf{u}^d + f\|^2_{\Omega_0} + \frac{C_{\alpha\beta}}{2} \|\operatorname{div} \tau + \psi + f\|^2_{\Omega_\psi} + \frac{a}{2} \|\psi - \mathsf{u}^d\|^2_{\Omega_\psi}. \tag{9.4.10}$$

Hence,

$$\inf_{\mathsf{v} \in \mathsf{U}} \overline{J}_1(\alpha, \beta; \eta, \tau, \mathsf{v}) = \widehat{J}_1(\alpha, \beta; \eta, \tau), \tag{9.4.11}$$

where

$$\widehat{J}_1(\alpha, \beta; \eta, \tau) = \frac{1+\alpha}{2} \|\nabla \eta - \sigma^d\|^2 + \frac{(1+\alpha)(1+\beta)}{2\alpha} \|\tau - \nabla \eta\|^2 + \widehat{\mathcal{M}}(\alpha, \beta; \tau, \mathsf{v}).$$

Then, we conclude that the problem of finding the sharpest upper bound for the cost functional can be formally reduced to the following minimization problem:

$$\inf_{\substack{\eta \in V_0, \tau \in H(\Omega, \operatorname{div}), \\ \alpha, \beta \in \mathbb{R}_+}} \widehat{J}_1(\alpha, \beta; \eta, \tau). \tag{9.4.12}$$

Lower bounds. Suppose $\sigma^d = \nabla \eta^d$, where $\eta^d \in V_0$. Then J_1 has the form

$$J_1(\eta, \mathsf{v}) := \frac{1}{2} \|\nabla(\eta - \eta^d)\|^2 + \frac{a}{2} \|\mathsf{v} - \mathsf{u}^d\|^2. \tag{9.4.13}$$

We note that if σ^d does not have such a form, then the optimization problem can be reduced to the above-considered case. Indeed, let $\widehat{\eta}^d$ be the projection of σ^d onto V_0, i.e.,

$$\int_\Omega (\nabla \widehat{\eta}^d - \sigma^d) \cdot \nabla w \, dx = 0, \qquad \forall w \in V_0.$$

Then

$$\|\nabla \eta - \sigma^d\|^2 = \|\nabla \eta - \nabla \widehat{\eta}^d\|^2 + \|\nabla \widehat{\eta}^d - \sigma^d\|^2$$

and

$$J(\eta, \mathsf{u}) = \frac{1}{2} \|\nabla \eta - \nabla \widehat{\eta}^d\|^2 + \frac{a}{2} \|\mathsf{u} - \mathsf{u}^d\|^2 + c,$$

where $c = \|\nabla \widehat{\eta}^d - \sigma^d\|^2$ is the distance from η^d to the set V_0. Thus, the cost functional can be reduced to the form (9.4.13).

We derive a lower bound for the functional $J_1(\eta_{\mathsf{v}}, \mathsf{v})$. For any $\eta \in V_0$, we have

$$\begin{aligned}
J_1(\eta_{\mathsf{v}}, \mathsf{v}) &:= \frac{1}{2}\|\nabla(\eta_{\mathsf{v}} - \eta)\|^2 + \frac{1}{2}\|\nabla(\eta - \eta^d)\|^2 \\
&\quad + \int_\Omega \nabla(\eta_{\mathsf{v}} - \eta) \cdot \nabla(\eta - \eta^d)\, dx + \frac{a}{2}\|\mathsf{v} - \mathsf{u}^d\|^2 \\
&= \frac{1}{2}\|\nabla(\eta_{\mathsf{v}} - \eta)\|^2 + \frac{1}{2}\|\nabla(\eta - \eta^d)\|^2 + \int_\Omega (f + \mathsf{v})(\eta - \eta^d)\, dx \\
&\quad - \int_\Omega \nabla\eta \cdot \nabla(\eta - \eta^d)\, dx + \frac{a}{2}\|\mathsf{v} - \mathsf{u}^d\|^2.
\end{aligned}$$

Hence,

$$\begin{aligned}
J_1(\eta_{\mathsf{u}}, \mathsf{u}) &= \inf_{\mathsf{v} \in \mathsf{U}} J_1(\eta_{\mathsf{v}}, \mathsf{v}) \\
&\geq \frac{1}{2}\|\nabla(\eta - \eta^d)\|^2 + \int_\Omega (f(\eta - \eta^d) - \nabla\eta \cdot \nabla(\eta - \eta^d))\, dx \\
&\quad + \inf_{\mathsf{v} \in \mathsf{U}} \left\{ \int_\Omega \mathsf{v}(\eta - \eta^d)\, dx + \frac{a}{2}\|\mathsf{v} - \mathsf{u}^d\|^2 \right\}.
\end{aligned} \tag{9.4.14}$$

Note that

$$\inf_{\mathsf{v} \in \mathsf{U}} \left\{ \int_\Omega g\mathsf{v}\, dx + \frac{a}{2}\|\mathsf{v} - \mathsf{u}^d\|^2 \right\} = \int_\Omega \mathcal{H}(a, \mathsf{u}^d, \psi, g)\, dx,$$

where

$$\mathcal{H}(a, \mathsf{u}^d, \psi, g)\, dx := \begin{cases} \mathsf{u}^d g - \dfrac{1}{2a} g^2 & \text{if } \widehat{\mathsf{v}} := \mathsf{u}^d - \dfrac{g}{a} \leq \psi, \\ \psi g + \dfrac{a}{2}(\psi - \mathsf{u}^d)^2 & \text{if } \widehat{\mathsf{v}} > \psi. \end{cases}$$

Now we obtain the following lower bound for the cost functional, which involves only known functions:

$$\begin{aligned}
J_1(\eta_{\mathsf{u}}, \mathsf{u}) \geq \underline{J}_1(\eta) &:= \frac{1}{2}\|\nabla(\eta - \eta^d)\|^2 + \int_\Omega (f(\eta - \eta^d) - \nabla\eta \cdot \nabla(\eta - \eta^d))\, dx \\
&\quad + \int_\Omega \mathcal{H}(a, \mathsf{u}^d, \psi, \eta - \eta^d)\, dx.
\end{aligned} \tag{9.4.15}$$

Assume that $\mathsf{U} = L^2(\Omega)$. Then, it is easy to show that the minorant is sharp. Indeed,

$$\begin{aligned}
\sup_{\eta \in V_0} \underline{J}_1(\eta) \geq \underline{J}_1(\eta_{\mathsf{u}}) &= \frac{1}{2}\|\nabla(\eta_{\mathsf{u}} - \eta^d)\|^2 + \int_\Omega (f(\eta_{\mathsf{u}} - \eta^d) - \nabla\eta_{\mathsf{u}} \cdot \nabla(\eta_{\mathsf{u}} - \eta^d))\, dx \\
&\quad + \int_\Omega \mathcal{H}(a, \mathsf{u}^d, \psi, \eta_{\mathsf{u}} - \eta^d)\, dx,
\end{aligned}$$

where

$$\mathcal{H}(a, \mathsf{u}^d, \psi, \eta_\mathsf{u} - \eta^d) = u^d(\eta_\mathsf{u} - \eta^d) - \frac{1}{2a}|\eta_\mathsf{u} - \eta^d|^2.$$

It is easy to prove (e.g., see [221]) that the corresponding solution of the optimal control problem satisfies the necessary condition

$$\mathsf{u} = \mathsf{u}^d + \frac{1}{a}(\eta^d - \eta_\mathsf{u}). \tag{9.4.16}$$

Therefore,

$$\begin{aligned}\int_\Omega (f(\eta_\mathsf{u} - \eta^d) - \nabla\eta_\mathsf{u} \cdot \nabla(\eta_\mathsf{u} - \eta^d))\,dx + \int_\Omega \mathcal{H}(a, \mathsf{u}^d, \psi, \eta_\mathsf{u} - \eta^d)\,dx \\ = \int_\Omega \left((\mathsf{u}^d - \mathsf{u})(\eta_\mathsf{u} - \eta^d) - \frac{1}{2a}(\eta_\mathsf{u} - \eta^d)^2\right) dx = \frac{a}{2}\|\mathsf{u} - \mathsf{u}^d\|^2.\end{aligned}$$

Thus, $J_1(\eta_\mathsf{u}, \mathsf{u}) = \underline{J}_1(\eta_\mathsf{u})$.

To find a more accurate lower bound for the cost functional, we estimate the first term on the right-hand side of (9.4.13) with the help of (3.1.13), which reads

$$\frac{1}{2}\|\nabla(\eta_\mathsf{v} - \eta)\|^2 \geq \int_\Omega \left(-\frac{1}{2}|\nabla w|^2 - \nabla w \cdot \nabla\eta + (\mathsf{v} + f)w\right) dx, \quad \forall w \in V_0. \tag{9.4.17}$$

This way results in a more complicated estimate:

$$\begin{aligned}J_1(\eta_\mathsf{u}, \mathsf{u}) \geq \underline{J}_1(\eta, w) := &-\frac{1}{2}\|\nabla w\|^2 + \frac{1}{2}\|\nabla(\eta - \eta^d)\|^2 \\ &+ \int_\Omega (f(w + \eta - \eta^d) - \nabla\eta \cdot \nabla(w + \eta - \eta^d))\,dx \\ &+ \inf_{\mathsf{v} \in \mathsf{U}} \left\{\int_\Omega \mathsf{v}(w + \eta - \eta^d)\,dx + \frac{a}{2}\|\mathsf{v} - \mathsf{u}^d\|^2\right\},\end{aligned} \tag{9.4.18}$$

where the last term is equal to $\int_\Omega \mathcal{H}(a, \mathsf{u}^d, \psi, w + \eta - \eta^d)\,dx$.

Remark 9.13. It should be noted that this estimate contains an additional function w, which makes computations of the lower more expensive with respect to (9.4.15). Numerical experiments performed (in part they are cited in [142, 143]) have shown that in most cases, specifications computed with the help of w are not very essential.

9.4.2 Estimates for state and control functions

Now our goal is to derive guaranteed upper bounds for the errors of v and η measured in terms of a *combined norm*

$$|[\mathsf{u} - \mathsf{v}]|^2 := \frac{1}{2}\|\nabla(\eta_\mathsf{u} - \eta_\mathsf{v})\|^2 + \frac{a}{2}\|\mathsf{u} - \mathsf{v}\|^2.$$

Our analysis is based upon the following result, which can be viewed as a generalization of the Mikhlin estimate (2.3.1) for the class of optimal control problems that we consider.

Theorem 9.14. *Let* $\mathsf{U} = L^2(\Omega)$. *For any control function* $\mathsf{v} \in \mathsf{U}$,

$$|[\mathsf{u}-\mathsf{v}]|^2 = J_1(\eta_{\mathsf{v}}, \mathsf{v}) - J_1(\eta_{\mathsf{u}}, \mathsf{u}). \tag{9.4.19}$$

Proof. We have

$$\begin{aligned} J(\eta_{\mathsf{v}}, \mathsf{v}) - J(\eta_{\mathsf{u}}, \mathsf{u}) &= \frac{1}{2}\|\nabla(\eta_{\mathsf{v}} - \eta_{\mathsf{u}})\|^2 + \frac{a}{2}\|\mathsf{v}-\mathsf{u}\|^2 \\ &\quad + \int_\Omega \nabla(\eta_{\mathsf{u}} - \eta^d)\cdot\nabla(\eta_{\mathsf{v}} - \eta_{\mathsf{u}})\,dx \\ &\quad + a\int_\Omega (\mathsf{u}-\mathsf{u}^d)(\mathsf{v}-\mathsf{u})\,dx. \end{aligned} \tag{9.4.20}$$

Note that

$$\int_\Omega \nabla(\eta_{\mathsf{v}} - \eta_{\mathsf{u}})\cdot\nabla(\eta_{\mathsf{u}} - \eta^d)\,dx = \int_\Omega (\mathsf{v}-\mathsf{u})(\eta_{\mathsf{u}} - \eta^d)\,dx.$$

By (9.4.16), we know that

$$(\eta_{\mathsf{u}} - \eta^d) + a(\mathsf{u}-\mathsf{u}^d) = 0. \tag{9.4.21}$$

In view of this relation,

$$\int_\Omega \nabla(\eta_{\mathsf{v}} - \eta_{\mathsf{u}})\cdot\nabla(\eta_{\mathsf{u}} - \eta^d)\,dx = -a\int_\Omega (\mathsf{v}-\mathsf{u})(\mathsf{u}-\mathsf{u}^d)\,dx$$

and the last two terms in (9.4.20) vanish. Hence, we arrive at (9.4.19). □

Corollary 9.15. *From (9.4.6), (9.4.14), and (9.4.19), it follows that*

$$|[\mathsf{v}-\mathsf{u}]|^2 \le \overline{\mathfrak{M}}_{\mathrm{opt}}(\alpha,\beta,\eta,\tau,\mathsf{v}) := \overline{J}_1(\alpha,\beta;\eta,\tau,\mathsf{v}) - \underline{J}_1(\eta), \tag{9.4.22}$$

where v *is an arbitrary control function in* $L^2(\Omega)$, $\tau \in H(\Omega, \mathrm{div})$, α *and* β *are arbitrary positive numbers, and*

$$\begin{aligned} \overline{\mathfrak{M}}_{\mathrm{opt}}(\alpha,\beta,\eta,\tau,\mathsf{v}) &:= \frac{\alpha}{2}\|\nabla(\eta-\eta^d)\|^2 + \frac{(1+\alpha)(1+\beta)}{2\alpha}\|\tau - \nabla\eta\|^2 \\ &\quad + \frac{(1+\alpha)(1+\beta)}{2\alpha\beta}C^2_{F\Omega}\,\|\mathrm{div}\,\tau + \mathsf{v} + f\|^2 + \frac{1}{2}\|\mathsf{v}-\mathsf{u}^d\|^2 \\ &\quad - \int_\Omega ((f+\mathsf{u}^d)(\eta-\eta^d) - \nabla\eta\cdot\nabla(\eta-\eta^d))\,dx \\ &\quad + \frac{1}{2a}\|\eta-\eta^d\|^2\,dx. \end{aligned}$$

Proposition 9.16. *The majorant* $\overline{\mathfrak{M}}_{\rm opt}(\alpha,\beta,\eta,\tau,\mathsf{v})$ *attains the exact lower bound on the exact solution of the optimal control problem, i.e.,*

$$\inf_{\alpha,\beta>0} \overline{\mathfrak{M}}_{\rm opt}(\alpha,\beta,\eta_{\mathsf{u}},\nabla\eta_{\mathsf{u}},\mathsf{u}) = 0.$$

Proof. In view of (9.4.21),

$$\int_\Omega ((f+\mathsf{u}^d)(\eta_{\mathsf{u}}-\eta^d) - \nabla\eta_{\mathsf{u}}\cdot\nabla(\eta_{\mathsf{u}}-\eta^d))\,dx$$
$$= \int_\Omega ((\mathsf{u}-\mathsf{u}^d)(\eta^d-\eta_{\mathsf{u}}) = \frac{1}{a}\|\eta^d-\eta_{\mathsf{u}}\|^2 \qquad (9.4.23)$$

and

$$\frac{1}{2}\|\mathsf{u}-\mathsf{u}^d\|^2 = \frac{1}{2a}\|\eta^d-\eta_{\mathsf{u}}\|^2.$$

For this reason, the last three terms of $\overline{\mathfrak{M}}_{\rm opt}(\alpha,\beta,\eta,\tau,\mathsf{v})$ vanish. The second term is also equal to zero, as well as the third one. Thus,

$$\overline{\mathfrak{M}}_{\rm opt}(\alpha,\beta,\eta_{\mathsf{u}},\nabla\eta_{\mathsf{u}},\mathsf{u}) = \frac{\alpha}{2}\|\nabla(\eta_{\mathsf{u}}-\eta^d)\|^2$$

and the result follows if we let α go to zero. □

Remark 9.17. Finally, we note that it may be useful to represent the basic problem in another (but equivalent) form. For a function $\zeta \in V_0$, consider the following problem: Minimize

$$J_{1,\zeta}(\eta,\mathsf{u}) := \frac{1}{2}\|\nabla(\eta-\zeta)\|^2 + \frac{a}{2}\|\mathsf{u}-\widetilde{u}^d(\zeta)\|^2 \qquad (9.4.24)$$

over $(\eta,\mathsf{u}) \in V_0 \times L^2(\Omega)$ such that

$$-\Delta\eta = \mathsf{u}+f \quad \text{a.e. in } \Omega, \qquad (9.4.25)$$

where $\widetilde{u}^d(\zeta) = \mathsf{u}^d + \frac{1}{a}(\eta^d-\zeta)$.

In [143], it is shown that for any $\eta \in V_0$,

$$J_1(\eta,\mathsf{u}) = J_{1,\zeta}(\eta,\mathsf{u}) + C_\zeta, \qquad (9.4.26)$$

where

$$C_\zeta := C(\zeta;\eta^d,f,\mathsf{u}^d) = \frac{1}{2}\|\nabla\zeta-\nabla\eta^d\|^2 + \frac{a}{2}\|\widetilde{u}^d(\zeta)-\mathsf{u}^d\|^2$$
$$+\left(\int_\Omega \nabla\zeta\cdot\nabla(\eta^d-\zeta)\,dx - \int_\Omega (\widetilde{u}^d(\zeta)+f)(\eta^d-\zeta)\,dx\right).$$

We outline that C_ζ depends only on ζ and known functions η^d, f, and u^d. This constant gives a quantity, which is contained in the upper bound of the cost functional, as well as in the lower one. In practice, it is convenient to reformulate the problem in such a way that this a priori known constant is sufficiently large. Then the estimates based on the comparison of upper and lower estimates of the cost functional would become more efficient (see [143] for details and numerical experiments).

9.4.3 Estimate in a combined norm

Introduce the following combined norm:

$$|[\mathsf{v}; q]|_\lambda^2 := |[\mathsf{v}]|^2 + \frac{1}{2}\|q\|^2 + \lambda\|\operatorname{div} q\|^2,$$

where $(\mathsf{v}, q) \in L^2(\Omega) \times H(\Omega, \operatorname{div})$ and $\lambda \in (0, a)$. This norm can be regarded as a full primal-dual norm associated with the problem under consideration.

Proposition 9.18.

$$|[(\mathsf{v} - \mathsf{u}); (\tau - p)]|_\lambda^2 \leq c_\oplus \overline{\mathfrak{M}}_{\text{opt}}(\alpha, \beta, \eta, \tau, \mathsf{v}), \tag{9.4.27}$$

where $p := \nabla\eta_{\mathsf{u}}$ *and*

$$c_\oplus := 3 + \tfrac{2\alpha}{(1+\alpha)(1+\beta)} \max\left\{6, \tfrac{\beta}{C_{F\Omega}^2}\left(4C_{F\Omega}^2 + \tfrac{\lambda a}{a-\lambda}\right)\right\}.$$

Proof. By the obvious inequality

$$\begin{aligned}\frac{1}{2}\|\tau - p\|^2 = \frac{1}{2}\|\tau - \nabla\eta_{\mathsf{u}}\|^2 &\leq \|\nabla(\eta_{\mathsf{v}} - \eta_{\mathsf{u}})\|^2 + \|\nabla\eta_{\mathsf{v}} - \tau\|^2\\ &= 2|[\mathsf{v} - \mathsf{u}]|^2 - a\|\mathsf{v} - \mathsf{u}\|^2 + \|\nabla\eta_{\mathsf{v}} - \tau\|^2\end{aligned} \tag{9.4.28}$$

and (9.4.22), we find that

$$\frac{1}{2}\|\tau - p\|^2 \leq 2\overline{\mathfrak{M}}_{\text{opt}}(\alpha, \beta, \eta, \tau, \mathsf{v}) - a\|\mathsf{v} - \mathsf{u}\|^2 + \|\nabla\eta_{\mathsf{v}} - \tau\|^2. \tag{9.4.29}$$

For any positive λ and μ, we have

$$\begin{aligned}\lambda\|\operatorname{div}\tau - \operatorname{div} p\| &= \lambda\|\operatorname{div}\tau + \mathsf{u} + f\|^2\\ &\leq \lambda\left((1+\mu)\|\operatorname{div}\tau + \mathsf{v} + f\|^2 + \frac{1+\mu}{\mu}\|\mathsf{v} - \mathsf{u}\|^2\right).\end{aligned} \tag{9.4.30}$$

Setting $\lambda\frac{1+\mu}{\mu} = a$, we find that $\mu = \frac{\lambda}{a-\lambda}$. Now (9.4.29) and (9.4.30) imply the estimate

$$\begin{aligned}\frac{1}{2}\|\tau - p\|^2 + \lambda\|\operatorname{div}(\tau - p)\|^2 &\leq 2\overline{\mathfrak{M}}_{\text{opt}}(\alpha, \beta, \eta, \tau, \mathsf{v})\\ &\quad + \frac{\lambda a}{a-\lambda}\|\operatorname{div}\tau + \mathsf{v} + f\|^2 + \|\nabla\eta_{\mathsf{v}} - \tau\|^2,\end{aligned}$$

which, together with (9.4.22), yields

$$|[(\mathsf{u}-\mathsf{u});(\tau-p)]|_\lambda^2 \le 3\overline{\mathfrak{M}}_{\mathrm{opt}}(\alpha,\beta,\eta,\tau,\mathsf{v}) + \frac{\lambda a}{a-\lambda}\|\operatorname{div}\tau+\mathsf{v}+f\|^2 + \|\nabla\eta_{\mathsf{v}}-\tau\|^2. \quad (9.4.31)$$

For the last term, we have

$$\begin{aligned}\|\nabla\eta_{\mathsf{v}}-\tau\|^2 &\le 2\|\nabla(\eta_{\mathsf{v}}-\eta)\|^2 + 2\|\nabla\eta-\tau\|^2\\ &\le 2(\|\nabla\eta-\tau\| + C_{F\Omega}\|\operatorname{div}\tau+\mathsf{v}+f\|)^2 + 2\|\nabla\eta-\tau\|^2\\ &\le 6\|\nabla\eta-\tau\|^2 + 4C_{F\Omega}^2\|\operatorname{div}\tau+\mathsf{v}+f\|^2.\end{aligned}$$

By (9.4.31), we obtain

$$|[(\mathsf{u}-\mathsf{u});(\tau-p)]|_\lambda^2 \le 3\overline{\mathfrak{M}}_{\mathrm{opt}}(\alpha,\beta,\eta,\tau,\mathsf{v}) + \left(4C_{F\Omega}^2 + \frac{\lambda a}{a-\lambda}\right)\|\operatorname{div}\tau+\mathsf{v}+f\|^2 + 6\|\nabla\eta-\tau\|^2.$$

Recalling the structure of $\overline{\mathfrak{M}}_{\mathrm{opt}}(\alpha,\beta,\eta,\tau,\mathsf{v})$, we arrive at (9.4.27). □

Remark 9.19. In the proof, we overestimated the right-hand side of inequalities several times. Therefore, in reality the constant $c_\oplus$ is essentially smaller than it is defined by Proposition 9.18.

Remark 9.20. With the help of similar arguments, it can be shown that

$$|[(\mathsf{v}-\mathsf{u});(\tau-p)]|_\lambda^2 \ge c_\ominus\overline{\mathfrak{M}}_{\mathrm{opt}}(\alpha,\beta,\eta,\tau,\mathsf{v}),$$

where $c_\ominus$ depends only on the data of the problem.

9.4.4 Generalizations

As we have seen, error majorants of the functional type allow to consider state equations of optimal control problems in the form of penalty functionals. This method yields computable upper bounds for cost functionals of many other optimal control problems (provided that the majorant for the state problem is known). Below we briefly consider several examples related to nonlinear problems.

1. Consider the problem with the cost functional J_1 in which the state function is defined not by (9.4.1) and (9.4.2) but by the *variational inequality* (see 8.1.1)

$$\int_\Omega \nabla\eta_{\mathsf{v}}\cdot\nabla(\eta-\eta_{\mathsf{v}})\,dx \ge \int_\Omega(\mathsf{v}+f)(\eta-\eta_{\mathsf{v}})\,dx, \qquad \forall\eta\in K, \quad (9.4.32)$$

where

$$K := \{\eta\in V_0 := \overset{\circ}{H}{}^1(\Omega) \mid \phi(x)\le\eta(x)\le\psi(x) \text{ a.e. in } \Omega\}.$$

Now the problem is to minimize $J_1(\eta_{\mathsf{v}}, \mathsf{v})$ on $(\eta_{\mathsf{v}}, \mathsf{v}) \in K \times \mathsf{U}$. We use (9.4.3) and estimate the term $\|\nabla(\eta_{\mathsf{v}} - \eta)\|$ by the majorant derived in Section 8.1. Instead of (9.4.4), we obtain

$$J_1(\eta_{\mathsf{v}}, \mathsf{v}) \leq \overline{J}_1(\eta, \tau, \mathsf{v}) := \frac{a}{2}\|\mathsf{v} - \mathsf{u}^d\|^2 + \frac{1}{2}\Big(\|\nabla\eta - \sigma^d\| + \|\tau - \nabla\eta\| + C_{F\Omega}\| \lessdot \operatorname{div} y + \mathsf{v} + f \gtrdot_\eta \|\Big)^2, \tag{9.4.33}$$

where $y \in H(\Omega, \operatorname{div})$. For any $(\eta_{\mathsf{v}}, \mathsf{v}) \in K \times \mathsf{U}$, the quantity $J_1(\eta_{\mathsf{v}}, \mathsf{v})$ is computable and gives a guaranteed upper bound for the cost functional.

Also, we can derive a computable lower bound. Suppose $\sigma^d = \nabla\eta^d$, where $\eta^d \in K$. Let $\mathsf{v} \in \mathsf{U}$ be an approximation of u and η be an approximation of η_{v}. Then,

$$\begin{aligned}
J_1(\eta_{\mathsf{v}}, \mathsf{v}) &:= \frac{1}{2}\|\nabla(\eta_{\mathsf{v}} - \eta)\|^2 + \frac{1}{2}\|\nabla(\eta - \eta^d)\|^2 \\
&\quad + \int_\Omega \nabla(\eta_{\mathsf{v}} - \eta)\cdot\nabla(\eta - \eta^d)\,dx + \frac{a}{2}\|\mathsf{v} - \mathsf{u}^d\|^2 \\
&\geq \frac{1}{2}\|\nabla(\eta - \eta^d)\|^2 + \frac{a}{2}\|\mathsf{v} - \mathsf{u}^d\|^2 - \|\nabla(\eta_{\mathsf{v}} - \eta)\|\|\nabla(\eta - \eta^d)\| \\
&\geq \frac{1-\beta}{2}\|\nabla(\eta - \eta^d)\|^2 + \frac{a}{2}\|\mathsf{v} - \mathsf{u}^d\|^2 \\
&\quad - \frac{1}{2\beta}\left(\|\tau - \nabla\eta\| + C_{F\Omega}\| \lessdot \operatorname{div}\tau + \mathsf{v} + f \gtrdot_\eta \|\right)^2.
\end{aligned} \tag{9.4.34}$$

The right-hand side of (9.4.34) can be maximized with respect to $\mathsf{v} \in \mathsf{U}$ by the same method, which we applied to the optimal control problem with linear state equation. Then we obtain a computable lower bound for the cost functional. The quality of this lower bound depends on the value of approximation error for the state function.

2. The method is extendable to optimal control problems with *convex cost functionals* and also to those that have the control function in the *main part* of the differential operator (in the latter case, the existence of u requires a special investigation). If the cost functional $J(\eta, v)$ is convex with respect to the first variable for any admissible v, them we have the relations

$$J(\eta_{\mathsf{v}}, \mathsf{v}) \leq \lambda J\left(\frac{\eta}{\lambda}, \mathsf{v}\right) + (1-\lambda)J\left(\frac{\eta_{\mathsf{v}} - \eta}{1-\lambda}, \mathsf{v}\right), \tag{9.4.35}$$

$$J\left(\frac{\eta_{\mathsf{v}}}{\lambda}, \mathsf{v}\right) \geq \frac{1}{\lambda}J(\eta, \mathsf{v}) - \frac{(1-\lambda)}{\lambda}J\left(\frac{\eta - \eta_{\mathsf{v}}}{1-\lambda}, \mathsf{v}\right), \tag{9.4.36}$$

where $\lambda \in (0, 1)$. Assume that J satisfies the growth condition

$$J(\eta, \mathsf{v}) \leq C(\mathsf{v})\|\eta\|_V^\alpha, \tag{9.4.37}$$

where $C > 0$ does not depend on η and V is the energy space of the boundary value problem (which defines the state function)

$$A(\mathsf{v})\eta_{\mathsf{v}} = f(\mathsf{v}) \tag{9.4.38}$$

generated by an elliptic operator A. If for the problem (9.4.38) we have a computable upper bound of the deviation from η_{v}

$$\|\eta_{\mathsf{v}} - \eta\|_V \leq \overline{\mathfrak{M}}\,(\eta, A(\mathsf{v}), f(\mathsf{v}), D), \tag{9.4.39}$$

then (9.4.35), (9.4.37), and (9.4.39) imply a computable upper bound for the cost functional, namely,

$$J(\eta_{\mathsf{v}}, \mathsf{v}) \leq \inf_{\lambda\in(0,1)} \left\{\lambda J\left(\frac{\eta}{\lambda}, \mathsf{v}\right) + (1-\lambda)^{1-\alpha} C(\mathsf{v})\overline{\mathfrak{M}}^{\alpha}(\eta, A(\mathsf{v}), f(\mathsf{v}), D)\right\}. \tag{9.4.40}$$

A lower bound follows from (9.4.36). However, in general, getting computable lower bounds for cost functionals is a more difficult task. Usually it requires a deeper analysis, which attracts specific properties of the problem considered.

9.4.5 Comments

For the reader interested in other approaches to a posteriori estimation for optimal control problems (and in the adaptive numerical methods developed for such problems), we recommend the papers by R. Becker and R. Rannacher [49], R. Becker, H. Kapp, and R. Rannacher [47], M. Hintermüller [173], R. H. W. Hoppe, Y. Iliash, C. Iyyunni, and N. H. Sweilam [177], A. Gaevskaya, R. H. W. Hoppe, Y. Iliash, and M. Kieweg [141], D. Meidner and B. Vexler [230], and R. Becker and B. Vexler [50]. These papers also contain an overview of the results in the area and many references.

9.5 Estimates for nonconforming approximations

Let Ω be divided into a collection of subdomains Ω_i, $i = 1, 2, \ldots, N$, and

$$\overline{\Omega} = \bigcup_i \overline{\Omega}_i.$$

We consider nonconforming approximations that may violate the continuity on the boundaries of subdomains Ω_i and the boundary conditions on Γ_1. The corresponding functions are marked by "hats" and form a *broken* Sobolev space

$$\widehat{H}^1 := \left\{\widehat{w} \in L^2(\Omega) | \widehat{w} \in H^1(\Omega_i), \quad i = 1, 2, \ldots, N\right\},$$

the norm of which is defined by the relation

$$|[w]|^2 = \sum_i \|w\|^2_{1,2,\Omega_i}.$$

Also, we define the following norms:

$$|||q|||^2_{\Omega_i} := \int_{\Omega_i} Aq \cdot q\,dx, \qquad |||q|||^2_{*,\Omega_i} := \int_{\Omega_i} A^{-1}q \cdot q\,dx,$$

$$|||[q]|||^2 := \sum_i |||q|||^2_{\Omega_i}, \qquad |||[q]|||^2_* := \sum_i |||q|||^2_{*,\Omega_i}.$$

Guaranteed bounds of approximation errors generated by nonconforming approximations can be derived by two methods. The first method projects a nonconforming approximation into the energy space (with the help of a suitable post-processing procedure) and applies the functional error majorant to the post-processed approximation. Since the majorant is valid for any conforming approximation, this procedure implies computable and guaranteed error bounds. The second method is based on the Helmholtz decomposition of the error. Below we briefly discuss both methods with the paradigm of the problem (4.1.1)–(4.1.3).

9.5.1 Estimates based on projecting to the energy space

Let $\widehat{v} \in \widehat{V}$ be a nonconforming approximation of $u \in u_0 + V_0$ (we recall that V_0 contains functions vanishing on Γ_1). Probably the simplest modus operandi is to define

$$\widetilde{v} := \mathsf{P}(\widehat{v}),$$

where $\mathsf{P} : \widehat{V} \to u_0 + V_0$ is a projection operator. In particular, one can use the orthogonal projection to a certain finite-dimensional subspace of $u_0 + V_0$. If P produces

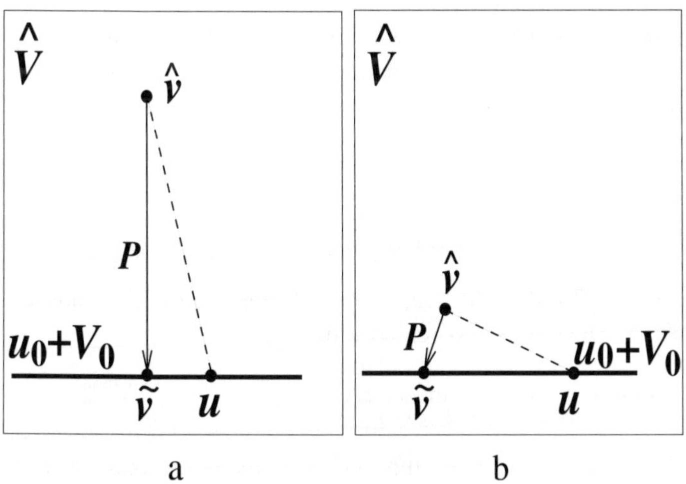

Figure 9.5.1 Projection to $u_0 + V_0$.

almost orthogonal projection (see Figure 9.5.1 a), then $\widetilde{v}$ is a better approximation of

u than $\widehat{v}$. In this case, it is logical to consider $\widetilde{v}$ as an approximate solution and use the estimates

$$\begin{aligned} &|||\nabla u - \nabla\widetilde{v}||| \leq \overline{\mathfrak{M}}_{\rm DF}(\widetilde{v}, y), && y \in H(\Omega, {\rm div}),\\ &|||\nabla u - \nabla\widetilde{v}||| \geq \underline{\mathfrak{M}}_{\rm DF}(\widetilde{v}, w), && w \in V_0, \end{aligned}$$

to evaluate the respective error. Similar estimates can easily be obtained for nonconforming approximations of other boundary value problems, where the corresponding functional error majorants are known.

However, orthogonal projection is equivalent to solving of an auxiliary problem, which may lead to essential expenditures. On the other hand, cheap post-processing procedures may destroy approximation properties of $\widehat{v}$, so that $\widetilde{v}$ is less accurate than $\widehat{v}$ (see Figure 9.5.1 b). In this case, we should estimate the error in terms of the "broken" norm $|||[\widehat{\nabla}\widehat{v} - \nabla u]|||$, in which $\widehat{\nabla}\widehat{v}$ is defined at almost all points of Ω by the relation

$$\widehat{\nabla}\widehat{v}(x) := \nabla\widehat{v}(x), \qquad x \in \Omega_i, \quad i = 1, 2, \ldots, N.$$

The simplest way to obtain such an estimate is as follows. For any $v \in u_0 + V_0$, we have the triangle inequality

$$|||[\nabla u - \widehat{\nabla}\widehat{v}]||| \leq |||[\nabla(u - v)]||| + |||[\widehat{\nabla}\widehat{v} - \nabla v]|||, \qquad \forall v \in u_0 + V_0. \tag{9.5.1}$$

Remark 9.21. It should be noted that $|||[\widehat{\nabla}\widehat{v}]|||$ is a seminorm on $\widehat{V}$. One can view this quantity as a factor norm defined for the class of functions that differ from each other by a constant, so that in (9.5.1) we can write $v \in u_0 + V_0 + \mathbb{R}$.

Consider the first term on the right-hand side of (9.5.1). Since $u - v \in V_0$, the broken norm $|||[\nabla(u - v)]|||$ coincides with the usual energy norm $|||\nabla(u - v)|||$, which is estimated from above by the error majorant. Moreover, the value of this norm does not change if we replace v by $v + c$ (where $c \in \mathbb{R}$). For this reason, we arrive at the estimate

$$|||[\nabla u - \widehat{\nabla}\widehat{v}]||| \leq \inf_{\substack{v \in u_0 + V_0\\ y \in H(\Omega, {\rm div})}} \left\{\overline{\mathfrak{M}}_{\rm DF}(v, y) + |||[\widehat{\nabla}\widehat{v} - \nabla v]|||\right\}. \tag{9.5.2}$$

If $v = u$ and $y = A\nabla u$, then $\overline{\mathfrak{M}}_{\rm DF}(v, y) = 0$. Therefore, (9.5.2) holds as the equality.

Let $\widetilde{v} := \mathsf{P}(\widehat{v})$. Then, (9.5.2) is replaced by

$$|||[\nabla u - \widehat{\nabla}\widehat{v}]||| \leq \inf_{y \in H(\Omega, {\rm div})} \overline{\mathfrak{M}}_{\rm DF}(\widetilde{v}, y) + |||[\widehat{\nabla}\widehat{v} - \nabla\widetilde{v}]|||, \tag{9.5.3}$$

where $y \in H(\Omega, {\rm div})$ should be used to minimize the right-hand side of (9.5.2).

Lower bounds of approximation errors are derived analogously. By the obvious inequality

$$|||[\nabla u - \widehat{\nabla}\widehat{v}]||| \geq |||[\nabla(u - v)]||| - |||[\widehat{\nabla}\widehat{v} - \nabla v]|||, \qquad \forall v \in u_0 + V_0, \tag{9.5.4}$$

we find that

$$|||[\nabla u - \widehat{\nabla}\widehat{v}]||| \geq \sup_{\substack{v \in u_0 + V_0 \\ w \in V_0}} \left\{ \underline{\mathfrak{M}}_{\mathrm{DF}}(v, w) - |||[\widehat{\nabla}\widehat{v} - \nabla v]||| \right\} \tag{9.5.5}$$

and, therefore,

$$|||[\nabla u - \widehat{\nabla}\widehat{v}]||| \geq \sup_{w \in V_0} \underline{\mathfrak{M}}_{\mathrm{DF}}(\widetilde{v}, w) - |||[\widehat{\nabla}\widehat{v} - \nabla\widetilde{v}]|||. \tag{9.5.6}$$

If the term $|||[\widehat{\nabla}\widehat{v} - \nabla\widetilde{v}]|||$ (related to the "nonconformity error") is large, then the estimates (9.5.3) and (9.5.6) may be not sharp. In this case, estimates of a somewhat different type can be helpful. They are obtained if instead of the triangle inequality we use another representation of the error.

9.5.2 Estimates based on the Helmholtz decomposition

Consider the error function

$$\eta := \nabla u - \widehat{\nabla}\widehat{v}.$$

Since $\widehat{\nabla}\widehat{v}$ is defined at almost all points of Ω and $\nabla\widehat{v} \in L^2(\Omega_i, \mathbb{R}^d)$ for $i = 1, \dots, N$, we can regard η as a vector-valued function in $L^2(\Omega, \mathbb{R}^d)$, for which the well-known Helmholtz decomposition takes place. For our purposes, it is more convenient to use a similar decomposition of $A\eta$, namely,

$$A\eta = A\nabla u_\eta + \tau_\eta, \tag{9.5.7}$$

where u_η is a function in V_0 and

$$\tau_\eta \in \mathrm{S}_{\Gamma_2} := \left\{ \tau \in L^2(\Omega, \mathbb{R}^d) \mid \int_\Omega \tau \cdot \nabla w \, dx = 0, \quad \forall w \in V_0 \right\}.$$

If $\tau \in \mathrm{S}_{\Gamma_2}$ then $\operatorname{div} \tau = 0$ and $\tau \cdot n = 0$ on Γ_2 (these relations should be understood in a generalized sense; they hold in the classical sense if τ is a sufficiently regular function).

The decomposition (9.5.7) is motivated by the problem: Find $u_\eta \in V_0$ such that

$$\int_\Omega A\nabla u_\eta \cdot \nabla w \, dx = \int_\Omega A\eta \cdot \nabla w \, dx, \qquad \forall w \in V_0, \tag{9.5.8}$$

which is uniquely solvable. From (9.5.8) it follows that

$$\int_\Omega \tau_\eta \cdot \nabla w \, dx = 0, \qquad \forall w \in V_0, \qquad \text{where } \tau_\eta := A(\eta - \nabla u_\eta), \tag{9.5.9}$$

which means that $\tau_\eta \in \mathrm{S}_{\Gamma_2}$ and we obtain (9.5.7).

By the definition of S_{Γ_2}, τ_η is orthogonal to ∇w for any $w \in V_0$. Therefore, we arrive at the important relation

$$
\begin{aligned}
|||[\eta]|||^2 &= \sum_i \int_{\Omega_i} A\eta \cdot \eta \, dx \\
&= \int_\Omega \left(A\nabla u_\eta \cdot \nabla u_\eta + A^{-1}\tau_\eta \cdot \tau_\eta - 2\nabla u_\eta \cdot \tau_\eta \right) dx \\
&= |||\nabla u_\eta|||^2 + |||\tau_\eta|||_*^2. \qquad (9.5.10)
\end{aligned}
$$

A posteriori estimates based on the Helmholtz type decomposition of η were studied in E. Dari, R. Duran, C. Padra, and V. Vampa [113] and M. Ainsworth [4, 5]. In those papers, τ_η is represented as curl $\boldsymbol{\psi}$ and $\boldsymbol{\psi}$ is defined by the relation

$$
\int_\Omega A^{-1} \operatorname{curl} \boldsymbol{\psi} \cdot \operatorname{curl} \boldsymbol{\phi} \, dx = \int_\Omega \eta \cdot \operatorname{curl} \boldsymbol{\phi} \, dx, \qquad \forall \boldsymbol{\phi} \in \mathcal{H},
$$

where $\mathcal{H}$ is the subspace of $H^1/\mathbb{R}$ that consists of functions having zero tangential derivatives on Γ_2. Hence, the overall error is represented by (9.5.10), where u_η and τ_η are defined as solutions of auxiliary boundary value problems.

Below we discuss a somewhat different modus operandi, which was suggested and numerically tested in S. Repin and S. Tomar [317]. First, we estimate u_η with the help of the same method as we used for deriving a posteriori error estimates (for conforming approximations) in previous chapters. This method provides computable two-sided bounds for the term $|||\nabla u_\eta|||$. After that, we show that the value of $|||\tau_\eta|||$ is estimated from above by the broken norm of the difference $\nabla v - \widehat{\nabla} \widehat{v}$ (i.e., by a "penalty for nonconformity"). The sum of these two estimates yields a directly computable bound of the error, expressed in terms of the broken energy norm.

Upper bound of $|||[\eta]|||$. First we find computable upper bounds for the norms $|||\nabla u_\eta|||$. Note that

$$
\begin{aligned}
\int_\Omega A\nabla u_\eta \cdot \nabla w \, dx &= \sum_i \int_{\Omega_i} A\eta \cdot \nabla w \, dx - \int_\Omega \tau_\eta \cdot \nabla w \, dx \\
&= \int_\Omega A\nabla u \cdot \nabla w \, dx - \sum_i \int_{\Omega_i} A\nabla \widehat{v} \cdot \nabla w \, dx \\
&= \int_\Omega f w \, dx + \int_{\Gamma_2} F w \, ds \\
&\quad - \sum_i \int_{\Omega_i} A\nabla \widehat{v} \cdot \nabla w \, dx, \qquad \forall w \in V_0. \qquad (9.5.11)
\end{aligned}
$$

Now, we rearrange the right-hand side of (9.5.11) by introducing a vector-valued function $y \in H(\Omega, \mathrm{div})$ and obtain

$$\int_\Omega A\nabla u_\eta \cdot \nabla w\, dx = \int_\Omega (\mathrm{div}\, y + f) w\, dx + \int_{\Gamma_2} (F - y \cdot n) w\, ds + \sum_i \int_{\Omega_i} (y - A\nabla\widehat{v}) \cdot \nabla w\, dx. \tag{9.5.12}$$

From (9.5.12), it follows that

$$|||\nabla u_\eta||| \le |||[y - A\widehat{\nabla}\widehat{v}]|||_* + \mathrm{C}\left(\|\mathrm{div}\, y + f\|^2 + \|y \cdot n - f\|^2_{\Gamma_2}\right)^{1/2}. \tag{9.5.13}$$

For the function τ_η, we have

$$\begin{aligned}\int_\Omega A^{-1}\tau_\eta \cdot \tau_\eta\, dx &= \sum_i \int_{\Omega_i} \eta \cdot \tau_\eta\, dx = \sum_i \int_{\Omega_i} (\nabla u - \nabla\widehat{v}) \cdot \tau_\eta\, dx \\ &= \sum_i \int_{\Omega_i} (\nabla v - \nabla\widehat{v}) \cdot \tau_\eta\, dx, \qquad \forall v \in u_0 + V_0.\end{aligned} \tag{9.5.14}$$

Hence,

$$|||\tau_\eta|||_*^2 \le \left(\sum_i \int_{\Omega_i} A(\nabla v - \nabla\widehat{v}) \cdot (\nabla v - \nabla\widehat{v})\, dx\right)^{1/2} |||\tau_\eta|||_*,$$

and we find that

$$|||\tau_\eta|||_* \le |||[\nabla v - \widehat{\nabla}\widehat{v}]|||. \tag{9.5.15}$$

By (9.5.10), (9.5.13), and (9.5.15), we deduce the estimate

$$\begin{aligned}|||[\nabla u - \widehat{\nabla}\widehat{v}]|||^2 \le \inf_{v \in u_0 + V_0} |||[\nabla v - \widehat{\nabla}\widehat{v}]|||^2 + \Big(|||[y - A\widehat{\nabla}\widehat{v}]|||_* \\ + \mathrm{C}\left(\|\mathrm{div}\, y + f\|^2 + \|y \cdot n - F\|^2_{\Gamma_2},\right)^{1/2}\Big)^2,\end{aligned} \tag{9.5.16}$$

which provides an upper bound of the error in terms of the broken energy norm.

Remark 9.22. The right-hand side of (9.5.16) presents a natural decomposition of the overall error into three terms: the error owing to nonconformity, the error in the duality relation for fluxes, and the error in the equilibrium equation and boundary condition for fluxes.

Remark 9.23. If y is subject to the boundary condition $y \cdot n = F$, then (9.5.16) takes a simplified form

$$|||[\eta]|||^2 \le |||[\nabla v - \widehat{\nabla}\widehat{v}]|||^2 + \left(|||[y - A\widehat{\nabla}\widehat{v}]||| + \mathrm{C}\,\|\mathrm{div}\, y + f\|\right)^2, \tag{9.5.17}$$

where v is an arbitrary function in $u_0 + V_0$. We set $v = \mathsf{P}(\widehat{v})$ (i.e., define v as a projection onto $u_0 + V_0$). Then,

$$|||[\eta]|||^2 \leq \eta_{\mathsf{P}}^2 + (1+\beta)|||[y - A\widehat{\nabla v}]|||^2 + \mathrm{C}^2\left(1 + \frac{1}{\beta}\right)\|\operatorname{div} y + f\|^2, \qquad (9.5.18)$$

where β is an arbitrary positive number and $\eta_{\mathsf{P}} := |||[\nabla \mathsf{P}(\widehat{v}) - \widehat{\nabla v}]|||$ is the projection error (which is directly computable). Minimization with respect to y is now reduced to a quadratic problem.

Finally, we consider a modification of (9.5.16). Let $\Gamma_{ij} = \partial\Omega_i \cap \partial\Omega_j$ and n_{ij} denote the unit normal vector to Γ_{ij} external to Ω_i if $i < j$. In (9.5.16), we assume that y belongs to $H(\Omega, \operatorname{div})$ (which means that $y \cdot n_{ij}$ is continuous on Γ_{ij}). However, $A\widehat{\nabla v}$ is a vector-valued function that may have jumps on Γ_{ij}. For this reason, it may be useful to have another form of the upper bound, which operates with y from a wider set. To deduce such an estimate, we transform the right-hand side of (9.5.11) as follows. Let

$$\widehat{y} \in \widehat{Y} := \left\{\widehat{y} = y^{(i)} \text{ in } \Omega_i, \quad y^{(i)} \in H(\Omega_i, \operatorname{div}),\ i = 1, \ldots, N\right\}.$$

Denote the "broken" divergence by $\widehat{\operatorname{div}}$ (i.e., $\widehat{\operatorname{div}}\, y^{(i)} = \operatorname{div} y^{(i)}$ in Ω_i).

For any $\sigma \in H(\Omega, \operatorname{div})$, we have

$$\begin{aligned}\int_\Omega (w\,\widehat{\operatorname{div}}\,\widehat{y} + \nabla w \cdot \widehat{y})\,dx &= \sum_i \int_{\Omega_i} (w \operatorname{div} y^{(i)} + \nabla w \cdot y^{(i)})\,dx \\ &= \int_\Omega (w\,\widehat{\operatorname{div}}\,(\widehat{y} - \sigma) + \nabla w \cdot (\widehat{y} - \sigma))\,dx + \int_{\Gamma_2} (\sigma \cdot n) w\,ds.\end{aligned}$$

Substitute this identity into (9.5.11). We have

$$\begin{aligned}\int_\Omega A\nabla u_\eta \cdot \nabla w\,dx = &\int_\Omega (f + \widehat{\operatorname{div}}\,\widehat{y}) w\,dx + \int_{\Gamma_2} (F - \sigma \cdot n) w\,ds \\ &+ \sum_i \int_{\Omega_i} (\widehat{y} - A\nabla\widehat{v}) \cdot \nabla w\,dx \\ &- \int_\Omega (w\,\widehat{\operatorname{div}}\,(\widehat{y} - \sigma) + \nabla w \cdot (\widehat{y} - \sigma))\,dx.\end{aligned}$$

Since

$$\left|\int_\Omega (w\,\widehat{\operatorname{div}}\,(\widehat{y} - \sigma) + \nabla w \cdot (\widehat{y} - \sigma))\,dx\right| \leq \left(C_{F\Gamma_1}\|\widehat{\operatorname{div}}\,(\widehat{y} - \sigma)\| + \|\widehat{y} - \sigma\|\right)\|\nabla w\|,$$

we find that

$$\begin{aligned}|||\nabla u_\eta||| \leq c_1^{-1}\big(&C_{F\Gamma_1}\|f + \widehat{\operatorname{div}}\,\widehat{y}\| + C_{T\Gamma_2}\|F - \sigma \cdot n\|_{\Gamma_2} \\ &+ C_{F\Gamma_1}\|\widehat{\operatorname{div}}\,(\widehat{y} - \sigma)\| + \|\widehat{y} - \sigma\|\big) + |||[\widehat{y} - A\widehat{\nabla v}]|||_*.\end{aligned}$$

Now, by (9.5.10) and (9.5.13) we obtain

$$
\begin{aligned}
|||[\nabla u - \widehat{\nabla}\widehat{v}]|||^2 \le & \inf_{v \in u_0 + V_0} |||[\nabla v - \widehat{\nabla}\widehat{v}]|||^2 \\
& + \Big(c_1^{-1} (C_{F\Gamma_1} \| f + \widehat{\operatorname{div}}\,\widehat{y} \| + C_{T\Gamma_2} \| F - \sigma \cdot n \|_{\Gamma_2} \\
& + C_{F\Gamma_1} \| \widehat{\operatorname{div}}\,(\widehat{y} - \sigma) \| + \| \widehat{y} - \sigma \|) + |||[\widehat{y} - A\widehat{\nabla}\widehat{v}]|||_* \Big)^2, \quad (9.5.19)
\end{aligned}
$$

where $\widehat{y}$ is any vector-valued function in $\widehat{Y}$.

From (9.5.19), we deduce the estimate

$$
\begin{aligned}
|||[\nabla u - \widehat{\nabla}\widehat{v}]|||^2 \le & \inf_{v \in u_0 + V_0} |||[\nabla v - \widehat{\nabla}\widehat{v}]|||^2 \\
& + \Big(c_1^{-1} C_{F\Gamma_1} \| f + \widehat{\operatorname{div}}\,\widehat{y} \| + |||[\widehat{y} - A\widehat{\nabla}\widehat{v}]|||_* \\
& + c_1^{-1} \inf_{\substack{\sigma \in H(\Omega, \operatorname{div}) \\ \sigma \cdot n = F \text{ on } \Gamma_2}} \Big(C_{F\Gamma_1} \| \widehat{\operatorname{div}}\,(\widehat{y} - \sigma) \| + \| \widehat{y} - \sigma \| \Big) \Big)^2, \quad (9.5.20)
\end{aligned}
$$

which operates with piecewise continuous approximations and fluxes. We note that the last term on the right-hand side of the above estimate can be viewed as a penalty for "nonconformity" of $\widehat{y}$.

Lower bound of $|||[\eta]|||$. To derive a lower bound of the error we again use (9.5.10). Now, our goal is to find lower bounds for the norms on the right-hand side of (9.5.10). To estimate the first term, we rewrite (9.5.11) in the form

$$
\int_\Omega A \nabla u_\eta \cdot \nabla w \, dx = \ell(w), \qquad (9.5.21)
$$

where $\ell : V_0 \to \mathbb{R}$ is a linear functional defined as

$$
\ell(w) = \int_{\Gamma_2} F w \, ds + \int_\Omega f w \, dx - \sum_i \int_{\Omega_i} A \nabla \widehat{v} \cdot \nabla w \, dx. \qquad (9.5.22)
$$

Note that (9.5.21) is the Euler's equation of the variational problem

$$
\min_{w \in V_0} J_\ell(w), \quad \text{where } J_\ell(w) = \frac{1}{2} |||\nabla w|||^2 - \ell(w). \qquad (9.5.23)
$$

From (9.5.21) it follows that

$$
|||\nabla u_\eta|||^2 = \langle \ell, u_\eta \rangle, \qquad (9.5.24)
$$

and, therefore,

$$
J_\ell(u_\eta) = -\frac{1}{2} |||\nabla u_\eta|||^2. \qquad (9.5.25)
$$

From (9.5.24) and (9.5.25), we obtain

$$\begin{aligned}|||\nabla u_\eta|||^2 &= -2J_\ell(u_\eta) = -2\inf_{w\in V_0}\frac{1}{2}\left\{|||\nabla w|||^2 - \ell(w)\right\}\\ &= \sup_{w\in V_0}\left\{2\ell(w) - |||\nabla w|||^2\right\}.\end{aligned}$$

Hence,

$$|||\nabla u_\eta|||^2 \geq 2\ell(w) - |||\nabla w|||^2, \qquad \forall w \in V_0. \tag{9.5.26}$$

For the second term in (9.5.10), we proceed analogously. In view of (9.5.7), τ_η meets the identity

$$\begin{aligned}\int_\Omega A^{-1}\tau_\eta\cdot\tau_0\,dx &= \int_\Omega (\eta - \nabla u_\eta)\cdot\tau_0\,dx\\ &= \int_\Omega (\nabla u - \widehat{\nabla}\widehat{v} - \nabla u_\eta)\cdot\tau_0\,dx, \qquad \forall \tau_0 \in \mathrm{S}_{\Gamma_2}.\end{aligned} \tag{9.5.27}$$

Introduce the functional $\mu : \mathrm{S}_{\Gamma_2} \to \mathbb{R}$,

$$\mu(\tau_0) := -\sum_i \int_{\Omega_i} (\nabla u_0 - \widehat{\nabla}\widehat{v})\cdot\tau_0\,dx. \tag{9.5.28}$$

Then, τ_η is a minimizer of the variational problem

$$\min_{\tau_0\in\mathrm{S}_{\Gamma_2}} I_\mu(\tau_0), \quad \text{where} \quad I_\mu(\tau_0) := \left\{\frac{1}{2}|||\tau_0|||_*^2 - \mu(\tau_0)\right\}. \tag{9.5.29}$$

By the arguments similar to those used before, we conclude that

$$\begin{aligned}|||\tau_\eta|||_*^2 &= -2I_\mu(\tau_\eta) = \sup_{\tau_0\in\mathrm{S}_{\Gamma_2}}\left\{2\mu(\tau_0) - |||\tau_0|||_*^2\right\}\\ &\geq 2\mu(\tau_0) - |||\tau_0|||_*^2, \qquad \forall \tau_0 \in \mathrm{S}_{\Gamma_2}.\end{aligned} \tag{9.5.30}$$

Combining (9.5.26) and (9.5.30), we obtain the following lower estimate of the error

$$|||[\eta]|||^2 \geq 2\ell(w) + 2\mu(\tau_0) - |||\nabla w|||^2 - |||\tau_0|||_*^2. \tag{9.5.31}$$

In (9.5.31), w and τ_0 are arbitrary functions in V_0 and S_{Γ_2}, respectively. Certainly, getting a realistic estimate requires a proper selection of these functions.

Remark 9.24. If $\widehat{v} \in u_0 + V_0$, then $\mu(\tau_0) \equiv 0$ and we should take $\tau_0 = 0$ to make the right-hand side of (9.5.31) maximal. In this case, (9.5.31) is transformed to the lower bound that was derived for conforming approximations in Chapter 4.

9.5.3 Accuracy of approximations obtained by the Trefftz method

Approximate solutions to boundary value problems can be constructed as series $\sum_i \alpha_i \phi_i(x)$, where the ϕ_i exactly satisfy the differential equation. The coefficients α_i are selected in order to approximate the Dirichlet boundary condition as accurately as possible (in the sense of least squares). Numerical methods of this type originate from the Trefftz method (e.g., see S. Mikhlin [232]). For the problem (4.1.1)–(4.1.3), the respective approximate solution $\widehat{u}$ satisfies the equation

$$\operatorname{div} A\nabla\widehat{u} + f = 0 \quad \text{in } \Omega,$$

but violates the condition $u = u_0$ on Γ_1. We can view such a function as a nonconforming approximation of u and apply (9.5.16). In this case,

$$\widehat{V} = H^1(\Omega) \quad \text{and} \quad |||[\nabla u - \widehat{\nabla}\widehat{u}]||| = |||\nabla(u - \widehat{u})|||.$$

Set $y = A\nabla\widehat{u}$. Then, (9.5.16) implies the estimate

$$|||\nabla u - \nabla\widehat{u}|||^2 \le \inf_{v \in u_0 + V_0} |||\nabla(v - \widehat{u})|||^2 + \mathrm{C}^2 \|A\nabla\widehat{u} \cdot n - f\|^2_{\Gamma_2}. \tag{9.5.32}$$

If $\Gamma_2 = \emptyset$, then we arrive at a simple projection estimate

$$|||\nabla u - \nabla\widehat{u}||| \le \inf_{v \in u_0 + V_0} |||\nabla(v - \widehat{u})||| \le |||\nabla(\mathrm{P}\widehat{u} - \widehat{u})|||. \tag{9.5.33}$$

Since

$$\ell(w) = \int_{\Gamma_2} Fw\,ds + \int_{\Omega} fw\,dx - \int_{\Omega} A\nabla\widehat{u} \cdot \nabla w\,dx = \int_{\Gamma_2} (F - A\nabla\widehat{u} \cdot n)w\,ds$$

and

$$\mu(\tau_0) = \int_{\Omega} (\nabla u_0 - \nabla\widehat{u}) \cdot \tau_0\,dx,$$

we find that

$$\begin{aligned} |||\nabla(u - \widehat{u})|||^2 \ge 2\int_{\Gamma_2} (F - A\nabla\widehat{u})w\,ds + 2\int_{\Gamma_1} (\widehat{u} - u_0)\tau_0 \cdot n\,ds \\ - |||\nabla w|||^2 - |||\tau_0|||_*^2, \qquad \forall w \in V_0,\ \tau_0 \in \mathrm{S}_{\Gamma_2}. \end{aligned} \tag{9.5.34}$$

If $\Gamma_2 = \emptyset$, then the right-hand side of (9.5.34) does not have the corresponding boundary integral. In this case, the best choice of w is the zero function. Then, the lower bound is given by a simple relation:

$$|||\nabla(u - \widehat{u})|||^2 \ge 2\int_{\Gamma} (\widehat{u} - u_0)\tau_0 \cdot n\,ds - |||\tau_0|||_*^2, \qquad \tau_0 \in H_0. \tag{9.5.35}$$

The vector-valued functions in S_{Γ_2} can be represented in the form $\alpha\bar{\tau}_0$, where α is a real number and $|||\bar{\tau}_0|||_* = 1$. Take

$$\alpha = \int_\Gamma (\widehat{u} - u_0)\bar{\tau}_0 \cdot n\, ds.$$

Then,

$$|||\nabla(u - \widehat{u})||| \geq \sup_{\substack{\bar{\tau}_0 \in H_0, \\ |||\bar{\tau}_0|||_* = 1}} \int_\Gamma (\widehat{u} - u_0)\bar{\tau}_0 \cdot n\, ds. \tag{9.5.36}$$

Remark 9.25. Consider the special case in which $\widehat{u} = u + c$, $c \in \mathbb{R}$. The left-hand sides of (9.5.33) and (9.5.36) are equal to zero. If we take $v = u$, then the right-hand side of (9.5.33) is also zero. On Γ we have $\widehat{u} - u_0 = \widehat{u} - u = c$. Since $\int_\Gamma \bar{\tau}_0 \cdot n\, ds = 0$, we conclude that the right-hand side of (9.5.36) is also equal to zero.

Similar estimates are easily obtained for different generalizations of the method. For example, the differential equation may be satisfied in subdomains Ω_i only approximately, i.e.,

$$\operatorname{div} A\nabla\widehat{u} + f = \epsilon_i(x), \qquad x \in \Omega_i, \tag{9.5.37}$$

where $\epsilon_i(x)$ is a small residual. For this case, we have the following upper bound of the error

$$|||[\eta]||| \leq \inf_{v \in V_0 + u_0} |||[\nabla v - \widehat{\nabla}\widehat{u}]|||^2 + C^2 \left(\sum_i \|\epsilon_i(x)\|^2_{\Omega_i} + \|A\nabla\widehat{u} \cdot n - f\|^2_{\Gamma_2} \right). \tag{9.5.38}$$

9.5.4 Comments

A rapidly developing group of methods is related to nonconforming finite element approximations. Here, a posteriori error estimation methods are much less developed than for classical (conforming) finite element methods. Concerning a posteriori estimates for *Discontinuous Galerkin* (DG) approximations of elliptic type equations, we refer to R. Becker, P. Hansbo, and M. G. Larsson [46] and R. Bustinza, G. N. Gatica, and B. Cockburn [83], where a modification of the residual based estimate for the energy norm of the error was suggested. In P. Castillo [101] a posteriori estimates in the L^2-norm were derived for the so-called "local DG method" applied to an elliptic boundary value problem. A posteriori error estimates for DG approximations were also obtained for other classes of problems. In particular, in S. Sun and M. F. Wheeler [346] time-dependent (transport) equations were considered and in P. Houston, I. Perugia, and D. Schotzau [181] the authors investigated elliptic problems of the Maxwell's type. The paper by J. Ma and H. Brunner [224] deals with a posteriori error estimates for DG approximations of integral equations. In A. Ern and J. Proft [127] a posteriori estimates were obtained for DG approximations of the convection-diffusion equation. In A. Ern, A.F. Stephensen, and M. Vohralik [128] advanced a posteriori estimates using constants in Poincaré inequalities for elements were derived for DG approximations. Error reduction and convergence analysis of an adaptive nonconforming finite element method was studied in C. Carstensen and R. H. W. Hoppe [94].

Finite Volume (FV) method is another nonconforming scheme, which is widely used in modern numerical analysis. Various approaches to a posteriori error control of FV approximations can be found in, e.g., Y. Achdou, C. Bernardi and F. Coquel [1], A. Bergam, Z. Mghazli, and R. Verfürth [51], C. Carstensen, R. Lazarov, and S. Tomov [98], R. Lazarov and S. Tomov [219], V. Jovanovic and C. Rohde [191], D. Kröner and M. Ohlberger [196], K. W. Morton and E. Süli [238], S. Nicaise [247], and M. Vohralík [365].

A consequent discussion of an a posteriori error estimation method based on (9.5.3) and (9.5.6) is presented in R. Lazarov, S. Repin and S. Tomar [218], where it is studied with the paradigm of the Discontinuous Galerkin method (also see [350], which includes a discussion of numerical experiments and practical efficiency of the error estimation method). In the context of the Finite Volume method, functional a posteriori estimates were studied in S. Cochez-Dhondt, S. Nicaise, and S. Repin [109].

9.6 Uncertain data

In practice, the data of a problem are always defined with certain indeterminacy. This fact should be taken into account in constructing approximation and error control methods that must be stable with respect to small variations in the coefficients and other data caused by indeterminacy. Also, it is necessary to compare the approximation errors and the errors caused by indeterminacy in the data. There are three major sources of the latter errors related to indeterminacy in

(a) coefficients of a PDE;
(b) boundary (initial) conditions;
(c) the configuration of Ω.

In what follows, we discuss the cases (a)–(c) and apply a posteriori estimates derived in previous chapters to the analysis of such errors. As everywhere, in this chapter we demonstrate new possibilities arising on this way, using linear diffusion problem as a basic example. Generalizations to other problems are rather transparent and can be done by the thoughtful reader without big difficulties. However, we begin with a concise introduction to general principles underlying error control theory in the case of uncertain data.

9.6.1 Introduction

Take a boundary value problem in abstract form

$$\mathcal{A}u = \ell, \tag{9.6.1}$$

where the operator $\mathcal{A}$ and the functional ℓ are *defined with indeterminacy*. In this case, instead of concrete $\mathcal{A}$ and ℓ, it is only known that

$$\mathcal{A} \in \mathcal{U}_{\mathcal{A}} \quad \text{and} \quad \ell \in \mathcal{U}_{\ell}, \tag{9.6.2}$$

where $\mathcal{U}_{\mathcal{A}}$ and $\mathcal{U}_{\ell}$ are certain (bounded) sets of "possible" data. In the simplest case, we assume that all pairs

$$(\mathcal{A}, \ell) \in \mathcal{U}_{\mathcal{A}} \times \mathcal{U}_{\ell}$$

are of equal probability. In order to guarantee the solvability of problems, we must assume that for any $\mathcal{A}$, the problem (9.6.1) is correct and has a unique solution. Moreover, we must assume that variations of $\mathcal{A}$ are sufficiently small, so that all problems belong to the same type (e.g., the operator remains elliptic and bounded).

Exact solutions of (9.6.1) with data satisfying (9.6.2) form a set in the respective energy space V, which we call the *set of possible solutions* and denote $\boldsymbol{S}$. Formally, this set is defined as follows:

$$\boldsymbol{S} := \left\{ u \in V \;\middle|\; \mathcal{A}u = \ell \;\text{ for some } \mathcal{A} \in \mathcal{U}_{\mathcal{A}} \text{ and } \ell \in \mathcal{U}_{\ell} \right\}.$$

Let $v \in V$ be an approximation of the exact solution (which is an unknown element of $\boldsymbol{S}$). Since the data are indeterminate, the error estimation problem takes two different forms. The first form is defined by the quantity

$$\underline{e}(v, \boldsymbol{S}) = \inf_{u \in \boldsymbol{S}} \|v - u\|_V \tag{9.6.3}$$

that measures the distance between v and the set $\boldsymbol{S}$ in terms of a certain norm $\|\cdot\|_V$ selected for this purpose. The quantity $\underline{e}(v, \boldsymbol{S})$ is equal to zero if $\mathcal{A}v = \ell$ for some pair $(\mathcal{A}, \ell) \in \mathcal{U}_{\mathcal{A}} \times \mathcal{U}_{\ell}$. This quantity provides the lowest possible bound of the true error or the error in the *best-case situation* (cf. Figure 9.6.1). Another important quantity is

$$\overline{e}(v, \boldsymbol{S}) = \sup_{u \in \boldsymbol{S}} \|v - u\|_V \,. \tag{9.6.4}$$

It compares v with the most remote element of $\boldsymbol{S}$ and yields the error *in the worst-case situation*. Obviously,

$$\overline{e}(v, \boldsymbol{S}) \geq \underline{e}(v, \boldsymbol{S}) \quad \text{and} \quad \overline{e}(v, \boldsymbol{S}) > 0. \tag{9.6.5}$$

If the problem (9.6.1) is uniquely solvable and the data are exactly determined, then $\boldsymbol{S}$ contains only one element, which is the exact solution u. In this case,

$$\underline{e}(v, \boldsymbol{S}) = \overline{e}(v, \boldsymbol{S}) = \|v - u\|_V \,,$$

and the accuracy of v can be evaluated by the error norm. However, if the data are not fully determined, then the error estimation problem is more complicated, and we need at least two different quantities to have an idea of the quality of v.

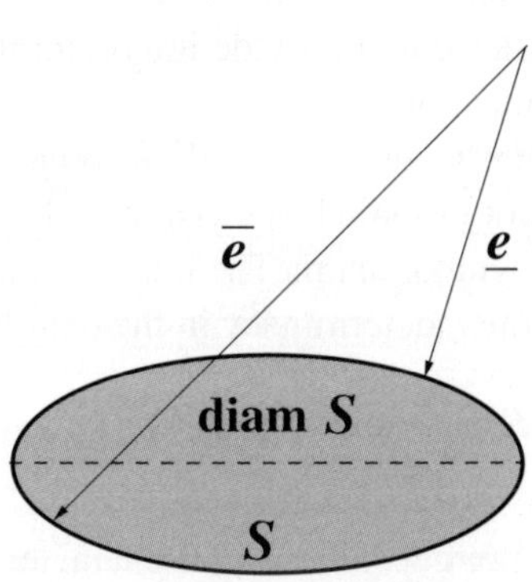

Figure 9.6.1 The set of solutions $\boldsymbol{S}$ and errors $\overline{e}$ and $\underline{e}$.

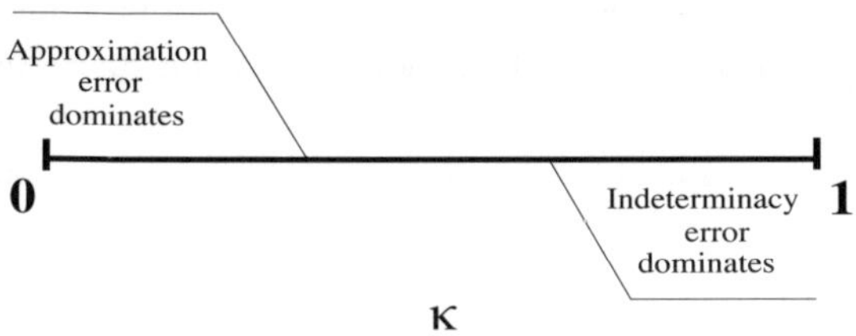

Figure 9.6.2 The indicator $\kappa(v, \boldsymbol{S})$.

In addition, we introduce the quantity

$$\kappa(v, \boldsymbol{S}) := \frac{\overline{e}(v, \boldsymbol{S}) - \underline{e}(v, \boldsymbol{S})}{\overline{e}(v, \boldsymbol{S}) + \underline{e}(v, \boldsymbol{S})} \in [0, 1], \tag{9.6.6}$$

which is easily computable provided that $\underline{e}(v, \boldsymbol{S})$ and $\overline{e}(v, \boldsymbol{S})$ (or bounds of them) are known.

This quantity could be used as an indicator that predicts the efficiency of expenditures spent on decreasing approximation errors. Indeed, if $\kappa(v, \boldsymbol{S})$ is close to zero, then approximation errors provide the major part of the overall error and the impact of indeterminacy is not significant. In the special case of fully determined data, $\kappa(v, \boldsymbol{S}) = 0$, and we have only the error of approximation. However, if $\kappa(v, \boldsymbol{S})$ is close to 1, then

$$\overline{e}(v, \boldsymbol{S}) \gg \underline{e}(v, \boldsymbol{S}),$$

which means that v is quite close to $\boldsymbol{S}$. If $\kappa(v, \boldsymbol{S}) = 1$, then $\underline{e}(v, \boldsymbol{S}) = 0$, i.e., $v \in \boldsymbol{S}$. In this case, further attempts at reducing approximation errors are obviously meaningless. These observations are schematically depicted in Figure 9.6.2.

Hence, *guaranteed and computable estimates of* $\underline{e}(v, S)$ *and* $\overline{e}(v, S)$, are indeed required in practice because they could provide important information for an efficient organization of the computational process.

If a boundary value problem is analyzed with account of the data indeterminacy, then one more important quantity should be determined. It is *the diameter* of S, which henceforth is denoted by the symbol diam. The quantity diam S gives an insight into the accuracy limit caused by the indeterminacy in the data. By definition,

$$\operatorname{diam} S := \sup_{u_1,u_2\in S} \|u_1 - u_2\|_V . \tag{9.6.7}$$

We outline that diam S does not depend on v. It characterizes a particular boundary value problem under the indeterminacy conditions imposed. The generation of approximate solutions that have approximation errors less than diam S has no practical sense.

In general, the exact values of $\underline{e}$, $\overline{e}$, and diam S cannot be found. However, functional a posteriori estimates supply a method for finding their computable bounds, which we shortly discuss below.

First, we note that a computable lower bound of diam S stems from the relation

$$\left\|u_i - u_j\right\|_V \geq \left\|v_i - v_j\right\|_V - \|u_i - v_i\|_V - \left\|u_j - v_j\right\|_V , \tag{9.6.8}$$

where v_i and v_j are conforming approximations of two exact solutions u_i and u_j (which belong to the set $\in S$), respectively. The functions v_i, $i = 1,\dots,m$, can be constructed by solving the boundary value problems with $\mathcal{A}_i \in \mathcal{U}_{\mathcal{A}}$ and $\ell_i \in \mathcal{U}_\ell$ with the help of some numerical method. Their accuracy is controlled by error majorants, which furnish relevant error bounds (e.g., see (7.1.20))

$$\|u_i - v_i\|_V \leq \epsilon_i := \overline{\mathfrak{M}}_{\mathcal{A}}(v_i, y),$$

where the functions y_i are properly selected to avoid significant overestimation. Then

$$\operatorname{diam} S \geq \sup_{i,j=1,2,\dots,m} \left\{\left\|v_i - v_j\right\|_V - \epsilon_i - \epsilon_j\right\}. \tag{9.6.9}$$

The right hand side of (9.6.9) is computable, because it contains only approximate solutions and known functions. The larger is m, the sharper lower bound of diam S is given by (9.6.9) (provided that the ϵ_i are sufficiently small).

However, such a straightforward procedure may be rather expensive. To simplify computations, we can use another method that we discuss with the paradigm of problem (7.1.1)–(7.1.3). Let $\mathcal{A}$ and $\ell \in \mathcal{V}$ be certain ("central") elements of the sets $\mathcal{U}_{\mathcal{A}}$ and $\mathcal{U}_\ell$, respectively. Consider the quantity

$$\rho(S)^2 := \sup_{\widetilde{u}\in S} |||\Lambda(u - \widetilde{u})|||^2,$$

where u is the solution generated by $\mathcal{A}$ and ℓ (we assume that it is either known or computed with a high accuracy confirmed by the the corresponding error majorant), $|||\cdot|||$ denotes the energy norm associated with $\mathcal{A}$, and $\widetilde{u}$ is the exact solution of the problem with $\widetilde{\mathcal{A}}$ and $\widetilde{\ell}$. It is easy to see that

$$\begin{aligned}\operatorname{diam} \boldsymbol{S} &= \sup_{u_1,u_2\in \boldsymbol{S}} |||\Lambda(u_1-u_2)||| \\ &\leq \sup_{u_1,u_2\in \boldsymbol{S}} \left(|||\Lambda(u_1-u)|||+|||\Lambda(u_2-u)|||\right)= 2\rho(\boldsymbol{S}).\end{aligned}$$

On the other hand, $\operatorname{diam} \boldsymbol{S} \geq \rho(\boldsymbol{S})$. Thus, the quantity $\rho(\boldsymbol{S})$ characterizes the diameter of $\boldsymbol{S}$. We can estimate $\rho(\boldsymbol{S})$ from below by (7.1.28), which reads as follows:

$$\begin{aligned}|||\Lambda(u-\widetilde{u})|||^2_{\widetilde{\mathcal{A}}} &\geq \sup_{w\in V_0} \left\{-(\widetilde{\mathcal{A}}\Lambda w,\Lambda w)-2(\widetilde{\mathcal{A}}\Lambda u,\Lambda w)-2\left\langle\widetilde{\ell},w\right\rangle\right\} \\ &= \sup_{w\in V_0} \left\{-(\widetilde{\mathcal{A}}\Lambda w,\Lambda w)-2((\widetilde{\mathcal{A}}-\mathcal{A})\Lambda u,\Lambda w)-2\left\langle\widetilde{\ell}-\ell,w\right\rangle\right\}.\end{aligned}$$

From this relation, it follows that

$$\begin{aligned}\rho(\boldsymbol{S})^2 &= \sup_{\widetilde{\mathcal{A}}\in\mathcal{U}_{\mathcal{A}},\widetilde{\ell}\in\mathcal{U}_{\ell}} \sup_{w\in V_0} |||\Lambda(u-\widetilde{u})|||^2 \\ &= \sup_{w\in V_0} \sup_{\widetilde{\mathcal{A}}\in\mathcal{U}_{\mathcal{A}},\widetilde{\ell}\in\mathcal{U}_{\ell}} \Big\{-(\widetilde{\mathcal{A}}\Lambda w,\Lambda w) \\ &\qquad -2((\widetilde{\mathcal{A}}-\mathcal{A})\Lambda u,\Lambda w)-2\left\langle\widetilde{\ell}-\ell,w\right\rangle\Big\}. \qquad (9.6.10)\end{aligned}$$

Computable lower bounds follow from (9.6.10) if V_0, $\mathcal{U}_{\mathcal{A}}$, and $\mathcal{U}_{\ell}$ are replaced by some finite-dimensional subsets.

Further simplifications can be made if we assume that

$$\lambda\mathcal{A} \in \mathcal{U}_{\mathcal{A}} \quad \text{for all} \quad \lambda \in [1-\mu_1, 1+\mu_2],$$

where μ_1 and μ_2 are some known nonnegative numbers, and there exists $\delta > 0$ such that

$$\mathcal{B}(\ell,\delta) := \left\{\widetilde{\ell} \in \mathcal{V} \mid \|\widetilde{\ell}-\ell\|_{\mathcal{V}} \leq \delta\right\} \subset \mathcal{U}_{\ell}.$$

Set $\widetilde{\mathcal{A}} = \widetilde{\mathcal{A}}_{\lambda} := \lambda\mathcal{A}$. Then

$$|||\Lambda(u-\widetilde{u})|||^2_{\widetilde{\mathcal{A}}_{\lambda}} = \lambda|||\Lambda(u-\widetilde{u})|||^2,$$

and we find that

$$\begin{aligned}\rho(S)^2 &= \sup_{\widetilde{\mathcal{A}},\widetilde{\ell}} |||\Lambda(u-\widetilde{u})|||^2 \geq \frac{1}{\lambda}\sup_{\widetilde{\ell}}\sup_{w\in V_0} |||\Lambda(u-\widetilde{u})|||^2_{\mathcal{A}_\lambda}\\ &\geq \sup_{w\in V_0}\Big\{-(\mathcal{A}\Lambda w,\Lambda w) - 2\left(1-\frac{1}{\lambda}\right)(\mathcal{A}\Lambda u,\Lambda w)\\ &\qquad + \frac{2}{\lambda}\sup_{\widetilde{\ell}\in\mathcal{U}_\ell}\left\langle\widetilde{\ell}-\ell, w\right\rangle\Big\}. \end{aligned} \tag{9.6.11}$$

Since

$$\sup_{\widetilde{\ell}\in\mathcal{U}_\ell}\left\langle\widetilde{\ell}-\ell, w\right\rangle \geq \sup_{\widetilde{\ell}\in\mathcal{B}(\ell,\delta)} (\widetilde{\ell}-\ell, w) = \delta\|w\|_{\mathcal{V}},$$

we have

$$\rho(S)^2 \geq \sup_{w\in V_0}\left\{-(\mathcal{A}\Lambda w,\Lambda w) - 2\left(1-\frac{1}{\lambda}\right)(\mathcal{A}\Lambda u,\Lambda w) + \frac{2\delta}{\lambda}\|w\|_{\mathcal{V}}\right\}, \tag{9.6.12}$$

where $\lambda \in [1-\mu_1, 1+\mu_2]$. Taking the supremum with respect to $w \in V_{0h} \subset V_0$ and λ, we obtain a lower bound of $\rho(S)$.

Finally, consider the special case in which $u_0 = 0$. Then we set $w = \alpha u$ and obtain

$$\begin{aligned}\rho(S)^2 &\geq \frac{1}{\lambda}\sup_{\alpha\in\mathbb{R}}\left\{(-\lambda\alpha^2 - 2\alpha\lambda + 2\alpha)|||\Lambda u|||^2 + 2|\alpha|\delta\|u\|_{\mathcal{V}}\right\}\\ &\geq \frac{(1-\lambda)^2}{\lambda^2}|||\Lambda u|||^2 + 2\frac{\delta}{\lambda^2}|1-\lambda|\|u\|_{\mathcal{V}}. \end{aligned} \tag{9.6.13}$$

If $\delta = 0$, then (9.6.13) implies the estimate

$$\rho(S) \geq \max_{\lambda\in[1-\mu_1,1+\mu_2]} \frac{|1-\lambda|}{\lambda}|||\Lambda u|||, \tag{9.6.14}$$

which (for this special case) can be derived in a simpler way.

9.6.2 Errors caused by indeterminacy in coefficients

Indeterminacy in coefficients of a boundary value problem that models a certain physical object is quite typical. For example, the coefficients of diffusion, elasticity constants, and viscosity are usually known only approximately. Functional a posteriori estimates provide a way for estimating these errors and comparing them with approximation errors. In this section, we consider this question by using again the problem (4.1.1)–(4.1.3) as a basic example. Assume that the indeterminacy in the coefficients of the differential equation (4.1.1) is described by fixing some "mean" elements

$$A_0 \in L^\infty(\Omega; M_s^{d\times d}), \qquad (f_0, F_0) \in L^2(\Omega) \times L^2(\Gamma_2),$$

and by defining bounds of possible variations. In this case, the elements of the sets $\mathcal{U}_{\mathcal{A}}$ and $\mathcal{U}_l$ are represented by elements of the following two functional sets:

$$\mathcal{U}_A := \left\{ A \in L^\infty(\Omega; M_s^{d\times d}) \mid A = A_0 + \epsilon E,\ E \in \mathcal{E} \right\},$$

$$\mathcal{U}_\ell := \left\{ (f, F) \mid f = f_0 + \delta_1 \varphi_1,\ F = F_0 + \delta_2 \varphi_2,\ \varphi_i \in F_i,\ i = 1, 2 \right\},$$

where

$$\mathcal{E} := \left\{ E \in L^\infty(\Omega; M_s^{d\times d}) \mid \| \, |E| \, \|_{\infty,\Omega} \leq 1 \right\},$$

$$F_1 := \left\{ \varphi_1 \in L^2(\Omega) \mid \|\varphi_1\|_{2,\Omega} \leq 1 \right\},$$

$$F_2 := \left\{ \varphi_2 \in L^2(\Gamma_2) \mid \|\varphi_2\|_{2,\Gamma_2} \leq 1 \right\}.$$

Here, ϵ and δ_i are small parameters characterizing the range of indeterminacy. Henceforth, we assume that $\epsilon < c_1^2$, where c_1 is the ellipticity constant for A_0:

$$c_1^2|\xi|^2 \leq A_0\xi \cdot \xi \leq c_2^2|\xi|^2, \qquad \forall \xi \in \mathbb{R}^d. \tag{9.6.15}$$

Since $|E\xi \cdot \xi| \leq |E|\,|\xi|^2$, we find that

$$A\xi \cdot \xi \geq A_0\xi \cdot \xi - \epsilon|E|\,|\xi|^2 \geq (c_1^2 - \epsilon)|\xi|^2, \tag{9.6.16}$$

$$A\xi \cdot \xi \leq A_0\xi \cdot \xi + \epsilon|E|\,|\xi|^2 \leq (c_2^2 + \epsilon)|\xi|^2 \tag{9.6.17}$$

and analogous estimates for the inverse matrix are

$$c_2^{-2}|\xi|^2 \leq A_0^{-1}\xi \cdot \xi \leq c_1^{-2}|\xi|^2, \tag{9.6.18}$$

$$(c_2^2 + \epsilon)^{-1}|\xi|^2 \leq A^{-1}\xi \cdot \xi \leq (c_1^2 - \epsilon)^{-1}|\xi|^2. \tag{9.6.19}$$

Let the energy norm $|||\nabla(u - v)|||_{A_0}$ generated by the mean matrix A_0 be selected as the energy norm. Our goal is to find estimates of the quantities

$$\underline{e}(v, \boldsymbol{S}) = \inf_{u \in \boldsymbol{S}} |||\nabla(v - u)|||_{A_0} \quad \text{and} \quad \overline{e}(v, \boldsymbol{S}) = \sup_{u \in \boldsymbol{S}} |||\nabla(v - u)|||_{A_0},$$

where

$$\boldsymbol{S}(\mathcal{U}_A, \mathcal{U}_\ell) := \Big\{ u \in V \Big| \int_\Omega A\nabla u \cdot \nabla w \, dx = \int_\Omega f w \, dx + \int_{\Gamma_2} F w \, ds,$$
$$\forall w \in V_0, \text{ for } A \in \mathcal{U}_A \text{ and } \ell \in \mathcal{U}_\ell \Big\}.$$

The idea behind the derivation of computable bounds of $\overline{e}(v, \boldsymbol{S})$ and $\underline{e}(v, \boldsymbol{S})$ is to use the majorants and minorants discussed in Chapter 4 that explicitly depend on the coefficients. Then, taking the supremum (infimum) over $u \in \boldsymbol{S}$ is reduced to a minimization (maximization) problem for the coefficients of a majorant (minorant).

By (4.1.12), we know that

$$
\begin{aligned}
|||\nabla(u-v)|||_{A_0}^2 &\le (1+c_\epsilon)|||\nabla(u-v)|||_A^2 \\
&\le \int_\Omega \big(A\nabla v\cdot\nabla v + A^{-1}y\cdot y - \nabla v\cdot y\big)\,dx \\
&\quad + \frac{1}{c_1}\Big(C_{F\Gamma_1}\|f+\operatorname{div} y\| + C_{T\Gamma_2}\|F-y\cdot n\|_{\Gamma_2}\Big), \qquad (9.6.20)
\end{aligned}
$$

where $c_\epsilon = \frac{\epsilon}{c_1^2-\epsilon}$. Analogously, by (4.1.13) we obtain

$$
\begin{aligned}
|||\nabla(u-v)|||_{A_0}^2 &\ge (1-c_\epsilon)|||\nabla(u-v)|||_A^2 \\
&\ge (1-c_\epsilon)\bigg(\int_\Omega (-A\nabla w\cdot\nabla w - 2A\nabla v\cdot\nabla w + 2fw)\,dx \\
&\quad + 2\int_{\Gamma_2} Fw\,ds\bigg). \qquad (9.6.21)
\end{aligned}
$$

Below we use (9.6.20) and (9.6.21) to find the desired error bounds.

Upper bound of $\overline{e}(v,S)$. Let $u_{(\mathcal{A},f,F)}$ denote the solution of the problem generated by A, f, and F. Then,

$$
\begin{aligned}
&|||\nabla(v-u_{(A,f,F)})|||_{A_0} \\
&\quad\le |||\nabla(v-u_{(A,f_0,F_0)})|||_{A_0} + |||\nabla(u_{(A,f_0,F_0)}-u_{(A,f,F)})|||_{A_0}. \qquad (9.6.22)
\end{aligned}
$$

Also, we have

$$
\int_\Omega A\nabla u_{(A,f,F)}\cdot\nabla w\,dx = \int_\Omega fw\,dx + \int_{\Gamma_2} Fw\,ds, \qquad \forall w\in V_0, \qquad (9.6.23)
$$

$$
\int_\Omega A\nabla u_{(A,f_0,F_0)}\cdot\nabla w\,dx = \int_\Omega f_0 w\,dx + \int_{\Gamma_2} F_0 w\,ds. \qquad (9.6.24)
$$

From (9.6.23) and (9.6.24), we conclude that

$$
\begin{aligned}
&|||\nabla(u_{(A,f_0,F_0)}-u_{(A,f,F)})|||_A^2 \\
&\le \big(C_{F\Gamma_1}\|f-f_0\|_\Omega + C_{T\Gamma_2}\|F-F_0\|_{\Gamma_2}\big)\|\nabla(u_{(A,f_0,F_0)}-u_{(A,f,F)})\|. \qquad (9.6.25)
\end{aligned}
$$

Hence,

$$
\begin{aligned}
&|||\nabla(u_{(A,f_0,F_0)}-u_{(A,f,F)})|||_{A_0} \\
&\quad\le \sqrt{1+c_\epsilon}\,|||\nabla(u_{(A,f_0,F_0)}-u_{(A,f,F)})|||_A \\
&\quad\le \frac{c_1}{c_1^2-\epsilon}\big(C_{F\Gamma_1}\delta_1\|\varphi_1\|_\Omega + C_{T\Gamma_2}\delta_2\|\varphi_2\|_{\Gamma_2}\big). \qquad (9.6.26)
\end{aligned}
$$

By (9.6.22) and (9.6.26), we obtain

$$\begin{aligned}&|||\nabla(v - u_{(A,f,F)})|||_{A_0}\\ &\qquad \le |||\nabla(v - u_{(A,f_0,F_0)})|||_{A_0} + \frac{c_1}{c_1^2 - \epsilon}\big(C_{F\Gamma_1}\delta_1 + C_{T\Gamma_2}\delta_2\big). \quad (9.6.27)\end{aligned}$$

Here,

$$\begin{aligned}|||\nabla(v - u_{(A,f_0,F_0)})|||_{A_0} &\le (1 + c_\epsilon)|||\nabla(v - u_{(A,f_0,F_0)})|||_A\\ &\le (1 + c_\epsilon)\big(|||A\nabla v - y|||_{A^{-1}} + C_{F\Gamma_1}\|\operatorname{div} y + f_0\|\\ &\qquad + C_{T\Gamma_2}\|y \cdot n - F_0\|_{\Gamma_2}\big). \quad (9.6.28)\end{aligned}$$

We must estimate the term

$$|||A\nabla v - y|||_{A^{-1}}^2 = (A\nabla v, \nabla v) + (A^{-1}y, y) - 2(\nabla v, y).$$

Since $A \in \mathcal{U}_{\mathcal{A}}$, we have

$$A^{-1} = \big(A_0(\mathbb{I} + \epsilon A_0^{-1}E)\big)^{-1} = (\mathbb{I} + \epsilon B)^{-1} A_0^{-1},$$

where $B = A_0^{-1}E$ and

$$\epsilon|B| \le \epsilon|A_0^{-1}|\,|E| \le \epsilon c_1^{-2} < 1.$$

Thus, the absolute values of all eigenvalues of the matrix ϵB are less than one and we can use the representation

$$(\mathbb{I} + \epsilon B)^{-1} = \mathbb{I} + \sum_{j=1}^{\infty}(-1)^j\epsilon^j B^j.$$

Hence,

$$\begin{aligned}A^{-1}y \cdot y &= (I + \epsilon B)^{-1}A_0^{-1}y \cdot y\\ &= A_0^{-1}y \cdot y - \epsilon BA_0^{-1}y \cdot y + \sum_{j=2}^{\infty}(-1)^j\epsilon^j B^j A_0^{-1}y \cdot y.\end{aligned}$$

Estimate the last term as follows:

$$B^j A_0^{-1}y \cdot y \le |B^j A_0^{-1}|\,|y|^2 \le |A_0^{-1}|^{j+1}\,|E|^j\,|y|^2 \le c_1^{-2(j+1)}|y|^2$$

and

$$\begin{aligned}\int_\Omega A^{-1}y \cdot y\,dx \le \int_\Omega (A_0^{-1}y \cdot y - \epsilon BA_0^{-1}y \cdot y)\,dx\\ + \left(\sum_{j=2}^{\infty}(-1)^j\epsilon^j c_1^{-2(j+1)}\right)\|y\|^2.\end{aligned}$$

In view of (9.6.1) and the relation

$$\int_\Omega A\nabla v\cdot\nabla v\,dx=\int_\Omega (A_0\nabla v\cdot\nabla v+\epsilon E\nabla v\cdot\nabla v)\,dx,$$

we find that

$$|\!|\!| A\nabla v-y|\!|\!|^2_{A^{-1}}=|\!|\!| A_0\nabla v-y|\!|\!|^2_{A_0^{-1}}+\epsilon\int_\Omega (E\nabla v\cdot\nabla v-BA_0^{-1}y\cdot y)\,dx$$
$$+\frac{1}{c_1^2}\|y\|^2\sum_{j=2}^{\infty}\left(-\frac{\epsilon}{c_1^2}\right)^j. \qquad (9.6.29)$$

Since A_0 is symmetric, we have

$$E\nabla v\cdot\nabla v-A_0^{-1}EA_0^{-1}y\cdot y=E(\nabla v-A_0^{-1}y)\cdot(\nabla v+A_0^{-1}y).$$

To find an upper bound of $\overline{e}(v,\boldsymbol{S})$, we note that

$$\sup_{E\in\mathcal{E}}\int_\Omega E:\left((\nabla v-A_0^{-1}y)\otimes(\nabla v+A_0^{-1}y)\right)dx\le I_{1;\mathcal{E},\epsilon},$$

where

$$I_{1;\mathcal{E},\epsilon}(v,y):=\epsilon\int_\Omega|\nabla v-A_0^{-1}y||\nabla v+A_0^{-1}y|\,dx.$$

Define

$$I_{2;\epsilon}:=\left(\frac{\epsilon}{c_1^2}\right)^2\frac{1}{(\epsilon+c_1^2)}\|y\|^2.$$

Then,

$$|\!|\!| A\nabla v-y|\!|\!|_{A^{-1}}\le|\!|\!| A_0\nabla v-y|\!|\!|_{A_0^{-1}}+\sqrt{I_{1;\mathcal{E},\epsilon}+I_{2;\epsilon}}, \qquad (9.6.30)$$

and by (9.6.28), we conclude that

$$|\!|\!|\nabla(v-u_{(A,f_0,F_0)})|\!|\!|_{A_0}\le\sqrt{1+c_\epsilon}\left(\overline{\mathfrak{M}}_{A_0,f_0,F_0}(v,y)+\sqrt{I_{1;\mathcal{E},\epsilon}+I_{2;\epsilon}}\right), \qquad (9.6.31)$$

where

$$\overline{\mathfrak{M}}_{A_0,f_0,F_0}(v,y):=|\!|\!| A_0\nabla v-y|\!|\!|_{A_0^{-1}}+C_{F\Gamma_1}\|\operatorname{div}y+f_0\|+C_{T\Gamma_2}\|y\cdot n-F_0\|_{\Gamma_2}.$$

Now (9.6.27) and (9.6.31) imply the estimate

$$|\!|\!|\nabla(v-u_{(A,f,F)})|\!|\!|_{A_0}\le\frac{c_1}{\sqrt{c_1^2-\epsilon}}\overline{\mathfrak{M}}_{A_0,f_0,F_0}(v,y)+M_{\mathcal{E},\epsilon}, \qquad (9.6.32)$$

where

$$M_{\delta,\epsilon} := \frac{c_1}{\sqrt{c_1^2 - \epsilon}} \sqrt{I_{1;\delta,\epsilon} + I_{2;\epsilon}} + \frac{1}{\sqrt{c_1^2 - \epsilon}} (C_{F\Gamma_1}\delta_1 + C_{T\Gamma_2}\delta_2).$$

We note that the right-hand side of (9.6.32) contains only known functions and parameters. It gives an upper bound of the error for any $u \in \boldsymbol{S}$. Hence,

$$\overline{e}(v, \boldsymbol{S}) \leq \frac{c_1}{\sqrt{c_1^2 - \epsilon}} \overline{\mathfrak{M}}_{\mathrm{A_0,f_0,F_0}}(v, y) + M_{\delta,\epsilon}. \tag{9.6.33}$$

In (9.6.33), the term $\overline{\mathfrak{M}}_{\mathrm{A_0,f_0,F_0}}(v, y)$ is related to the approximation error of a solution to the "mean" problem and $M_{\delta,\epsilon}$ reflects the impact of data indeterminacy.

Lower bound of $\underline{e}(v, \boldsymbol{S})$. In view of (9.6.16), we have

$$\begin{aligned}
&|||\nabla(u - v)|||_{A_0}^2 \geq (1 - c_\epsilon)|||\nabla(u - v)|||_A^2 \\
&\geq (1 - c_\epsilon)\left(\int_\Omega (-A\nabla w \cdot \nabla w - 2A\nabla v \cdot \nabla w + 2fw)\, dx + 2\int_{\Gamma_2} Fw\, ds\right) \\
&= (1 - c_\epsilon)\underline{\mathfrak{M}}^2_{(\mathrm{A_0,f_0,F_0})}(v, w) + (1 - c_\epsilon)\epsilon \int_\Omega (-E\nabla w \cdot \nabla w - 2E\nabla v \cdot \nabla w)\, dx \\
&\quad + 2(1 - c_\epsilon)\int_\Omega (f - f_0)w\, dx + 2(1 - c_\epsilon)\int_{\Gamma_2} (F - F_0)w\, ds,
\end{aligned} \tag{9.6.34}$$

where w is an arbitrary function in V_0 and

$$\underline{\mathfrak{M}}^2_{(\mathrm{A_0,f_0,F_0})}(v, w) := \int_\Omega (-A_0\nabla w \cdot \nabla w - 2A_0\nabla v \cdot \nabla w + 2f_0 w)\, dx + 2\int_{\Gamma_2} F_0 w\, ds.$$

Therefore,

$$\begin{aligned}
\underline{e}^2(v, \boldsymbol{S}) &= \inf_{u \in \boldsymbol{S}} |||\nabla(v - u)|||_{A_0}^2 \\
&\geq (1 - c_\epsilon) \inf_{\substack{E \in \mathcal{E}, \\ \phi_1 \in F_1, \phi_2 \in F_2}} \sup_{w \in V_0} \Big(\underline{\mathfrak{M}}^2_{(\mathrm{A_0,f_0,F_0})}(v, w) \\
&\quad - \epsilon \int_\Omega (E\nabla w \cdot \nabla w + 2E\nabla v \cdot \nabla w)\, dx \\
&\quad + 2\int_\Omega \delta_1\phi_1 w\, dx + 2\int_{\Gamma_2} \delta_2\phi_2 w\, ds\Big) \\
&\geq (1 - c_\epsilon) \sup_{w \in V_0} \Big(\underline{\mathfrak{M}}^2_{(\mathrm{A_0,f_0,F_0})}(v, w) \\
&\quad + \epsilon \inf_{E \in \mathcal{E}} \int_\Omega (E\nabla w \cdot \nabla w + 2E\nabla v \cdot \nabla w)\, dx - 2\delta_1\|w\| - 2\delta_2\|w\|_{\Gamma_2}\Big).
\end{aligned}$$

Here,

$$\inf_{E\in\mathcal{E}} \int_\Omega (E\nabla w\cdot\nabla w + 2E\nabla v\cdot\nabla w)\,dx = -\int_\Omega \Big|(\nabla w + 2\nabla v)\otimes\nabla w\Big|\,dx.$$

Now, we obtain

$$\underline{e}^2(v,\boldsymbol{S}) \geq (1-c_\epsilon)\sup_{w\in V_0}\Big(\mathfrak{M}^2_{(A_0,f_0,F_0)}(v,w) - S_{\epsilon,\delta}(w)\Big), \qquad (9.6.35)$$

where the term

$$S_{\epsilon,\delta}(w) := \epsilon\int_\Omega \Big|(\nabla w + 2\nabla v)\otimes\nabla w\Big|\,dx + 2\delta_1\|w\| + 2\delta_2\|w\|_{\Gamma_2}$$

depends on the indeterminacy parameters.

To obtain a computable lower bound, we replace V_0 by a finite-dimensional subspace V_{0k} and solve a finite-dimensional maximization problem

$$\sup_{w\in V_{0k}}\Big(\mathfrak{M}^2_{(A_0,f_0,F_0)}(v,w) - S_{\epsilon,\delta}(w)\Big).$$

9.6.3 Errors owing to uncertain Ω

In practice, the domain Ω may be also not completely defined. For example, it may be only known that

$$\Omega_- \subset \Omega \subset \Omega_+, \qquad (9.6.36)$$

where Ω_- and Ω_+ are given domains with Lipschitz continuous boundaries Γ_- and Γ_+, respectively. Henceforth, for the sake of simplicity, we assume that Ω is a simply connected domain.

In general, accounting of this type uncertainty imposes a more complicated task. One way to solve it is to make a proper parametrization that maps Ω to an etalon domain $\widehat{\Omega}$. Then the uncertainty is transformed to the coefficients of a modified differential problem related to the new coordinate system. Then the uncertainty is transformed to the coefficients of a modified differential problem related to the new coordinate system. In this case, the influence of uncertain geometry can be estimated by the above-discussed methods.

Below we consider another (simpler) way the applicability of which is, however, restricted to the case of Dirichlet boundary conditions.

Consider the problem

$$\operatorname{div} A\nabla u + f = 0 \qquad \text{in } \Omega$$

with homogeneous Dirichlet boundary conditions. We assume that $f \in L^2(\Omega_+)$ and the matrix A is defined and positive definite in Ω_+. Let us denote the exact solutions of the boundary value problems in Ω_+ and Ω_- by u_+ and u_-, respectively.

A function $v \in \overset{\circ}{H}{}^1(\Omega)$ can be extended to $\overset{\circ}{H}{}^1(\Omega_+)$ if we set $v = 0$ in $\Omega_+ \setminus \Omega$. Similarly, u_- can be viewed as a function in $\overset{\circ}{H}{}^1(\Omega_+)$ or $\overset{\circ}{H}{}^1(\Omega)$ with zero values outside Ω_-.

By analogy with the estimate (2.3.1), we deduce the relation

$$\frac{1}{2}|||\nabla(u_+ - u)|||^2_{\Omega_+} = J_{\Omega_+}(u) - J_{\Omega_+}(u_+), \tag{9.6.37}$$

where

$$J_{\Omega_+}(v) := \int_{\Omega_+} (A\nabla v \cdot \nabla v - f v)\, dx.$$

Since

$$J_{\Omega_+}(u) = J_\Omega(u) \leq J_\Omega(u_-)$$

we find that

$$\begin{aligned}\frac{1}{2}|||\nabla(u_+ - u)|||^2_{\Omega_+} &\leq J_\Omega(u_-) - J_{\Omega_+}(u_+) = J_{\Omega_+}(u_-) - J_{\Omega_+}(u_+) \\ &= \frac{1}{2}|||\nabla(u_- - u_+)|||^2_{\Omega_+},\end{aligned}$$

On the other hand,

$$\begin{aligned}\frac{1}{2}|||\nabla(u_- - u)|||^2_{\Omega} &= J_\Omega(u_-) - J_\Omega(u) = J_{\Omega_+}(u_-) - J_{\Omega_+}(u) \\ &\leq J_{\Omega_+}(u_-) - J_{\Omega_+}(u_+) = \frac{1}{2}|||\nabla(u_- - u_+)|||^2_{\Omega_+},\end{aligned}$$

Hence, we conclude that

$$|||\nabla(u_\pm - u)|||_{\Omega_+} \leq |||\nabla(u_+ - u_-)|||_{\Omega_+}. \tag{9.6.38}$$

This estimate leads to an upper bound of

$$\operatorname{diam} \boldsymbol{S} := \sup_{u_1, u_2} |||\nabla(u_1 - u_2)|||_{\Omega_+},$$

where u_1 and u_2 are the exact solutions related to Ω_1 and Ω_2, respectively (it is assumed that $\Omega_- \subset \Omega_i \subset \Omega_+$ for $i = 1, 2$). Then,

$$\begin{aligned}|||\nabla(u_1 - u_2)|||_{\Omega_+} &\leq |||\nabla(u_1 - u_-)|||_{\Omega_+} + |||\nabla(u_- - u_+)|||_{\Omega_+} + |||\nabla(u_+ - u_2)|||_{\Omega_+} \\ &\leq 3|||\nabla(u_- - u_+)|||_{\Omega_+}\end{aligned}$$

and we find that

$$|||\nabla(u_- - u_+)|||_{\Omega_+} \leq \operatorname{diam} \boldsymbol{S} \leq 3|||\nabla(u_- - u_+)|||_{\Omega_+}. \tag{9.6.39}$$

Let $v_- \in \overset{\circ}{H}{}^1(\Omega_-)$ be an approximation of u_- extended by zero to Ω_+. Then

$$\begin{aligned}\frac{1}{2}|||\nabla(u_- - u_+)|||^2_{\Omega_+} &= J_{\Omega_+}(u_-) - J_{\Omega_+}(u_+) \\ &\le J_{\Omega_+}(v_-) - J_{\Omega_+}(u_+) = \frac{1}{2}|||\nabla(v_- - u_+)|||^2_{\Omega_+}\end{aligned}$$

and by (9.6.39) we conclude that

$$\operatorname{diam} \boldsymbol{S} \le 3|||\nabla(v_- - u_+)|||_{\Omega_+}. \tag{9.6.40}$$

Apply the majorant to the right-hand side of (9.6.40). We have

$$\begin{aligned}|||\nabla(u_+ - v_-)|||_{\Omega_+} &\le |||A\nabla v_- - y|||_{*,\Omega_+} + C_{F\Omega_+}\|\operatorname{div} y + f\|_{\Omega_+} \\ &=: \overline{\mathfrak{M}}_{\Omega_+}(v_-, y),\end{aligned} \tag{9.6.41}$$

where $y \in H(\Omega_+, \operatorname{div})$.

Remark 9.26. By the same arguments, we can obtain computable estimates of the error arising from the uncertainty of the Dirichlet part of the boundary in many other problems, for which the principal relation (9.6.37) holds.

9.6.4 Comments

In the fully reliable mathematical modeling, the errors caused by uncertainties in the problem data must be measured, as well as the approximation errors. However, this question has only recently started receiving serious attention. For example, effects associated with uncertainty in the boundary conditions was considered in the papers by I. Babuška and J. Chleboun [23, 24]. Also, we refer to the papers by I. Babuška, F. Nobile, and R. Tempone [29] and J. T. Oden, I. Babuška, F. Nobile, Y. Feng, and R. Tempone [250]. A study of mathematical models generated by uncertain input data and the worst scenario method is the main subject of the book by I. Hlaváček, J. Chleboun, and I. Babuška [175].

In [290] (see also P. Neittaanmäki and S. Repin [244]), a posteriori estimates discussed in Chapter 4 were used to evaluate errors induced by indeterminacy in coefficients of elliptic problems. Subsequent investigations of this problem and results of numerical tests are presented in O. Mali and S. Repin [225]. Recently, estimates of diam $\boldsymbol{S}$ for linear diffusion problems has been derived by a different method (see O. Mali and S. Repin [226, 227]).

It should be noted that the influence of various uncertainties (e.g., roundoff errors) affecting the accuracy of computations can also be investigated in terms of the so-called *interval analysis*. In it, the operations are performed for intervals instead of numbers. This theory is beyond the scope of the present book. The reader interested in it is referred to, e.g., G. Alefeld and J. Herzberger [11], B. S. Dobronets and V. V. Shaĭdurov [115], Y. V. Matijasevich [228], R. E. Moore [237] and to the literature cited therein.

9.7 Error estimates in terms of functionals and nonenergy norms

A posteriori estimates are intended to present a computable measure of the difference between an exact solution of a certain boundary value problem and an approximate one obtained by some numerical technology. In most cases, a posteriori estimates are derived in global (e.g., energy) norms and, therefore, justify the overall accuracy of an approximation considered. However, such information may be not sufficient because numerical analysts are often interested in local errors and in errors expressed in terms of special problem-oriented functionals. In other words, it may be desirable to have estimates of the type

$$\Phi(u - v) \leq \overline{\mathfrak{M}}_{\Phi}(v, y, \mathcal{D}),$$

where $\Phi : V \to \mathbb{R}_+$ is a given functional and $\mathbb{R}_+$ denotes the set of nonnegative real numbers. In this section, we derive computable estimates in terms of local norms and other quantities, which may be used to obtain a comprehensive presentation on the error structure.

9.7.1 General framework

It is often required to measure the accuracy of approximate solutions in terms of functionals other than the energy norm. If the basic *error control problem* stated at the beginning of Section 4 is solved and we have certain $\underline{\mathfrak{M}}$ and $\overline{\mathfrak{M}}$ satisfying (8.5.5), then on can to construct computable and sharp error estimates for any error functional subject to the energy norm.

Assume that we are interested in the value of $\Phi(v - u)$, where $\Phi : V \to \mathbb{R}$ is a given functional such that

$$|\Phi(v_1 + v_2)| \leq \mu_1 \left(|\Phi(v_1)| + |\Phi(v_2)|\right), \qquad \forall v_1, v_2 \in V, \tag{9.7.1}$$

$$|\Phi(v_1)| \leq \mu_2 |||v_1|||, \tag{9.7.2}$$

where $\mu_1 \geq 1$ and $\mu_2 > 0$.

Set $v_1 = \phi$ and $v_2 = u - v - \phi$, where $\phi \in V_0$. From (9.7.1) it follows that

$$\begin{aligned} |\Phi(u - v)| &\leq \mu_1 |\Phi(\phi)| + \mu_1 |\Phi(u - v - \phi)| \\ &\leq \mu_1 |\Phi(\phi)| + \mu_1 \mu_2 |||u - v - \phi|||. \end{aligned} \tag{9.7.3}$$

Consider $v + \phi \in u_0 + V_0$ as an approximation of u and apply the majorant to the second term. We have

$$|\Phi(u - v)| \leq \mu_1 |\Phi(\phi)| + \mu_1 \mu_2 \overline{\mathfrak{M}}(v + \phi, \mathcal{D}), \tag{9.7.4}$$

where $\mathcal{D}$ stands for the problem data and free functions.

Set $v_1 = u - v$ and $v_1 + v_2 = \phi$. Then,

$$|\Phi(u-v)| \geq \frac{1}{\mu_1}\,\Phi(\phi) - \mu_2 \overline{\mathfrak{M}}\,(v+\phi, \mathcal{D}). \tag{9.7.5}$$

Since $\overline{\mathfrak{M}}\,(u, \mathcal{D}) = 0$, we observe that, by choosing $\phi = u - v$ the right-hand side of (9.7.4) is equal to $|\Phi(u-v)|$. If $\mu_1 = 1$, then the lower bound (9.7.5) is also sharp.

Also, from (9.7.4) we observe that an upper bound of $|\Phi(u-v)|$ can be obtained if $\overline{\mathfrak{M}}$ *is additionally minimized on the kernel of* Φ, i.e.,

$$|\Phi(u-v)| \leq \mu_1\mu_2 \inf_{\phi \in \operatorname{Ker}\Phi} \left\{\overline{\mathfrak{M}}\,(v+\phi, \mathcal{D})\right\}. \tag{9.7.6}$$

9.7.2 Estimates in local norms

Consider the diffusion problem (4.1.1)–(4.1.3). Let ω be a subdomain of Ω with Lipschitz continuous boundary $\partial\omega$. Set

$$\Phi(u-v) = |||\nabla(u-v)|||_\omega := \left(\int_\omega A\nabla(u-v)\cdot(u-v)\,dx\right)^{1/2}.$$

In this case, $\mu_1 = \mu_2 = 1$ and we have the following estimates:

$$|||\nabla(u-v)|||_\omega \leq |||\nabla\phi|||_\omega + \overline{\mathfrak{M}}_{\mathrm{DF}}(v+\phi, y) =: \overline{\mathfrak{M}}_{\mathrm{DF}\omega}(v, \phi, y), \tag{9.7.7}$$

$$|||\nabla(u-v)|||_\omega \geq |||\nabla\phi|||_\omega - \overline{\mathfrak{M}}_{\mathrm{DF}}(v+\phi, y) =: \underline{\mathfrak{M}}_{\mathrm{DF}\omega}(v, \phi, y), \tag{9.7.8}$$

where $\phi \in V_0$. It is easy to see that the estimates (9.7.7) and (9.7.8) have no gaps (if $y = A\nabla u$ and $\phi = u - v$, they hold as equalities).

In particular, from (9.7.7) it follows that an upper bound of $|||\nabla(u-v)|||_\omega$ is obtained if $\overline{\mathfrak{M}}_{\mathrm{DF}}(v+\phi, y)$ is minimized over functions ϕ such that $\nabla\phi = 0$ in ω (see [291, 295, 294]).

Square both parts of (9.7.7) and use (4.1.14). Then, we obtain

$$\begin{aligned} |||\nabla(u-v)|||_\omega^2 \leq (1+\gamma)|||\nabla\phi|||_\omega^2 &+ \frac{(1+\beta)(1+\gamma)}{\gamma}|||A\nabla(v-\phi) - y|||_*^2 \\ &+ \frac{(1+\beta)(1+\gamma)}{\beta\gamma}\mathrm{C}\left(\|\operatorname{div} y + f\|^2 + \|F - y\cdot n\|^2\right), \end{aligned} \tag{9.7.9}$$

where β and γ are positive numbers, $y \in H(\Omega, \operatorname{div})$, and ϕ is an arbitrary function from $V_{0\omega}$. If y, β, and γ are defined, then optimization of the upper bound is reduced to minimization of the quadratic functional

$$J_\omega(\phi) := \kappa|||\nabla\phi|||^2 + |||\nabla\phi|||_\omega^2 - 2\kappa\int_\Omega \nabla\phi\cdot(A\nabla v - y)\,dx \tag{9.7.10}$$

on the set V_0. In (9.7.10), $\kappa = \frac{1+\beta}{\gamma}$. This problem can be approximately solved if J_ω is minimized on a sufficiently reach subspace $V_{0h} \in V_0$.

It should be noted that the last term on the right-hand side of (9.7.9) does not depend on ϕ, so that the quality of y states an accuracy limit for the local norm. Therefore, an efficient evaluation of $|||\nabla(u-v)|||^2_\omega$ requires not only finding ϕ but also needs a vector-valued function y such that $\|\operatorname{div} y + f\|$ is sufficiently small. However, such difficulties are quite predictable and it is natural to await that getting guaranteed and accurate estimates for local norms (which provide a more detailed information on the quality of an approximate solution) should be more expensive than for the global energy norm.

Remark 9.27. If y is equilibrated, then we can set $\phi = w$, where w is obtained by maximization of the minorant $\underline{\mathfrak{M}}_{\mathrm{DF}}(v, w)$. The same choice is valid for any error functional Φ that satisfies (9.7.1)–(9.7.2). Thus, we conclude that if the global upper and lower bounds computed by the majorant and minorant are sufficiently close, then the corresponding w provides a good image of $u-v$ and, therefore, it can also be used for estimation of local errors and other quantities of interest without noticeable increasing in the computational cost. In fact, these observations mean that the expenditures spent for a careful control of global errors can be utilized for other purposes.

9.7.3 Estimates in terms of linear functionals

In Section 2.6.4, we discussed some error estimation methods used to evaluate the quantity

$$\mathcal{E}_\ell(u-v) :=| \langle \ell, u-v \rangle |,$$

where $\ell \in V_0^*$ is a given linear functional and $\langle \cdot, \cdot \rangle$ denotes the duality pairing of the spaces V_0^* and V_0. The functional ℓ is usually taken in such a way that its value characterizes some especially important properties of a solution. It is necessary to note that $\mathcal{E}_\ell$ is *only a seminorm* on V, so that small values of this indicator do not guarantee small values of the error. Indeed, if $u \neq v$, but $e := u - v \in \operatorname{Ker} \ell$, then $\mathcal{E}_\ell(u-v) = 0$. Thus, indicators of this type are meaningful only if e contains a significant component orthogonal to $\operatorname{Ker} \ell$. Despite of this fact, error indicators of such a type are often used in error control methods in combination with global error estimates. Guaranteed upper bounds of $|\langle \ell, u-v \rangle|$ can be derived in several different ways discussed below.

1. The simplest way is to use the inequality

$$|\langle \ell, u-v \rangle| \leq \boldsymbol{|} \ell \boldsymbol{|} \, |||\nabla(u-v)|||, \tag{9.7.11}$$

which implies the estimate

$$|\langle \ell, u-v \rangle| \leq \boldsymbol{|} \ell \boldsymbol{|} \, \overline{\mathfrak{M}}_{\mathrm{DF}}(v, y), \tag{9.7.12}$$

where

$$|\!|\ell|\!| := \sup_{w\in V_0} \frac{|\langle \ell, w\rangle|}{|\!|\!|\nabla w|\!|\!|}.$$

However, the right-hand side of (9.7.11) may essentially overestimate the left-hand one and, in general, the upper bound given by (9.7.12) may be very coarse.

2. Another upper bound follows from (2.6.34). Since

$$E_2(u, u_h, u_\ell, u_{\ell\tau}) \le |\!|\!|\nabla(u_h - u)|\!|\!|\,|\!|\!|\nabla(u_{\ell\tau} - u_\ell)|\!|\!|,$$

we find that

$$|\langle \ell, u - u_h\rangle| \le \Big| \int_\Omega (f u_{\ell\tau} - A\nabla u_h \cdot \nabla u_{\ell\tau})\, dx \\ + \overline{\mathfrak{M}}_{\mathrm{f,DF}}(u_h, y_1)\, \overline{\mathfrak{M}}_{\ell,\mathrm{DF}}(u_{\ell\tau}, y_2)\Big|, \tag{9.7.13}$$

where y_1 and y_2 are two (different) functions in $H(\Omega, \mathrm{div})$. By $\overline{\mathfrak{M}}_{\mathrm{f,DF}}$ and $\overline{\mathfrak{M}}_{\ell,\mathrm{DF}}$ we denote the majorants related to the diffusion problems with source terms f and ℓ, respectively. This upper bound is sharper than in (9.7.12). Regrettably, its computation is rather expensive, because, in addition to $u_{\ell\tau}$, it requires finding y_1 and y_2 sufficiently close to exact fluxes of solutions to the original and adjoint problems, respectively.

3. Another method is based upon the relation

$$|\langle \ell, u - v\rangle| = |\langle \ell, u - v - \phi\rangle|, \qquad \forall \phi \in V_{0\ell}(\Omega), \tag{9.7.14}$$

where

$$V_{0\ell}(\Omega) := \{\phi \in V_0(\Omega) \mid \langle \ell, \phi\rangle = 0\}.$$

By (9.7.14) we obtain

$$|\langle \ell, u - v\rangle| \le |\!|\ell|\!| \inf_{\phi\in V_{0\ell}} |\!|\!|u - v - \phi|\!|\!|. \tag{9.7.15}$$

We show that (9.7.15) holds as equality. Indeed, let $u_\ell \in V_0$ be a function such that

$$\int_\Omega A\nabla u_\ell \cdot \nabla w\, dx = \langle \ell, w\rangle, \qquad \forall w \in V_0. \tag{9.7.16}$$

Since ℓ is a bounded linear functional (on V_0), the solution u_ℓ exists, is unique, and satisfies the relation $|\!|\!|\nabla u_\ell|\!|\!| = |\!|\ell|\!|$.

Set

$$\bar\phi = u - v - \frac{\langle \ell, u - v\rangle}{|\!|\ell|\!|^2}\, u_\ell.$$

Then,

$$\langle \ell, \bar{\phi} \rangle = \langle \ell, u - v \rangle \left(1 - \frac{\langle \ell, u_\ell \rangle}{|\ell|^2} \right) = 0,$$

so that $\bar{\phi} \in V_{0\ell}$. Hence,

$$\begin{aligned} \inf_{\phi \in V_{0\ell}} |||u - v - \phi||| &\leq |||u - v - \bar{\phi}||| \\ &= |\langle \ell, u - v \rangle| \frac{|||\nabla u_\ell|||}{|\ell|^2} = \frac{1}{|\ell|} |\langle \ell, u - v \rangle|, \end{aligned}$$

and we see that the left-hand side of (9.7.15) is equal to the right-hand one.

Estimate the right-hand side of (9.7.15) by the error majorant. Then, we find that

$$|\langle \ell, u - v \rangle| \leq |\ell| \, \overline{\mathfrak{M}}_{\mathrm{DF}}(v + \phi, y). \tag{9.7.17}$$

It is easy to see that

$$\begin{aligned} |\langle \ell, u - v \rangle| &\leq |\ell| \inf_{\substack{y \in H_{\Gamma_2}(\Omega, \mathrm{div}) \\ \phi \in V_{0\ell}}} \overline{\mathfrak{M}}_{\mathrm{DF}}(v + \phi, y) \leq |\ell| \, \overline{\mathfrak{M}}_{\mathrm{DF}}(v + \bar{\phi}, A\nabla u) \\ &= |\ell| \, |||A\nabla(u - v - \bar{\phi})|||_* = |\ell| \, |||u - v - \bar{\phi}||| = |\langle \ell, u - v \rangle|. \end{aligned}$$

Thus, additional minimization of the majorant $\overline{\mathfrak{M}}_{\mathrm{DF}}$ over the set $V_{0\ell}$ is sufficient to get a sharp upper bound of $|\langle \ell, u - v \rangle|$.

4. An estimate of the quantity $|\langle \ell, u - v \rangle|$ follows from (9.7.4) if we set

$$\Phi(u - v) = \ell(u - v).$$

In this case, $\mu_1 = 1$ and $\mu_2 = |\ell|$. Therefore, we have the estimate

$$|\langle \ell, u - v \rangle| \leq |\langle \ell, \phi \rangle| + |\ell| \, \overline{\mathfrak{M}}_{\mathrm{DF}}(v + \phi, y), \tag{9.7.18}$$

where y and ϕ are arbitrary functions in $H_{\Gamma_2}(\Omega, \mathrm{div})$ and V_0, respectively. Usually, the value of $|\ell|$ is not difficult to estimate. For example, if

$$\langle \ell, u - v \rangle = \int_\Omega \lambda (u - v) \, dx, \qquad \text{where} \quad \lambda \in L^2(\Omega), \tag{9.7.19}$$

then $|\ell| \leq \|\lambda\| \frac{C_{F\Omega}}{c_1}$. If

$$\langle \ell, u - v \rangle = \int_\Omega \tau \cdot \nabla(u - v) \, dx, \tag{9.7.20}$$

where $\tau \in L^2(\Omega, \mathbb{R}^d)$ is a given vector-valued function, then

$$\left| \int_\Omega \tau \cdot \nabla w \, dx \right| \leq |||\tau|||_* |||\nabla w|||$$

and $\| \ell \| \leq |||\tau|||_*$.

In another frequently encountered case, the functional is defined by the integral

$$\langle \ell, u - v \rangle = \int_{\partial\omega} \gamma \cdot (u - v) \, ds, \tag{9.7.21}$$

where $\gamma \in L^2(\partial\omega)$ is a given weight function. In this case, the value of $\| \ell \|$ is estimated by the constant in the trace inequality associated with $\partial\omega$.

9.7.4 Estimates based on the Poincaré inequality

In Sections 3.5.3, 4.1.3, and 4.2 we used the Poincaré inequality for deriving error estimates in terms of energy norms. Below we shortly discuss one other application of this inequality.

Let ω be again a connected subset of Ω with a Lipschitz boundary. In view of the Poincaré inequality, we have

$$\|u - v\|_{2,\omega}^2 \leq C_{P\omega} \left(\|\nabla(u - v)\|_{2,\omega}^2 + \left(\int_\omega (u - v) \, dx \right)^2 \right), \tag{9.7.22}$$

where u is the exact solution of (4.1.1)–(4.1.3) and $v \in u_0 + V_0$ is an approximation. For some subdomains (simplexes, circles, squares, cubes, etc.), the constants $C_{P\omega}$ can be evaluated analytically. For example, if

$$\omega = \Pi_l := \{x \mid 0 < x_i < l_i, \; i = 1, 2, \ldots, d\},$$

then (see (1.4.31))

$$\|u - v\|_{\Pi_l}^2 \leq \frac{1}{|\Pi_l|} \left(\int_{\Pi_l} (u - v) \, dx \right)^2 + \frac{d}{2} \max_i \{l_i^2\} \, \|\nabla(u - v)\|_{\Pi_l}^2 . \tag{9.7.23}$$

Thus, if the error in the local energy norm is evaluated, then adding the error in terms of the linear functional $\int_{\Pi_l} (u - v) \, dx$ provides an upper bound of the local error in L^2. Also, (9.7.23) implies a special characteristic of the error. Introduce the quantity

$$\mathrm{Os}_\omega(u - v) := \inf_{\alpha \in \mathbb{R}} \|u - v - \alpha\|_\omega ,$$

which characterizes the oscillatory part of the error $u - v$ related to ω. Obviously

$$\mathrm{Os}_\omega(u - v) \leq \|u - v - \bar{\alpha}\|_\omega ,$$

where $\bar{\alpha} = \{u - v\}_\omega$. Therefore, (9.7.23) implies the estimate

$$\mathrm{Os}_{\Pi_l}(u - v) \leq \sqrt{\frac{d}{2}} \max_i \{l_i\} \, \|\nabla(u - v)\|_{\Pi_l} . \tag{9.7.24}$$

Hence, by (9.7.7) we can also evaluate certain part of the local error.

9.7.5 Estimates based on multiplicative inequalities

To derive estimates in L^p norms, we can use well-known embedding inequalities analogous to (9.7.22). However, these estimates involve new constants (in place of $C_{P\omega}$). The multiplicative inequality (e.g., see O. A. Ladyzhenskaya and N. N. Uraltseva [214]) opens a way for avoiding the necessity of computing new constants. It has the form

$$\|w\|_{p,\omega} \leq C_{p,\alpha} \|\nabla w\|_{2,\omega}^{\alpha} \|w\|_{2,\omega}^{1-\alpha}, \qquad \forall w \in \overset{\circ}{H}{}^1(\omega),\ p > 2, \tag{9.7.25}$$

where

$$\alpha = \left(\tfrac{1}{2} - \tfrac{1}{p}\right)\left(\tfrac{1}{2} - \tfrac{d-2}{2d}\right)^{-1}.$$

If $d = 2$, then we can take $p \in [2, +\infty)$, and the constant is defined by the relation $C_{p,\alpha} = \max\left\{\frac{p}{2}, 2\right\}^{\alpha}$. If $d = 3$, then $p \in [2, 6]$ and $C_{p,\alpha} = 4^{\alpha}$.

Let $\phi \in W^{1,\infty}(\omega)$ be a function such that

$$0 \leq \phi(x) \leq 1, \quad \operatorname{supp}\phi \subset \omega, \quad \text{and} \quad \phi(x) = 1 \text{ in } \omega' \subset \omega.$$

By (9.7.25), we conclude that

$$\|\phi(u-v)\|_{p,\omega} \leq C_{p,\alpha} \|\nabla(\phi(u-v))\|_{2,\omega}^{\alpha} \|\phi(u-v)\|_{2,\omega}^{1-\alpha}.$$

Since

$$\|\nabla(\phi(u-v))\|_{2,\omega}^2 \leq (1+\delta)\,\|\phi\,|\nabla(u-v)|\|_{2,\omega}^2 + \frac{1+\delta}{\delta}\,\||\nabla\phi|\,(u-v)\|_{2,\omega}^2,$$

where δ is an arbitrary positive number, we obtain

$$\begin{aligned}\|u-v\|_{p,\omega'} &\leq \|\phi(u-v)\|_{p,\omega} \\ &\leq C_{p,\alpha}\Big((1+\delta)\,\|\nabla(u-v)\|_{2,\omega}^2 \\ &\qquad + \frac{1+\delta}{\delta}\left\||\nabla\phi|^2\right\|_{\infty,\omega} \|u-v\|_{2,\omega}^2\Big)^{\alpha/2} \|u-v\|_{2,\omega}^{1-\alpha}.\end{aligned} \tag{9.7.26}$$

Thus, having estimates of the local errors in terms of L^2-norms, we obtain an upper bound of the error in any L^p-norm without computing new global constants.

9.7.6 Estimates based on the maximum principle

Pointwise estimates of approximation errors can be derived with the help of known estimates for partial differential equations, which follow from the *maximum principle* (e.g., see D. Gilbarg and N. S. Trudinger [151] and O. A. Ladyzhenskaya and N. N. Uraltseva [214]). In the simplest case, it reads as follows:

Theorem 9.28. *Let $\mathcal{A}$ be a uniformly elliptic operator of the second order, which is defined in a bounded domain Ω with Lipschitz boundary Γ. Assume that $u^+ \in C^2(\Omega) \bigcap C^0(\Omega)$ and*

$$\mathcal{A}u^+ \geq 0. \tag{9.7.27}$$

Then, the function u^+ attains its maximum on Γ, i.e.,

$$\sup_\Omega u^+ = \sup_\Gamma u^+. \tag{9.7.28}$$

This principle holds for many elliptic operators. In particular, it holds for the operator $Av := \operatorname{div} A\nabla v + b \cdot \nabla v + cv$ provided that $c \leq 0$, the coefficients are bounded, and $|b|$ is small with respect to the ellipticity constant c_1 (so that the ellipticity condition is satisfied).

The function u^+ in Theorem 9.28 is called a *sub-solution* associated with $\mathcal{A}$. A function $u^- \in C^2(\Omega) \bigcap C^0(\Omega)$ that satisfies the condition $\mathcal{A}u^- \leq 0$ is called a *super-solution*. If $\mathcal{A}$ is the operator Δ, then sub- and super-solutions are presented by *sub-* and *super-harmonic* functions, respectively.

Consider the problem $Au = f$ in Ω with homogeneous Dirichlet boundary conditions. Assume that we have an approximate solution $\widetilde{u} \in C(\bar{\Omega}) \cap C^2(\Omega)$ that satisfies the condition

$$A\widetilde{u} \geq f \quad \text{in} \quad \Omega.$$

Then $A(\widetilde{u} - u) \geq 0$ and by the maximum principle we conclude that

$$\sup_\Omega (\widetilde{u} - u) \leq \sup_\Gamma (\widetilde{u} - u) = \sup_\Gamma \widetilde{u}. \tag{9.7.29}$$

The estimate (9.7.29) shows that the (pointwise) error is bounded by the quantity $\sup_\Gamma \widetilde{u}$. If $\widetilde{u}$ satisfies the relation $A\widetilde{u} \leq f$, then we find that $\inf_\Omega (\widetilde{u} - u) \geq \inf_\Gamma \widetilde{u}$.

Also, one can apply a more sophisticated estimate that based upon the following theorem (e.g., see [151]):

Theorem 9.29. *Let $A\widetilde{u} \geq f$ and $\widetilde{u} \in C(\bar{\Omega}) \cap C^2(\Omega)$. Then*

$$\sup_\Omega \widetilde{u} \leq \sup_\Gamma (\widetilde{u})_+ + \mu \sup_\Omega |(f)_-|/\lambda. \tag{9.7.30}$$

If $A\widetilde{u} = f$, then

$$\sup_\Omega |\widetilde{u}| \leq \sup_\Gamma |(\widetilde{u})_+| + \mu \sup_\Omega |f|/\lambda, \tag{9.7.31}$$

where $\lambda(x)$ is the lowest eigenvalue of $A(x)$ and the constant μ depends only on $\operatorname{diam}\Omega$ *and on the ratio $\beta = \sup|b|/\lambda$. In particular, if Ω lies between two parallel surfaces and d is the distance between them, then* $\mu = \exp\{(\beta + 1)d\} - 1$.

Theorem 9.29 implies estimates of the difference between u and any approximation $\widetilde{u} \in C(\bar{\Omega}) \cap C^2(\Omega)$. Set $e = u - \widetilde{u}$. We have

$$A e = Au - A\widetilde{u} = f - A\widetilde{u} =: \widetilde{f}.$$

By (9.7.31), we obtain

$$\sup_{\Omega} |e| \leq \sup_{\Gamma} |(u - \widetilde{u})_+| + \mu \sup_{\Omega} |\widetilde{f}|/\lambda, \tag{9.7.32}$$

where we can estimate λ by (4.1.4). If the residual $\widetilde{f} = Av - f$ is of constant sign in Ω, then we can apply (9.7.30). The maximum principle and the estimate (9.7.30) can be extended to a wider class of functions. However, these functions must have second generalized derivatives summable in any subdomain of Ω, so that the regularity conditions that we must impose on $\widetilde{u}$ are rather strong. This fact may lead to certain technical difficulties in practical applications.

9.7.7 Estimates in weighted norms

Estimates in weighted norms can be useful if the significance of errors in different parts of Ω is different. We consider a way of deriving such estimates with the paradigm of the problem

$$\Delta u + f = 0 \qquad \text{in } \Omega, \tag{9.7.33}$$

$$u = u_0 \qquad \text{on } \Gamma. \tag{9.7.34}$$

Let ϕ be a smooth (or piecewise smooth) positive weight function. We wish to measure the error in terms of the norm

$$\|\nabla(u - v)\|^2_{[\phi]} := \int_{\Omega} \phi |\nabla(u - v)|^2 \, dx.$$

From the corresponding integral identity, we find that

$$\int_{\Omega} \nabla(u - v) \cdot \nabla(\phi w) \, dx = \int_{\Omega} (f \phi w - \nabla v \cdot \nabla(\phi w)) \, dx, \qquad \forall w \in V_0, \tag{9.7.35}$$

where $V_0 = \mathring{H}^1(\Omega)$ and $v \in u_0 + V_0$ is an approximate solution. Set $w = u - v$ and rewrite (9.7.35) in the form

$$\int_{\Omega} \phi |\nabla(u - v)|^2 \, dx + \frac{1}{2} \int_{\Omega} \nabla \phi \cdot \nabla((u - v)^2) \, dx$$
$$= \int_{\Omega} ((f \phi - \nabla \phi \cdot \nabla v)(u - v) - \phi \nabla v \cdot \nabla(u - v)) \, dx. \tag{9.7.36}$$

By the identity

$$\int_\Omega (\phi y \cdot \nabla(u-v) + (u-v) y \cdot \nabla\phi + \phi(u-v)\operatorname{div} y)\,dx = 0,$$

we obtain

$$\begin{aligned}\int_\Omega \phi|\nabla(u-v)|^2\,dx + \frac{1}{2}\int_\Omega \nabla\phi\cdot\nabla((u-v)^2)\,dx \\ = \int_\Omega \big((f+\operatorname{div} y)\phi + \nabla\phi\cdot(y-\nabla v))(u-v) \\ + \phi(y-\nabla v)\cdot\nabla(u-v)\big)\,dx.\end{aligned} \tag{9.7.37}$$

If ϕ is sufficiently regular, then we integrate by parts in the second term and deduce the estimate

$$\begin{aligned}\int_\Omega \phi|\nabla(u-v)|^2\,dx - \frac{1}{2}\int_\Omega \Delta\phi\,(u-v)^2\,dx \\ = \int_\Omega \big((f+\operatorname{div} y)\phi + \nabla\phi\cdot(y-\nabla v))(u-v) \\ + \phi(y-\nabla v)\cdot\nabla(u-v)\big)\,dx.\end{aligned} \tag{9.7.38}$$

If ϕ is a harmonic (or superharmonic) function, then (9.7.38) implies the estimate

$$\begin{aligned}&\int_\Omega \phi|\nabla(u-v)|^2\,dx \\ &\quad\le \int_\Omega ((f+\operatorname{div} y)\phi + \nabla\phi\cdot(y-\nabla v))(u-v) + \phi(y-\nabla v)\cdot\nabla(u-v))\,dx.\end{aligned}$$

It is easy to see that a constant $C_{\phi\Omega}$ exists such that

$$\|w\| \le C_{\phi\Omega}\|\nabla w\|_{[\phi]}, \qquad \forall w \in V_0.$$

In particular, one can set $C_{\phi\Omega} = C_{F\Omega}\,(\phi_0)^{-1/2}$, where $\phi_0 = \min_{x\in\Omega}\{\phi(x)\}$. Since

$$\begin{aligned}&\int_\Omega ((f+\operatorname{div} y)\phi + \nabla\phi\cdot(y-\nabla v))(u-v)\,dx \\ &\qquad \le C_{\phi\Omega}\|(f+\operatorname{div} y)\phi + \nabla\phi\cdot(y-\nabla v)\|\|\nabla(u-v)\|_{[\phi]},\end{aligned}$$

we arrive at the estimate

$$\|\nabla(u-v)\|_{[\phi]} \le C_{\phi\Omega}\|(f+\operatorname{div} y)\phi + \nabla\phi\cdot(y-\nabla v)\| + \|\phi^{1/2}(y-\nabla v)\|. \tag{9.7.39}$$

Remark 9.30. Let $\psi(x) \in C^2(\Omega)$ be a superharmonic function and $b \in \mathbb{R}^d$ be a vector independent of x. Set $\phi(x) = \psi(x) + b \cdot x + \epsilon$, where

$$\epsilon > |\min_{x\in\Omega}\{\psi(x) + b \cdot x\}|.$$

Since $\Delta\phi = \Delta\psi$, the function ϕ is a nonnegative superharmonic function, which can be used in (9.7.39).

We can rearrange (9.7.37) without assuming that ϕ is superharmonic. For this reason, we shift the term with $\nabla((u-v)^2$ to the right-hand side of (9.7.38). Let $(\Delta\phi)_+$ denote the positive part of $\Delta\phi$. Note that

$$\begin{aligned}\frac{1}{2}\left|\int_\Omega \nabla\phi \cdot \nabla((u-v)^2)\,dx\right| &= \frac{1}{2}\left|\int_\Omega \Delta\phi\,(u-v)^2\,dx\right| \\ &\leq \frac{1}{2}\int_\Omega (\Delta\phi)_+\,(u-v)^2\,dx \\ &\leq \|\,(\Delta\phi)_+\,\|_\infty \|u-v\|^2 \\ &\leq \|\,(\Delta\phi)_+\,\|_\infty C^2_{\phi\Omega}\|\nabla(u-v)\|_{[\phi]}.\end{aligned}$$

Hence, (9.7.37) implies the estimate

$$\begin{aligned}&\left(1 - C^2_{\phi\Omega}\|\,(\Delta\phi)_+\,\|_\infty\right)\|\nabla(u-v)\|_{[\phi]} \\ &\qquad \leq C_{\phi\Omega}\|(f + \operatorname{div} y)\phi + \nabla\phi\cdot(y - \nabla v)\| + \|\phi^{1/2}(y - \nabla v)\|, \qquad (9.7.40)\end{aligned}$$

which has a meaning only if $\|\,(\Delta\phi)_+\,\|_\infty$ is sufficiently small.

Another estimate follows from the relation

$$\begin{aligned}\int_\Omega \phi|\nabla(u-v)|^2\,dx &= \int_\Omega (\phi(f + \operatorname{div} y) + \nabla\phi\cdot(y - \nabla v))\,(u-v)\,dx \\ &\quad + \int_\Omega \phi(y - \nabla v)\cdot\nabla(u-v)\,dx \\ &\quad + \frac{1}{2}\int_\Omega (q - \nabla\phi)\cdot\nabla((u-v)^2)\,dx,\end{aligned}$$

where q is an arbitrary function in $L^\infty(\Omega, \mathbb{R}^d) \cap \mathrm{S}(\Omega)$. Since

$$\begin{aligned}&\frac{1}{2}\int_\Omega (q - \nabla\phi)\cdot\nabla((u-v)^2)\,dx \\ &\quad \leq \|q - \nabla\phi\|_\infty \int_\Omega |u-v|\,|\nabla(u-v)|\,dx \\ &\quad \leq \|q - \nabla\phi\|_\infty\|\nabla(u-v)\|_{[\phi]}\left(\int_\Omega \phi^{-1}|u-v|^2\,dx\right)^{1/2} \\ &\quad \leq \frac{C_{F\Omega}}{\phi_0}\|q - \nabla\phi\|_\infty\|\nabla(u-v)\|^2_{[\phi]}\end{aligned}$$

we find that

$$\left(1-\frac{C_{F\Omega}}{\phi_0}\|q-\nabla\phi\|_\infty\right)\|\nabla(u-v)\|_{[\phi]}$$
$$\leq C_{F\Omega}\|\phi^{1/2}(f+\operatorname{div} y)+\nabla\phi\cdot(y-\nabla v))\|+\|\phi^{1/2}(y-\nabla v)\|. \quad (9.7.41)$$

Certainly, this estimate makes a sense only if the quantity in parentheses is positive. If ϕ is a harmonic function, then its gradient is a divergence free vector-valued function and the term $\|q-\nabla\phi\|_\infty$ vanishes.

Finally, we note that a posteriori estimates in weighted norms can be derived for other elliptic problems, using transformations of integral identities quite similar to those we applied to the problem (9.7.33)–(9.7.34) (see [302]).

Bibliography

[1] Y. Achdou, C. Bernardi, and F. Coquel, *A priori and a posteriori analysis of finite volume discretizations of Darcy's equations*, Numer. Math. 96 (2003), pp. 17–42.

[2] A. Agouzal, *On the saturation assumption and hierarchical a posteriori error estimator,* Comput. Meth. Appl. Math. 1 (2001), pp. 125–131.

[3] M. Ainsworth, *A posteriori error estimation for fully discrete hierarchic models of elliptic boundary value problems in thin domains,* Numer. Math. 80 (1998), pp. 325–362.

[4] ———, *Robust a posteriori error estimation for nonconforming finite element approximation*, SIAM J. Numer. Anal. 42 (2005), pp. 2320–2341.

[5] ———, *A posteriori error estimation for discontinuous Galerkin finite element approximation*, SIAM J. Numer. Anal. 45 (2007), 4, pp. 1777–1798.

[6] M. Ainsworth and J. T. Oden, *A posteriori error estimation in the finite element method,* Numer. Math. 60 (1992), pp. 429–463.

[7] ———, *A unified approach to a posteriori error estimation using element residual methods,* Numer. Math. 65 (1993), pp. 23–50.

[8] ———, *A posteriori error estimation in finite element analysis*, Wiley and Sons, New York, 2000.

[9] M. Ainsworth, J. T. Oden, and C. Y. Lee, *Local a posteriori error estimators for variational inequalities*, Numer. Methods for PDE 9 (1993), pp. 23–33.

[10] M. Ainsworth, J. Z. Zhu, A. W. Craig, and O. C. Zienkiewicz, *Analysis of the Zienkiewicz–Zhu a posteriori error estimator in the finite element method*, Int. J. Numer. Methods Engrg. 28 (1989), pp. 2161–2174.

[11] G. Alefeld and J. Herzberger, *Introduction to interval computations,* Academic Press, New York, 1983.

[12] A. Alonso, *Error estimators for a mixed method*, Numer. Math. 74 (1996), pp. 385–395.

[13] M. Amara, M. Ben Younes, and C. Bernardi, *Error indicators for the Navier–Stokes equations in stream function and vorticity formulation*, Numer. Math. 80 (1998), pp. 181–206.

[14] A. M. Arthurth, *Complementary variational principles,* Clarendon Press, Oxford, 1980.

[15] S. N. Atluri (ed.), *Computational methods in the mechanics of fracture*, North-Holland, New York, 1986.

[16] J. P. Aubin, *Behavior of the error of the approximate solutions of boundary-value problems for linear elliptic operators by Galerkin's and finite difference methods*, Ann, Scuola Norm. Sup. Pisa 21 (1967), pp. 599–637.

[17] G. Auchmuty, *A posteriori error estimates for linear equations,* Numer. Math. 61 (1992), pp. 1–6.

[18] A. Babin, A. Mahalov, and B. Nicolaenko, *Global splitting, integrability and regularity of 3D Euler and Navier–Stokes equations for uniformly rotating fluids*, European. J. Mech. 15 (1996), pp. 291–300.

[19] ———, *Global regularity of 3D rotating Navier–Stokes equations for resonant domains*, Indiana Univ. Math. J. 48 (1999), pp. 1133–1176.

[20] I. Babuška, *Courant element: before and after*, Fifty Years of Courant Element, edited by M. Křížek, P. Neittaanmäki, and R. Stenberg, Marcel Dekker, 1994, pp. 37–51.

[21] ———, *The finite element method with Lagrangian multipliers,* Numer. Math. 20 (1973), pp. 179–192.

[22] I. Babuška and A. K. Aziz, *Surway lectures on the mathematical foundations of the finite element method*, The mathematical formulations of the finite element method with applications to partial differential equations, Academic Press, New York, 1972, pp. 5–359.

[23] I. Babuška and J. Chleboun, *Effects of uncertainties in the domain on the solution of Neumann boundary-value problems in two spatial dimensions,* Math. Comput. 71 (2001), pp. 1339–1370.

[24] ———, *Effects of uncertainties in the domain on the solution of Drichlet boundary-value problems*, Numer. Math. 93 (2003), pp. 583–610.

[25] I. Babuška, R. Duran and R. Rodriguez, *Analysis of the efficiency of a posteriori error estimator for linear triangular elements,* SIAM J. Numer. Anal. 29 (1992), pp. 947–964.

[26] I. Babuška and G. N. Gatica, *On the mixed finite element method with Lagrange multipliers*, Numer. Meth. Partial Diff. Eq. 19 (2003), pp. 192–210.

[27] I. Babuška, F. Ihlenburg, T. Strouboulis, and S. K. Gangaraj, *A posteriori error estimation for finite element solutions on Helmholtz' equation – Part II: estimation of the pollution error,* Int. J. Numer. Meth. Engrg. 40 (1997), pp. 3883–3900.

[28] I. Babuška, I. Lee, and C. Schwab, *On the a posteriori estimation of the modeling error for the heat conduction in a plate and its use for adaptive hierarchical modeling,* Appl. Numer. Math. 14 (1994), pp. 5–21.

[29] I. Babuška, F. Nobile, and R. Tempone, *Worst case scenarion analysis for elliptic problems with uncertainty,* Numer. Math. 101 (2005), pp. 185–219.

[30] I. Babuška and J. E. Osborn, *Can a finite element method perform arbitrarily badly?,* Math. Comput. 69 (2000), pp. 443–462.

[31] I. Babuška and W. C. Rheinboldt, *A-posteriori error estimates for the finite element method,* Internat. J. Numer. Meth. Engrg. 12 (1978), pp. 1597–1615.

[32] ———, *Error estimates for adaptive finite element computations,* SIAM J. Numer. Anal. 15 (1978), pp. 736–754

[33] I. Babuška and R. Rodriguez, *The problem of the selection of an a posteriori error indicator based on smoothing techniques,* Internat. J. Numer. Meth. Engrg. 36 (1993), pp. 539–567.

[34] I. Babuška and C. Schwab, *A posteriori error estimation for hierarchic models of elliptic boundary value problems on thin domains,* SIAM J. Numer. Anal. 33 (1996), pp. 221–246.

[35] I. Babuška and T. Strouboulis, *The finite element method and its reliability*, Oxford University Press, New York, 2001.

[36] I. Babuška, T. Strouboulis, S. K. Gangaraj, and C. S. Upadhyay, *Pollution-error in the h-version of the FEM and the local quality of recovered derivatives,* Comput. Methods Appl. Mech. Engrg. 140 (1997), pp. 1–37.

[37] I. Babuška, T. Strouboulis, A. Mathur, and C. S. Upadhyay, *Pollution error in the h-version of the FEM and the local quality of a posteriori error estimators,* Finite Elem. Anal. and Des. 17 (1994), pp. 273–321.

[38] I. Babuška, T. Strouboulis, and C. S. Upadhyay, *A model study of the quality of a posteriori error estimators for linear elliptic problems. Error estimation in the interior of patchwise uniform grids of triangles,* Comp. Meth. Appl. Mech. Engrg. 114 (1994), pp. 307–378.

[39] I. Babuška, T. Strouboulis, C. S. Upadhyay, and S. K. Gangaraj, *A posteriori estimation and adaptive control of the pollution-error in the h-version of the FEM,* Int. J. Numer. Meth. Engrg. 38 (1995), pp. 4207–4235.

[40] W. Bangerth and R. Rannacher, *Adaptive finite element methods for differential equations,* Birkhäuser-Verlag, Berlin, 2003.

[41] R. E. Bank and R. K. Smith, *Mesh smoothing using a posteriori error estimates,* SIAM J. Numer. Anal. 34 (1997), pp. 979–997.

[42] R. E. Bank and A. Weiser, *Some a posteriori error estimators for elliptic partial differential equations,* Math. Comput. 44 (1985), pp. 283–301.

[43] R. E. Bank and B. D. Welfert, *A posteriori error estimates for the Stokes problem,* SIAM J. Numer. Anal. 28 (1991), pp. 591–623.

[44] S. Bartels and C. Carstensen, *Averaging techniques yield reliable a posteriori finite element error control for obstacle problems,* Numer. Math. 99 (2004), pp. 225–249.

[45] R. Beck, R. Hiptmair, R. Hoppe, and B. Wohlmuth, *Residual based a posteriori error estimators for eddy current computation,* RAIRO – Math. Model. Numer. Anal. 34 (2000), 159–182.

[46] R. Becker, P. Hansbo, and M. G. Larsson, *Energy norm a posteriori error estimation for discontinuous Galerkin methods,* Comput. Methods Appl. Mech. Engrg. 192 (2003), pp. 723–733.

[47] R. Becker, H. Kapp, and R. Rannacher, *Adaptive finite element methods for optimal control of partial differential equations: Basic concept,* SIAM J. Control Optim. 39 (2000), pp. 113–132.

[48] R. Becker and R. Rannacher, *A feed-back approach to error control in finite element methods: Basic approach and examples,* East–West J. Numer. Math. 4 (1996), pp. 237–264.

[49] ———, *An optimal control approach to error estimation and mesh adaptation,* Acta Numerica 2000, edited by A. Iserles, Cambridge University Press, 2001, pp. 1–102.

[50] R. Becker and B. Vexler, *A posteriori error estimation for finite element discretization of parameter identification problems,* Numer. Math. 96 (2004), pp. 435–459.

[51] A. Bergam, Z. Mghazli, and R. Verfürth, *A posteriori estimators for the finite volume discretization of an elliptic problem,* Numer. Math. 95 (2003), pp. 599–624.

[52] M. Bergounioux, M. Haddou, M. Hintermüller, and K. Kunisch, *A comparison of a Moreau–Yosida-Based Active Set Strategy and Interior Point Methods for Constrained Optimal Control Problems,* SIAM J. Optim. 11 (2000), pp. 495–521.

[53] A. Bermúdez, R. Durán, and R. Rodríguez, *Finite element analysis of compressible and incompressible fluid-solid systems,* Math. Comput. 67 (1998), pp. 111–136.

[54] C. Bernardi and V. Girault, *A local regularization operator for triangular and quadrilateral finite elements,* SIAM J. Numer. Anal. 35 (1998) pp. 1893–1916.

[55] C. Bernardi and R. Verfürth, *A posteriori error analysis of the fully discretized time-dependent Stokes equations,* ECIAM:M2NL 38 (2004), pp. 437–455.

[56] M. Bildhauer and M. Fuchs, *Error estimates for obstacle problems of higher order,* Zap. Nauchn, Sem. S.-Peterburg, Otdel. Mat. Inst. Steklov. (POMI) 326 (2007), pp. 5–18.

[57] M. Bildhauer, M. Fuchs, and S. Repin, *A posteriori error estimates for stationary slow flows of power-law fluids,* J. Non-Newtonian Fluid Mech. 142 (2007), pp. 112–122.

[58] ———, *A functional type a posteriori error analysis for Ramberg–Osgood Model,* Z. Angew. Math. Mech. (ZAMM) 87 (2007), pp. 860–876.

[59] ———, *The elasto-plastic torsion problem: a posteriori estimates for approximate solutions,* Universität des Saarlandes, 2007, preprint.

[60] ———, *Duality based a posteriori error estimates for higher order variational inequalities with power growth functionals,* Annales Academiæ Scentiarum Fennicæ, Mathematics 33 (2008), pp. 475–490.

[61] M. Bildhauer and S. Repin, *Estimates for the deviation from exact solutions of variational problems with power growth functionals,* Zapiski Nauchn, Semin V. A. Steklov Mathematical Institute in St.-Petersburg (POMI) 336 (2006), pp. 5–24.

[62] J. Bonvin, M. Picasso, and R. Stenberg, *GLS and EVSS methods for a three-field Stokes problem arising from viscoelastic flows,* Comput. Methods Appl. Mech. Engrg. 190 (2001), pp. 3893–3914.

[63] B. Boroomand and O. C. Zienkiewicz, *Recovery by equilibrium in patches (REP),* Int. J. Numer. Meth. Engrg. 40 (1997), pp. 137–164.

[64] ———, *An improved REP recovery and the effectivity robustness test,* Int. J. Numer. Meth. Engrg. 40 (1997), pp. 3247–3277.

[65] M.-H. Bossel, *Membranes élastiquement liées: extension du théorème de Rayleigh–Faber–Krahn et de linéqualité de Cheeger,* C. R. Acad. Sci., Paris, Sér.I Math. 302 (1986), pp. 47-50.

[66] S. W. Brady and A. R. Elcrat, *Some results on a posteriori error estimation for approximate solution of second order elliptic problems,* SIAM J. Numer. Anal. 16 (1979), pp. 877–889.

[67] D. Braess, *Finite elements,* Cambridge University Press, Cambridge, 1997.

[68] ———, *A posteriori error estimators for obstacle problems – another look,* Numer. Math. 101 (2005), pp. 415–421.

[69] D. Braess and J. Schöberl, *Equilibrated residual error estimator for Maxwell's equation,* RICAM Preprint, 2006.

[70] D. Braess and R. Verfürth, *A posteriori error estimators for the Raviart–Thomas element,* SIAM J. Numer. Anal. 33 (1996), pp. 2431–2444.

[71] J. H. Bramble, *A proof of the inf-sup condition for the Stokes equations on Lipschitz domains,* Math. Models Methods Appl. Sci 13 (2003), pp. 361–371.

[72] J. H. Bramble, R. D. Lazarov, and J. E. Pasciak, *Least-squares for second-order elliptic problems,* Comput. Methods Appl. Mech. Engrg. 152 (1998), pp. 195–210.

[73] J. H. Bramble and A. H. Schatz, *Higher order local accuracy by averaging in the finite element method,* Math. Comput. 31 (1977), pp. 94–111.

[74] J. H. Brandts, *Superconvergence and a posteriori error estimation for triangular mixed finite elements,* Numer. Math. 68 (1994), pp. 311–324.

[75] K. Braune, *A posteriori error bounds in the computation of inverse standard functions with the aid of the Newton method,* Z. Angew. Math. Mech. 70 (1990), pp. T579–T581 (in German).

[76] S. Brenner and R. L. Scott, *The mathematical theory of finite element methods*, Springer-Verlag, Berlin, Heidelberg, New York, 1994.

[77] F. Brezzi, *On the existence, uniqueness and approximation of saddle-point problems arising from Lagrange multipliers,* R.A.I.R.O., Annal. Numer. 8 (1974), R-2, pp. 129–151.

[78] F. Brezzi, J. J. Douglas, R. Duran, and M. Fortin, *Mixed finite elements for second order elliptic problems in three variables,* Numer. Math. 51 (1987), pp. 237–250.

[79] F. Brezzi and M. Fortin, *Mixed and hybrid finite element methods*, Springer Series in Computational Mathematics, vol. 15, New York, 1991.

[80] F. Brezzi, M. Fortin, and R. Stenberg, *Error analysis of mixed-interpolated elements for Reissner-Midlin plates,* Math. Models and Methods in Applied Sciences 1 (1991), pp. 125–151.

[81] F. Brezzi, C. Johnson, and B. Mercier, *Analysis of a mixed finite element method for elasto-plastic plates,* Mathematics of Computation 31 (1977), pp. 809–817.

[82] H. Buss and S. Repin, *A posteriori error estimates for boundary-value problems with obstacles,* Numerical mathematics and advanced applications (Jyväskylä, 1999), World Sci. Publishing, River Edge, New York, 2000, pp. 162–170.

[83] R. Bustinza, G. N. Gatica, and B. Cockburn, *An a posteriori error estimate for the local Discontinuous Galerkin method applied to linear and nonlinear diffusion problems,* Journal of Scientific Computing 22/23 (2005), pp. 147–185.

[84] G. F. Carey and D. L. Humphrey, *Mesh refinement and iterative solution methods for finite element computations,* Int. J. Numer. Meth. Engrg. 17 (1981), pp. 1717–1734.

[85] G. F. Carey and A. I. Pehlivanov, *Local error estimation and adaptive remeshing scheme for least-squares mixed finite elements,* Comput. Methods Appl. Mech. Engrg. 150 (1997), pp. 125–131.

[86] C. Carstensen, *A posteriori estimate for the mixed finite element method,* Math. Comput. 66 (1997), pp. 218, 465–476.

[87] ———, Quasi-interpolation and a posteriori error analysis of finite element methods, Mathematical Modelling in Numerical Analysis 33 (1999), pp. 1187–1202.

[88] ———, *A posteriori error estimate for the mixed finite element method,* Math. Comput. 66 (1997), pp. 465–476.

[89] ———, *An a posteriori error estimate for first-kind integral equation,* Math. Comput. 66 (1997), pp. 139–155.

[90] C. Carstensen and S. Bartels, Each averaging technique yields reliable a posteriori error control in FEM on unstructured grids. I: Low order conforming, nonconforming, and mixed FEM, Math. Comput. 71 (2002), pp. 945–969.

[91] C. Carstensen and G. Dolzmann, *A posteriori error estimates for mixed FEM in elasticity,* Numer. Math. 81 (1998), pp. 187–209.

[92] C. Carstensen and S. A. Funken, *Fully reliable localized error control in the FEM,* SIAM J. Sci. Comput. 21 (2000), pp. 1465–1484.

[93] ———, *Constants in Clément-interpolation error and residual based a posteriori error estimates in Finite Element Methods,* East–West J. Numer. Math. 8 (2000), pp. 153–175.

[94] C. Carstensen and R. H. W. Hoppe, *Error reduction and convergence for an adaptive mixed finite element method,* Math. Comput. 75 (2006), pp. 1033–1042

[95] ———, *Convergence analysis of an adaptive nonconforming finite element method,* Numer. Math. 103 (2006), pp. 251–266.

[96] C. Carstensen, J. Hu, and A. Orlando, *Framework for the A Posteriori Error Analysis of Nonconforming Finite Elements*, SIAM J. Numer. Anal. 45 (2007), pp. 68–82.

[97] C. Carstensen, R. Klose, and A. Orlando, *Reliable and efficient equilibrated a posteriori finite element error control in elastoplasticity and elastoviscoplasticity with hardening,* Comput.Methods Appl. Mech. Engrg. 195 (2006), pp. 25742598.

[98] C. Carstensen, R. Lazarov, and S. Tomov, *Explicit and averaging a posteriori error estimates for adaptive finite volume methods,* SIAM J. Numer. Anal. 42 (2005), pp. 2496–2521.

[99] C. Carstensen and S. Sauter, *A posteriori error analysis for elliptic PDEs on domains with complicated structures,* Numer. Math. 96 (2004), pp. 691–721.

[100] C. Carstensen and R. Verfürth, *Edge residuals dominate a posteriori error estimates for low order finite element methods,* SIAM J. Numer. Anal. 36 (1999), pp. 1571–1587.

[101] P. Castillo, *An a posteriori error estimate for the local Discontinuous Galerkin method,* Journal of Scientific Computing 22/23 (2005), pp. 187–2004.

[102] A. Charbonneau, K. Dossou, and R. Pierre, *A residual-based a posteriori error estimator for Ciarlet–Raviart formulation of the first biharmonic problem,* Numer. Meth. for PDE's 13 (1997), pp. 93–111.

[103] Z. Chen and R. H. Nochetto, *Rezidual type a posteriori error estimates for elliptic obstacle problems,* Numer. Math. 84 (2000), pp. 527–548.

[104] X. Cheng, W. Han, and H. Huang, *Analysis of some mixed elements for the Stokes problem,* Journal of Computational and Applied Mathematics 85 (1997), pp. 19–35.

[105] E. Chizhonkov and M. Olshanskii, *On the domain geometry dependence of the LBB condition,* Mathematical Modelling and Numerical Analys 34 (2000), pp. 935–951.

[106] S.-S. Chow, G. F. Carey, and R. D. Lazarov, *Natural and post-processed superconvergence in semilinear problems,* Numer. Meth. PDE 7 (1991), pp. 245–259.

[107] P. G. Ciarlet, *The finite element method for elliptic problems*, North Holland, New York, Oxford, 1978.

[108] Ph. Clément, *Approximations by finite element functions using local regularization,* RAIRO Anal. Numér. 9 (1975), pp. 77–84.

[109] S. Cochez-Dhondt, S. Nicaise, and S. Repin, *A posteriori error estimates for finite volume approximations*, Technical report, Universitet Valenciennes, 2008.

[110] L. Collatz, *Funktionanalysis und Numerische Mathematik*, Springer-Verlag, Berlin, 1964.

[111] P. Coorevits, J.-P. Dumeau, and J.-P. Pelle, *Error estimator and adaptivity for three-dimensional finite element analysis*, Advances in Adaptive Computational Methods in Mechanics, edited by P. Ladevéze and J. T. Oden, Elsevier, New York, 1998, pp. 443–457.

[112] R. Courant, *Variational methods for some problems of equilibrium and vibration,* Bulletin of AMS 49 (1943), pp. 1–23.

[113] E. Dari, R. Duran, C. Padra, and V. Vampa, A posteriori error estimators for nonconforming finite element methods, M2AN Math. Model. Numer. Anal. 30 (1996), pp. 385–400.

[114] P. Destuynder and B. Métivet, *Explicit error bounds of a conforming finite element method,* Math. Comput. 68 (1999), pp. 1379–1396.

[115] B. S. Dobronets and V. V. Shaĭdurov, *Two-sided numerical methods*, Nauka Sibirsk. Otdel., Novosibirsk, 1990 (in Russian).

[116] M. Dobrowolski, *On the LBB constant on stretched domains,* Math. Nachr. 254–255 (2003), pp. 64–67.

[117] W. Dörfler and M. Rumpf, *An adaptive strategy for elliptic problems including a posteriori controlled boundary approximation,* Math. Comput. 67 (1998), 224, pp. 1361–1362.

[118] R. Duran, M. A. Muschietti, and R. Rodriguez, *On the asymptotic exactness of error estimators for linear triangle elements,* Numer. Math. 59 (1991), pp. 107–127.

[119] R. Duran and R. Rodriguez, *On the asymptotic exactness of Bank–Weiser's estimator,* Numer. Math. 62 (1992), pp. 297–303.

[120] G. Duvaut and J.-L. Lions, *Les inequations en mecanique et en physique*, Dunod, Paris, 1972.

[121] I. Ekeland and R. Temam, *Convex analysis and variational problems*, North-Holland, Amsterdam, 1976.

[122] B. Engelmann, R. H. W. Hoppe, B. Wohlmuth, Yu. Kuznetsov, Yu. Iliash, and Yu. Vasilevskii, *Adaptive macro-hybrid finite element methods,* Proc. of the 2nd European Conference on Numerical Mathematics and Advanced Applications, Heidelberg, 1997, World Sci., River Edge, New York, 1998, pp. 294–302.

[123] H. W. Engl and O. Scherzer, *Convergence rates results for iterative methods for solving nonlinear ill-posed problems,* Surveys on solution methods for inverse problems, Springer-Verlag, Vienna, 2000, pp. 7–34.

[124] K. Eriksson, D. Estep, P. Hansbo, and C. Johnson, *Computational differential equations*, Cambridge University Press, Cambridge, 1996.

[125] K. Eriksson and C. Johnson, *An adaptive finite element method for linear elliptic problems,* Math. Comput. 50 (1988), pp. 361–383.

[126] ———, *Adaptive finite element methods for parabolic problems. I. A linear model problem,* SIAM J. Numer. Anal. 28 (1991), pp. 43–77.

[127] A. Ern and J. Proft, *A posteriori Discontinuous Galerkin error estimates for transient convection-diffusion equations,* Appl. Math. Lett. 18 (2005), pp. 833–841.

[128] A. Ern, A. F. Stephansen, and M. Vohralik, *Improved energy norm a posteriori error estimation based on flux reconstruction for discontinuous Galerkin methods,* manuscript, 2007.

[129] R. E. Ewing, *A posteriori error estimation,* Comput. Meth. Appl. Mech. Engrg. 82 (1990), pp. 59–72.

[130] R. E. Ewing, R. D. Lazarov, and J. Wang, *Superconvergence of the velocity along the Gauss lines in mixed finite element methods,* SIAM J. Numer. Anal. 18 (1991), pp. 1015–1029.

[131] M. Feistauer, *Mathematical methods in fluid dynamics,* Longman, Harlow, 1993.

[132] W. Fenchel, *On the conjugate convex functions,* Canad. J. Math. 1 (1949), pp. 73–77.

[133] A. Friedman, *Variational principles and free-boundary problems*, Wiley and Sons, New York, 1982.

[134] M. Frolov, *Reliable control over approximation errors by functional type a posteriori error estimates,* Phd thesis, Jyväskylä Studies in Computing 44, University of Jyväskulä, 2004.

[135] ———, *On the efficiency of the dual majorant method for the quality estimation of approximate solutions of fourth-order elliptic boundary-value problems,* Russian J. Numer. Anal. Math. Modelling 19 (2004), pp. 407–418.

[136] M. Frolov, P. Neittaanmäki, and S. Repin, *On the reliability, effectivity and robustness of a posteriori error estimation methods*, Numerical Methods for Scientific Computing, Variational problems and applications, CIMNE, Barcelona, 2003, pp. 153–175.

[137] ———, *On computational properties of a posteriori error estimates based upon the method of duality error majorants*, Numerical mathematics and advanced applications, Springer-Verlag, Berlin, 2004, pp. 346–357.

[138] ———, *Guaranteed functional error estimates for the Reissner-Mindlin plate problem,* J. Math. Sci (New York) 132 (2006), pp. 553–561.

[139] M. Fuchs and S. Repin, *Estimates for the deviation from the exact solutions of variational problems modeling certain classes of generalized Newtonian fluids,* Math. Meth. Appl. Sci. 29 (2006), pp. 2225–2244.

[140] M. Fuchs and G. Seregin, *Variational methods for problems from plasticity theory and for generalized Newtonian fluids*, Lecture Notes in Math., vol. 1749, Springer-Verlag, Berlin, 2000.

[141] A. Gaevskaya, R. H. W. Hoppe, Y. Iliash, and M. Kieweg, *Convergence analysis of an adaptive finite element method for distributed control problems with control constraints,* Control of coupled partial differential equations, Internat. Ser. Numer. Math., 155 (2007), Birkhäuser-Verlag, Basel, pp. 47–68.

[142] A. Gaevskaya, R. H. W. Hoppe, and S. Repin, *A posteriori estimates for cost functionals of optimal control problems,* Proceedings of the 6th European Conference on Numerical Mathematics and Advanced Applications, Santiago de Compostela, 2005, Springer-Verlag, Berlin, 2006, pp. 291–298.

[143] ———, *A posteriori error estimation for elliptic optimal control problems with distributed control,* Journal of Mathematical Sciences 144 (2007), pp. 1–14.

[144] A. Gaevskaya and S. Repin, *A posteriori error estimates for approximate solutions of linear parabolic problems,* Differential Equations, 41 (2005), 7, pp. 970–983 (in Russian); translation in Differ. Equ. 41 (2005), 7, pp. 970–983.

[145] H. Gajewski, K. Gröger, and K. Zacharias, *Nichtlineare Operatorgleichungen and Operatordifferentialgleichungen,* Akademie-Verlag, Berlin, 1974.

[146] G. Galdi, *An introduction to the mathematical theory of the Navier–Stokes equations,* Springer-Verlag, New York, 1994.

[147] B. G. Galerkin, *Beams and plates. Series in some questions of elastic equilibrium of beams and plates,* Vestnik Ingenerov 19 (1915), pp. 897–908 (in Russian).

[148] L. Gallimard, P. Ladevéze, and J.-P. Pelle, *Error estimation and adaptivity in elastoplasticity,* Int. J. Numer. Meth. Engrg. 39 (1996), pp. 189–217.

[149] G. Gatica, *A note on the efficiency of residual-based a-posteriori error estimators for some mixed finite element methods,* Electron.Trans.Numer.Anal. 17 (2004), pp. 218233.

[150] E. Geisler, A. Tal, and D. Garg, *On a posteriori error bounds for the solution of ordinary nonlinear differential equations,* Computers and Mathematics with Applications, Pergamon, Oxford, 1976, pp. 407–416.

[151] D. Gilbarg and N. S. Trudinger, *Elliptic partial differential equations of second order*, Springer-Verlag, Berlin, 1977.

[152] V. Girault and P. A. Raviart, *Finite element approximation of the Navier–Stokes equations*, Springer-Verlag, Berlin, 1986.

[153] R. Glowinski, *Numerical methods for nonlinear variational problems*, Springer-Verlag, New-York, 1982.

[154] ———, *Finite element methods for incompressible viscous flow*, Handbook of numerical analysis, Vol. IX, 3–1176, North-Holland, Amsterdam, 2003.

[155] R. Glowinski, J.-L. Lions, R. Trémolierès, *Analyse numérique des inéquations variationnelles*, Dunod, Paris, 1976.

[156] R. Glowinski, T.-W. Pan, and J. Periaux, *A fictitious domain method for Dirichlet problem and applications,* Comput. Methods Appl. Mech. Engrg. 111 (1994), pp. 283–303.

[157] R. Glowinski and O. Pironneau, *Numerical Methods for the First Biharmonic Equation and for the Two-Dimensional Stokes Problem,* SIAM Review 21 (1999), pp. 167–212.

[158] E. Gorshkova, *A posteriori error estimates and adaptive methods for incompressible viscous flow problems*, Phd thesis, Jyväskylä Studies in Computing 86, University of Jyväskulä, 2007.

[159] E. Gorshkova, A. Mahalov, P. Neittaanmaki, and S. Repin, *Functional a posteriori estimates for viscous flow problems with rotation,* J. Math. Sci. New York 142 (2007), pp. 927–935.

[160] E. Gorshkova and P. Neittaanmaki. Mesh-adaptive methods for viscous flow problems with rotation, In "Adaptive Modeling and Simulation 2007", edited by K. Runesson and P. Diez, CIMNE, Barcelona, 2007, pp. 109–111.

[161] E. Gorshkova, P. Neittaanmaki, and S. Repin, *Comparative study of a posteriori error estimates for the Stokes problem,* Numerical Mathematics and Advanced Applications (ENUMATH 2005), Springer-Verlag, Berlin, 2006, pp. 255–262.

[162] ———, *A posteriori error estimates for semi-discrete approximations of the evolutionary Stokes problem,* Technical Report, University of Jyväskylä (2007).

[163] E. Gorshkova and S. Repin, *Error control of the approximate solution to the Stokes equation using a posteriori error estimates of functional type,* European Congress on Computational Methods in Applied Sciences and Engineering, ECCOMAS 2004, Jyväskylä, 24–28 July, 2004, edited by P. Neittaanmäki, T. Rossi, K. Majava, and O. Pironeau, O. Nevanlinna and R. Rannacher (assoc. eds.), (electronic).

[164] G. Haase, M. Kuhn, and U. Langer, *Parallel multigrid 3D Maxwell solvers*, Parallel Computing 27 (2001), pp. 761–775.

[165] L. Halpern, *Spectral methods in polar coordinates for the Stokes problem, Application to computation in unbounded domains,* Math. Comput. 65 (1996), pp. 507–531.

[166] W. Han, *Finite element analysis of a holonomic elasto-plastic problem,* Numer. Math. 60 (1992), pp. 493–508.

[167] ———, *A posteriori error analysis for linearization of nonlinear elliptic problems and their discretization,* Math. Meth. Appl. Sci. 17 (1994), pp. 487–508.

[168] A. Hanukainen, *Functional type a posteriori error estimates for the Maxwell's equations,* ENUMATH 2007, September 10–14 2007, book of abstracts, p. 114.

[169] Y. Hayashi, *On a posteriori error estimation in the numerical solution of systems of ordinary differential equations,* Hiroshima Math. J 9 (1979), pp. 201–243.

[170] B.-O. Heimsund, X.-C. Tai, and J. Wang, *Superconvergence for the gradient of finite element approximations by L^2 projections,* SIAM J. Numer. Anal. 40 (2002), pp. 1263–1280.

[171] J. G. Heywood, W. Nagata, and W. Xie, *A numerically based existence theorem for the Navier–Stokes equations,* J. Math. Fluid. Mech. 1 (1999), pp. 5–23.

[172] J. G. Heywood and R. Rannacher, *Finite element approximations of the nonstationary Navier–Stokes problem, Part II: Stability of solutions and error estimates uniform in time,* SIAM J. Numer. Anal. 23 (1986), pp. 750–777.

[173] M. Hintermüller, *A primal-dual active set algorithm for bilaterally control constrained optimal control problems,* Quarterly of Applied Mathematics LXI (2003), pp. 131–161.

[174] R. Hiptmair, *Multigrid method for Maxwell's equations,* SIAM J. Numer. Anal. 36 (1999), pp. 204–225.

[175] I. Hlaváček, J. Chleboun, and I. Babuška, *Uncertain input data problems and the worst scenario method*, Elsevier, Amsterdam, 2004.

[176] I. Hlaváček and M. Křížek, *On a superconvergence finite element scheme for elliptic systems. I. Dirichlet boundary conditions,* Aplikace Matematiky, 32 (1987), pp. 131–154.

[177] R. H. W. Hoppe, Y. Iliash, C. Iyyunni, and N. H. Sweilam, *A posteriori error estimates for adaptive finite element discretizations of boundary control problems,* J. Numer. Math. 14 (2006), pp. 57–82

[178] R. H. W. Hoppe and R. Kornhuber, *Adaptive multilevel methods for obstacle problems,* SIAM J. Numer. Anal. 31 (1994), pp. 301–323.

[179] C. O. Horgan. *Korn's inequalities and their applications in continuum mechanics,* SIAM Review 37 (1995), pp. 491–511.

[180] P. Houston, J. A. Mackenzie, E. Suli, and G. Warnecke, *A posteriori analysis for numerical approximations of Friederichs systems,* Numer. Math. 82 (1999), pp. 433–470.

[181] P. Houston, I. Perugia, and D. Schotzau, *A posteriori error estimation for discontinuous Galerkin discretizations of* $H(curl)$*elliptic partial differential equations,* IMA Journal of Numerical Analysis (2005), pp. 1–25.

[182] P. Houston, R. Rannacher, E. Süli, *A posteriori error analysis for stabilised finite element approximations of transport problems,* Comput. Methods Appl. Mech. Engrg. 190 (2000), pp. 1483–1508.

[183] A. A. Iljushin, *Plasticite. (Deformations elastico-plastiques)*, Eyrolles, Paris, 1956.

[184] Jin Qi-nian, *Error estimates of some Newton-type methods for solving nonlinear inverse problems in Hilbert scales,* Inverse Problems 16 (2000), pp. 187–197.

[185] C. Johnson, *Numerical Solution of Partial Differential Equations by the Finite Element Methods*, Cambridge University Press, Cambridge, 1987.

[186] ———, *Adaptive finite element methods for diffusion and convection problems,* Comput. Methods Appl. Mech. Engrg. 82 (1990), pp. 301–322.

[187] C. Johnson and P. Hansbo, *Adaptive finite elements in computational mechanics,* Comput. Methods Appl. Mech. Engrg. 101 (1992), pp. 143–181.

[188] C. Johnson and B. Mercier, *Some equilibrium finite element methods two-dimensional elasticity problems,* Numer. Math. 30 (1978), pp. 101–116.

[189] C. Johnson, R. Rannacher, and M. Boman, *Numerics in hydrodynamic stability. Towards error control in CFD,* SIAM J. Numer. Anal. 32 (1995), pp. 1058–1079.

[190] C. Johnson and A. Szepessy, *Adaptive finite element methods for conservation laws based on a posteriori error estimates,* Commun, Pure and Appl. Math. XLVIII (1995), pp. 199–234.

[191] V. Jovanovic and C. Rohde, *Finite-volume schemes for Friedrichs systems in multiple space dimensions: a priori and a posteriori error estimates,* Numer. Methods Partial Differ. Equations 21 (2005), pp. 104–131.

[192] T. Kato, *On some approximation methods concerning the operator TT^*, Math. Ann.* 126 (1953), pp. 253–262.

[193] D. Kay and D. Silvester, *A-posteriori error estimation for stabilised mixed approximations of the Stokes equations,* SIAM J. Sci. Comput. 21 (2000), pp. 1321–1336.

[194] D. W. Kelly, *The self equilibration of residuals and complementary error estimates in the finite element method,* Internat. J. Numer. Meth. Engrg. 20 (1984) pp. 1491–1506.

[195] D. W. Kelly, J. R. Gago, O. C. Zienkiewicz, and I. Babuška, *A posteriori error analysis and adaptive processes in the finite element method. Part I – error analysis,* Internat. J. Numer. Meth. Engrg. 19 (1983), pp. 1593–1619.

[196] D. Kröner and M. Ohlberger, *A posteriori error estimates for upwind finite volume schemes for nonlinear conservation laws in multi dimensions,* Math. Comp. 69 (1999), pp. 25–39.

[197] A. N. Kolmogorov and S. V. Fomin, *Introductory real analysis*, Dover Publications, Inc., New York, 1975.

[198] R. Kornhuber, *A posteriori error estimates for elliptic variational inequalities,* Comput. Math. Appl. 31 (1996), pp. 49–60.

[199] S. Korotov, P. Neittaanmaki, and S. Repin, *A posteriori error estimation of goal-oriented quantities by the superconvergence patch recovery,* J. Numer. Math. 11 (2003), pp. 33–59.

[200] M. Křížek, P. Neittaanmäki, and R. Stenberg (eds.), *Finite Element Methods. Superconvergence, Post-Processing and A Posteriori Error Estimates*, Lecture Notes in Pure and Applied Mathematics, Vol. 196, Marcel Dekker, New York, 1998.

[201] M. Křížek and P. Neittaanmäki, *Finite Element Approximations of Variational Problems and Applications*. Wiley and Sons, New York, 1990.

[202] ———, *Superconvergence phenomenon in the finite element method arising from averaging of gradients,* Numer. Math. 45 (1984), pp. 105–116.

[203] G. Kunert, *An a posteriori residual error estimator for the finite element method on anisotropic tetrahedral meshes,* Numer. Math. 86 (2000), pp.471–490.

[204] ———, *Towards anisotropic mesh construction and error estimation in the finite element method,* Numer. Meth. PDE 18 (2002), pp. 625–648.

[205] Yu. Kuznetsov and S. Repin, *New mixed finite element method on polygonal and polyhedral meshes,* Russian J. Numer. Anal. Math. Modeling 18 (2003), pp. 261–278.

[206] ———, *Convergence analysis and error estimates for mixed finite element method on distorted meshes,* Journal of Numerical Mathematics 13 (2005), pp. 22–51.

[207] P. Ladevéze and D. Leguillon, *Error estimate procedure in the finite element method and applications,* SIAM J. Numer. Anal. 20 (1983), pp. 485–509.

[208] P. Ladevéze, J.-P. Pelle, and Ph. Rougeot, *Error estimation and mesh optimization for classical finite elements,* Engineering Computations 8 (1991), pp. 69–80.

[209] P. Ladevéze and Ph. Rougeot, *New advances on a posteriori error on constitutive relation in f.e. analysis,* Comput. Methods Appl. Mech. Engrg. 150 (1997), pp. 239–249.

[210] O. A. Ladyzhenskaya, *The mathematical theory of viscous incompressible flow*, Gos. Izdat. Fiz.-Mat. Moscow, 1961 (in Russian); English transl. Gordon & Breach, New York-London, 1963.

[211] ———, *The boundary value problems of mathematical physics*, Springer-Verlag, New York, 1985.

[212] ———, *On mosified Navier–Stokes equations for large velocity gradients,* Zapiski Nauchnych Seminarov LOMI 7 (1968), pp. 126–154.

[213] O. A. Ladyzenskaja and V. A. Solonnikov, *Some problems of vector analysis, and generalized formulations of boundary value problems for the Navier–Stokes equation,* Zap. Nauchn, Sem. Leningrad. Otdel. Mat. Inst. Steklov. (LOMI) 59 (1976), pp. 81–116 (in Russian).

[214] O. A. Ladyzhenskaya and N. N. Uraltseva, *Linear and Quasilinear Elliptic equations*, Academic Press, New York, 1968.

[215] U. Langer and W. Queck, *On the convergence factor of Uzawa's algorithm,* J. Comput. Appl. Math. 15 (1986), pp. 191–202.

[216] M. G. Larson and A. Målquist, *Adaptive variational multiscale methods based on a posteriori error estimation: Energy norm estimates for elliptic problems,* Comput. Methods. Appl. Math. Engrg. 196 (2007), pp. 2313–2334.

[217] R. Lazarov, *Superconvergence of the gradient for triangular and tetrahedral finite elements of a solution of linear problems in elasticity theory,* Computational Processes and Systems 6 (1988), pp. 180–191 (in Russian).

[218] R. Lazarov, S. Repin, and S. Tomar, *Functional a posteriori error estimates for Discontinuous Galerkin method,* Radon Institute of Computational and Applied Mathematics, Preprint 2006-40, Linz.

[219] R. Lazarov and S. Tomov, *A posteriori error estimates for finite volume approximations of convection-diffusion-reaction equations,* Comput. Geosci. 6 (2002), pp. 483–503.

[220] A. S. Leonov, *Some a posteriori stopping rules for iterative methods for solving linear ill-posed problems,* Comput. Math. Math. Phys. 34 (1994), pp. 121–126.

[221] J.-L. Lions, *Optimal Control of Systems Governed by Partial Differential Equations,* Springer-Verlag, Berlin, Heidelberg, New York, 1971.

[222] J.-L. Lions and E. Magenes, *Problḿes aux limites non homogénes et applications,* Dunod, Paris, 1968.

[223] M. Lonsing and R. Verfürth, *A posteriori error estimators for mixed finite element methods in linear elasticity,* Numer. Math. 97 (2004), pp. 757–778.

[224] J. Ma and H. Brunner, *A posteriori error estimates of Discontinuous Galerkin methods for non-standard Volterra integro-differential equations,* IMA J. Numer. Anal. 26 (2006), pp. 78–95.

[225] O. Mali and S. Repin, *Estimates of the indeterminacy set for elliptic boundary-value problems with uncertain data,* In "Adaptive Modeling and Simulation 2007", edited by K. Runesson and P. Diez, CIMNE, Barcelona, 2007, pp. 158–161.

[226] ———, *Estimates of the indeterminacy set for elliptic boundary-value problems with uncertain data, J. Math. Sci.* 150 (2008), pp. 1869–1874.

[227] ———, *Computable estimates of the solution set for the reaction-diffusion problem with uncertain data* (to appear).

[228] Y. V. Matijasevich, *A posteriori interval analysis, Lecture Notes in Computer Sciences,* 204 (1985), pp. 328–334.

[229] J. Medina, M. Picasso, and J. Rappaz, *Error estimates and adaptive finite elements for nonlinear diffusion-convection problems*, Ecole Polytechnique Federale de Lausanne, Preprint CH-1015, 1995.

[230] D. Meidner and B. Vexler, *Adaptive space-time finite element methods for parabolic optimization problems*. SIAM J. Control Optim. 46 (2007), pp. 116–142.

[231] P. Meyer, *A unifying theorem on Newton's method,* Numer. Funct. Anal. Optim. 13 (1992), pp. 463–473.

[232] S. G. Mikhlin, *Variational methods in mathematical physics*, Pergamon, Oxford, 1964.

[233] ———, *Constants in some inequalities of analysis*, Wiley and Sons, Chiester–New York, 1986.

[234] ———, Error Analysis in Numerical Processes, Wiley and Sons, Chiester–New York, 1991.

[235] M. Möller and D. Kuzmin, *Adaptive mesh refinement for high-resolution finite element schemes,* Int. J. Numer. Meth. Fluids 52 (2006), pp. 545–569.

[236] P. Monk, *Finite Element Methods for Maxwell's Equations,* Clarendon Press, Oxford, 2003.

[237] R. E. Moore, *Interval analysis,* Prentice-Hall, Englewood Cliffs, New Jersey, 1966.

[238] K. W. Morton and E. Süli, *A posteriori and a priori error analysis of finite volume methods,* in *The Mathematics of Finite Elements and Applications*, edited by J. R. Whiteman, Wiley, 1994, pp. 267–288.

[239] P. Mosolov and V. Myasnikov, *Mechanics of rigid plastic bodies*, Nauka, Moscow, 1981 (in Russian).

[240] A. Muzalevsky and S. Repin, *On two-sided error estimates for approximate solutions of problems in the linear theory of elasticity,* Russian J. Numer. Anal. Math. Modelling 18 (2003), pp. 65–85.

[241] ———, *On error estimates for approximate solutions in problems of the linear theory of thermoelasticity,* Russian Math. (Iz. VUZ) 49 (2005), pp. 60–68.

[242] P. Neittaanmäki and M. Křížek, *On $O(h^4)$ superconvergence of piecewise bilinear FE-approximations,* Teubner-Texte Math., vol. 107, Teubner-Verlag, 1988, pp. 250–255.

[243] P. Neittaanmäki and S. Repin, *A posteriori error estimates for boundary-value problems related to the biharmonic operator,* East-West J. Numer. Math. 9 (2001), pp. 157–178.

[244] ———, *Reliable methods for computer simulation, Error control and a posteriori estimates*, Elsevier, New York, 2004.

[245] ———, *Guaranteed estimates of approximation errors for numerical solutions of the Maxwell equation* (to appear).

[246] P. Neittaanmäki, S. Repin, and P. Turchyn, *New A Posteriori Error Indicators in Terms of Linear Functionals for Linear Elliptic Problems,* Russian J. Numer. Anal. Math. Modelling 23 (2008), pp. 77–87.

[247] S. Nicaise, *A posteriori error estimation of some cell centered finite volume methods for diffusion-convection-reaction problems,* SIAM J. Numer. Anal. 44 (2006), pp. 949–978.

[248] S. Nicaise and S. Repin, *Functional a posteriori error estimates for the reaction-convection-diffusion problem,* Zapiski Nauchn, Semin V. A. Steklov Mathematical Institute in St.-Petersburg (POMI) 326 (2007), pp. 1–2.

[249] J. A. Nitsche, *Ein Kriterium für die Quasi-Optimalität des Ritzschen Verfahrens,* Numer. Math. 11 (1968), pp. 346–348.

[250] J. T. Oden, I. Babuška, F. Nobile, Y. Feng, and R. Tempone, *Theory and methodology for estimation and control of errors due to modeling, approximation, and uncertainty,* Comput. Methods Appl. Mech. Engrg. 194 (2005), pp 195–204.

[251] J. T. Oden, L. Demkowicz, W. Rachowicz, and T. A. Westermann, *Towards a universal $h - p$ adaptive finite element strategy. Part 2. A posteriori error estimation,* Comput. Methods Appl. Mech. Engrg. 77 (1989), pp. 113–180.

[252] J. T. Oden, L. Demkowicz, T. Strouboulis, and P. Devloo, *Adaptive methods for problems in solid and fluid mechanics,* Accuracy Estimates and Adaptive Refinements in Finite Element Computations (Lisbon, 1984), edited by I. Babuška, O. C. Zienkiewicz, J. Gago, and E. R. de Oliveira, Wiley, Chichester, 1996, pp. 249–280.

[253] J. T. Oden and S. Prudhomme, *Goal-oriented error estimation and adaptivity for the finite element method,* Comput. Math. Appl. 41 (2001), pp. 735–756.

[254] J. T. Oden and J. N. Reddy, *Mathematical theory of finite elements*, Wiley and Sons, 1976.

[255] J. T. Oden, W. Wu, and M. Ainsworth, *An a posteriori error estimate for finite element approximations of the Navier–Stokes equations,* Comput. Methods. Appl. Mech. Engrg. 111 (1994), pp. 185–202.

[256] L. A. Oganesjan and L. A. Ruchovec, *Study of the rate of convergence of variational difference scheme for second order elliptic equations in two-dimensional region with a smooth boundary,* U.S.S.R. Comput. Math. and Math. Phy. 9 (1969), pp. 158–183.

[257] A. Ostrowski, *Les estimations des erreurs a posteriori dans les procédés itératifs,* C. R. Acad. Sci, Paris, Sér. A–B 275 (1972), pp. A275–A278.

[258] C. Padra, *A posteriori error estimators for nonconforming approximation of some quasi-newtonian flows,* SIAM J. Numer. Anal. 34 (1997), pp. 1600–1615.

[259] P. Panagiotopoulos, *Inequality problems in mechanics and applications*, Birkhäuser-Verlag, Boston, 1985.

[260] J. Peraire and A. T. Patera, *Bounds for linear-functional outputs of coercive partial differential equations: Local indicators and adaptive refinement,* Advances in Adaptive Computational Methods in Mechanics, edited by P. Ladevéze and J. T. Oden, Elsevier, New York, 1998, pp. 199–228.

[261] T. Plewa, T. Linde, and W. G. Weirs (eds.), *Adaptive mesh refinement – theory and applications.* Proceedings of Chicago workshop on adaptive mesh refinement methods, Chicago, IL, USA, September 3–5, 2003, Springer-Verlag, Berlin, 2005.

[262] M. Plum, *Computer-assisted enclosure methods for elliptic differential equations,* Linear Algebra and its Applications 324 (2001) pp. 147187.

[263] F. Potra, *Sharp error bounds for a class of Newton-like methods,* Libertas Math. 5 (1985), pp. 71–84.

[264] J. Pousin and J. Rappaz, *Consistance, stabilité, erreurs a priori et a posteriori pour des probl emes non linéaries,* C. R. Acad. Sci. Paris 312 (1991), pp. 699–703.

[265] ——, *Consistency, stability, a priori and a posteriori errors for Petrov–Galerkin methods applied to nonlinear problems,* Numer. Math. 69 (1994), pp. 213–231.

[266] W. Prager and J. L. Synge, *Approximation in elasticity based on the concept of function space,* Quart. Appl. Math. 5 (1947), pp. 241–269.

[267] President's Information Technology Advisory Committee, *Computational Science: Ensuring America's Competitiveness*, Report to the President, June 2005, `http://www.nitrd.gov/pitac/reports`.

[268] Jin Qi-nian and Hou Zong-yi, *On a posteriori parameter choice strategy for Tikhonov regularization of nonlinear ill-posed problems,* Numer. Math. 83 (1999), pp 139–159.

[269] Lin Qun and Sci Jan, *Iterative correctins and a posteriori error estimate for integral equations,* Journal of Computational Math. 11 (1993), pp. 297–300.

[270] R. Rannacher, *The dual-weighted-residual method for error control and mesh adaptation in finite element methods,* The mathematics of finite elements and applications X, MAFELAP 1999. Proceedings of the 10th conference, Brunel Univ., Uxbridge, Middlesex, GB, June 22–25, 1999, edited by J. Whiteman, Elsevier, Amsterdam, 2000, pp. 97–116.

[271] ———, *Finite element methods for the incompressible Navier–Stokes equations,* Fundamental directions in mathematical fluid mechanics, edited by G. P. Galdi, J. H. Heywood, and R. Rannacher, Birkhauser-Verlag, Basel, 2000, pp. 191–293.

[272] R. Rannacher and F. T. Suttmeier, *A feed-back approach to error control in finite element methods: application to linear elasticity,* IWR, Preprint 96–42 (SFB 359), Heidelberg 1996.

[273] ———, *A posteriori error control and mesh adaptation for F.E. models in elasticity and elasto-plasticity,* Advances in Adaptive Computational Methods in Mechanics, edited by P. Ladevéze and J. T. Oden, Elsevier, New York, 1998, pp. 275–292.

[274] R. Rannacher and S. Turek, *A simple nonconforming quadrilateral Stokes element,* Numer. Meth. Part. Diff. Eq. 8 (1992), pp. 97–111.

[275] P.-A. Raviart and J.-M. Thomas, *A mixed finite element for second order elliptic problems,* Mathematical Aspects of Finite Element Methods, edited by I. Galligani and E. Magenes, Springer-Verlag, Berlin, 1977, pp. 292–315.

[276] S. Repin, *A posteriori estimates for approximate solutions of variational problems with strongly convex functionals,* Problems of Mathematical Analysis 17 (1997), pp. 199–226. (In Russian, translated in J. Math. Sci, (New York) 97 (1999), no. 4, pp. 4311–4328).

[277] ———, *A posteriori error estimation for nonlinear variational problems by duality theory,* Zapiski Nauchn, Semin, V. A. Steklov Mathematical Institute in St.-Petersburg (POMI) 243 (1997), pp. 201–214.

[278] ———, *A posteriori error estimation for variational problems with power growth functionals based on duality theory,* Zapiski Nauchn, Semin, V. A. Steklov Mathematical Institute in St.-Petersburg (POMI) 249 (1997), pp. 244–255.

[279] ———, *A posteriori error estimates for approximate solutions of variational problems,* Proc. 2nd European Conference on Numerical Mathematics and Advanced Applications, (Heidelberg) 1997, World Sci, Publishing, River Edge, New York, 1998, pp. 524–531.

[280] ———, *A unified approach to a posteriori error estimation based on duality error majorants,* Mathematics and Computers in Simulation 50 (1999), pp. 313–329.

[281] ———, *A posteriori estimates of the accuracy of variational methods for problems with nonconvex functionals,* Algebra i Analiz, 11 (1999), pp. 151–182 (in Russian, translated in St.-Petersburg Mathematical Journal 11 (2000), pp. 651–672).

[282] ———, *A posteriori error estimates for variational problems with uniformly convex functionals,* Math. Comput. 69 (230) (2000), pp. 481–500.

[283] ———, *Estimates of deviations from exact solutions of elliptic variational inequalities, Zapiski Nauchn. Semin. V. A. Steklov Mathematical Institute in St.-Petersburg (POMI)* 271 (2000), pp. 188–203.

[284] ——–, *Estimates for errors in two-dimensional models of elasticity theory,* J. Math. Sci, (New York) 106 (2001), pp. 3027–3041.

[285] ——–, *A posteriori estimates of the accuracy of dimensional reduction models in 3-D elasticity theory,* Finite element methods (Jyväskylä), 2000, GAKUTO Internat. Ser. Math. Sci, Appl., 15, Gakkotosho, Tokyo, 2001, pp. 240–253.

[286] ——–, *Two-sided estimates of deviation from exact solutions of uniformly elliptic equations,* Proc. St. Petersburg Math. Society IX (2001), pp. 143–171, translation in Amer. Math. Soc. Transl. Ser. 2, 209, Amer. Math. Soc., Providence, RI, 2003.

[287] ——–, *Estimates of deviation from exact solutions of initial-boundary value problems for the heat equation,* Rend. Mat. Acc. Lincei 13 (2002), pp. 121–133.

[288] ——–, *Aposteriori estimates for the Stokes problem,* J. Math. Sci, (New York) 109 (2002), pp. 1950–1964.

[289] ——–, *Estimates of deviations for generalized Newtonian fluids,* Zapiski Nauchn. Semin. V. A. Steklov Mathematical Institute in St.-Petersburg (POMI) 288 (2002), pp. 178–203.

[290] ——–, *A posteriori error estimates taking into account indeterminacy of the problem data,* Russian J. Numer. Anal. Math. Modelling 18 (2003), pp. 507–519.

[291] ——–, *A posteriori estimates in local norms. Problems in mathematical analysis,* J. Math. Sci, (New York) 124 (2004), pp. 5026–5035.

[292] ——–, *Functional type a posteriori error estimates for approximate solutions of problems with incompressibility condition,* European Congress on Computational Methods in Applied Sciences and Engineering, ECCOMAS 2004, Jyväskylä, 24–28 July, 2004, edited by P.Neittaanmäki, T. Rossi, K. Majava, and O. Pironeau, O. Nevanlinna and R. Rannacher (assoc. eds.), (electronic).

[293] ——–, *Estimates of deviations from exact solutions for some boundary-value problems with incompressibility condition, Algebra and Analiz* 16 (2004), pp. 124–161.

[294] ——–, *Local a posteriori estimates for the Stokes problem,* Zap. Nauchn, Sem. S.-Peterburg Otdel. Mat. Inst. Steklov. (POMI) 318 (2004), pp. 233–245.

[295] ——–, *Functional Approach to Locally Based A Posteriori Error Estimates for Elliptic and Parabolic Problems,* Proc. 6th European Conference on Numerical Mathematics and Advanced Applications (Santiago de Compostela) 2005, Springer-Verlag, Berlin, 2006, pp. 133–148.

[296] ——–, *On the derivation of functional a posteriori estimates from integral identities,* Advances in numerical mathematics. Proceedings of International Conference on the occasion of the 60th birthday of Yu. A. Kuznetsov Moscow, September 16–17, 2005, W. Fitzgibbon, R. Hoppe, J. Periaux, O. Pironneau, and Y. Vassilevski, Institute of Numerical Mathematics Russian Academy of sciences and Department of Mathematics University of Houston, USA, Moscow 2006, pp. 217–242.

[297] ——–, *Functional a posteriori estimates for the Maxwell's problem,* J. Math. Sci. New York 142 (2007), pp. 1821–1827.

[298] ——–, *Estimates of deviations from exact solutions of elliptic variational inequalities,* Zap. Nauchn, Sem. S.-Peterburg, Otdel. Mat. Inst. Steklov. (POMI) 326 (2007), pp. 147–164.

[299] ——–, *On a posteriori error estimates for the stationary Navier–Stokes problem, J. Math. Sci.* 150 (2008), pp. 1885–1889.

[300] ———, *A posteriori error estimation methods for partial differential equations,* In *Lectures on Advanced Computational Methods in Mechanics*, edited by J. Kraus and U. Langer, Walter de Gruyter, Berlin, 2007, pp. 161–226.

[301] ———, *A posteriori error estimates for evolutionary reaction-convection-diffusion problem,* Zap. Nauchn, Sem. St. Peterburg, Otdel. Mat. Inst. Steklov. (POMI) (2008), to appear.

[302] ———, *Advanced forms of functional a posteriori estimates for elliptic problems,* Russian J. Numer. Anal. Math. Modelling, to appear.

[303] S. Repin and M. Frolov, *A posteriori error estimates for approximate solutions of elliptic boundary value problems,* Zh. Vychisl. Mat. Mat. Fiz. 42 (2002), pp. 1774–1787 (in Russian).

[304] ———, *An estimate for deviations from the exact solution of the Reissner–Mindlin plate problem,* J. Math. Sci, (New York) 132 (2006), pp. 331–338.

[305] S. Repin and S. Sauter, *Functional A Posteriori Estimates for the Reaction-Diffusion Problem,* C. R. Acad. Sci. Paris, Ser. 1 343 (2006), pp. 349–354.

[306] ———, *Computable estimates of the modeling error related to Kirchhoff–Love plate model,* Preprint of the University of Zürich, 2008.

[307] S. Repin, S. Sauter, and A. Smolianski, *A posteriori error estimation for the Dirichlet problem with account of the error in the approximation of boundary conditions,* Computing 70 (2003), pp. 205–233.

[308] ———, *Duality Based A Posteriori Error estimator for the Dirichlet Problem,* Proc.Appl. Math. Mech. 2 (2003), pp. 513–514.

[309] ———, *A posteriori estimation of dimension reduction errors for elliptic problems in thin domains, SIAM J. Numer. Anal.* 42 (2004), pp. 1435–1451.

[310] ———, *A Posteriori Control of Dimension Reduction Errors on Long Domains,* Proceedings in Applied Mathematics and Mechanics 4 (2004), pp. 714–715.

[311] ———, *A posteriori error estimation for the Poisson equation with mixed Dirichlet/Neumann boundary conditions,* J. Comput. Appl. Math. 164/165 (2004), pp. 601–612.

[312] ———, *A posteriori estimation of dimension reduction errors,* Proc. 5th European Conference on Numerical Mathematics and applications (Prague) 2004, Springer-Verlag, pp. 717–725.

[313] ———, *Two-Sided A Posteriori Error Estimates for Mixed Formulations of Elliptic Problems,* SIAM J. Numer. Anal. 45 (2007), pp. 928–945.

[314] S. Repin and A. Smolianski, *Functional-type a posteriori error estimates for mixed finite element methods, Russian J. Numer. Anal. Math. Modelling* 20 (2005), pp. 365–382.

[315] S. Repin and R. Stenberg, *Two-sided a posteriori estimates for the generalized Stokes problem,* Helsinki University of Technology, 2006, Preprint A516.

[316] ———, *A posteriori error estimates for the generalized Stokes problem,* J. Math. Sci. New Jork 142 (2007), pp. 1828–1843.

[317] S. Repin and S. Tomar, *A posteriori error estimates for nonconforming approximations of elliptic problems based on Helmholtz decomposition of the error,* Radon Institute of Computational and Applied Mathematics, Preprint 2007-40.

[318] S. Repin and J. Valdman, *Functional a posteriori estimates for incremental models in elasto-plasticity*, Technical Report, Radon Institute of Computational and Applied Mathematics (RICAM), Preprint 2007-41.

[319] ———, *A posteriori error estimates for problems with nonlinear boundary conditions,* J. Numer. Math. 16 (2008), pp. 51–81.

[320] S. Repin and L. S. Xanthis, *A posteriori error estimation for elasto-plastic problems based on duality theory,* Comput. Methods Appl. Mech. Engrg. 138 (1996), pp. 317–339.

[321] ———, *A posteriori error estimation for nonlinear variational problems,* Comptes Rendus de l'Académie des Sciences, Mathématique 324 (1997), pp. 1169–1174.

[322] W. C. Rheinboldt, *On a theory of mesh-refinement processes,* SIAM J. Numer. Anal. 17 (1980), pp. 766–778.

[323] W. Ritz, *Über eine neue Methode zur Lösung gewisser Variationsprobleme der Mathematischen Physik,* J. Reine Angew. Math. 135 (1909), pp. 1–61.

[324] R. T. Rockafellar, *Convex analysis*, Princeton Univ. Press, 1970.

[325] J. E. Roberts and J.-M. Thomas, *Mixed and hybrid methods,* Handbook of numerical analysis, Vol. II, Handb. Numer. Anal., II, North-Holland, Amsterdam, 1991, pp. 523–639.

[326] R. Rodrigues, *Some remarks on Zienkiewicz–Zhu estimator,* Numer. Methods for PDE 10 (1994), pp. 625–635.

[327] J. Saranen, *On an inequality of Friedrichs,* Math. Scand. 51 (1982), pp. 310–322.

[328] O. Scherzer, H. W. Engl, and K. Kunisch, *Optimal a posteriori parameter choice for Tikhonov regularization for solving nonlinear ill-posed problems,* SIAM J. Numer. Anal. 30 (1993), pp. 1796–1838.

[329] J. Schröder, *Operator Inequalities,* Academic Press, New York, 1980.

[330] M. Schultz and O. Steinbach, *A new a posteriori error estimator in adaptive direct boundary element methods: the Dirichlet problem,* Calcolo 37 (2000), pp. 79–96.

[331] M. Schulz and W. L. Wendland, *A general approach to a posteriori error estimates for strictly monotone and Lipschitz continuous nonlinear operators illustrated in elasto-plasticity,* Proc. 2nd European Conference on Numerical Mathematics and Advanced Applications 1997 (Heidelberg), World Sci, Publishing, River Edge, New York, 1998, pp. 572–579.

[332] C. Schwab, *A-posteriori modelling error estimation for hierarchic plate models,* Numer. Math. 74 (1996), pp. 221–259.

[333] S. Shaw and J. R. Whiteman, *Negative norm error control for second-kind convolution Volterra equations,* Numer. Math. 85 (2000), pp. 329–341.

[334] K. G. Siebert, *An a posteriori error estimation for anisotropic refinement,* Numer. Math. 73 (1996), pp. 373–398.

[335] M. G. Slobodyanskii, *Estimates of the error of an approximate solution in linear problems reducing to variational ones, and their application to the determination of two-sided approximations in static problems of the theory of elasticity,* Akad. Nauk SSSR. Prikl. Mat. Meh. 16 (1952), pp. 449–464.

[336] S. L. Sobolev, *Some Applications of Functional Analysis in Mathematical Physics*, Izdt. Leningrad. Gos. Univ., Leningrad, 1955 (in Russian, translated in Translation of Mathematical Monographs, Volume 90, American Mathematical Society, Providence, RI, 1991).

[337] E. Stein, F. J. Bartold, S. Ohnimus, and M. Schmidt, *Adaptive finite elements in elastoplasticity with mechanical error indicators and Neumann-type estimators,* Advances in Adaptive Computational Methods in Mechanics, edited by P. Ladevéze and J. T. Oden, Elsevier, New York, 1998, pp. 81–100.

[338] E. Stein and S. Ohnimus, *Coupled model- and solution-adaptivity in the finite element method,* Comput. Methods Appl. Mech. Engrg. 150 (1997), pp. 327–350.

[339] ——–, *Coupled anisotropic discretization and modelerror estimation in solid mechanics by local Neumann problems,* Comput. Methods Appl. Mech. Engrg. 176 (1999), pp. 363–385.

[340] E. Stein, M. Rüter, and S. Ohnimus, *Error-controlled adaptive goal-oriented modeling and finite element approximations in elasticity,* Comput. Methods Appl. Mech. Engrg. 196 (2007), pp. 3598–3613.

[341] R. Stenberg, *Some new families of finite elements for the Stokes equations,* Numer. Math. 56 (1990), pp. 827–838.

[342] ——–, *Error analysis of some finite element methods for the Stokes problem,* Math. Comput. 54 (1990), pp. 495–508.

[343] G. Stoyan, *Towards Discrete Velte Decompositions and Narrow Bownds for Inf-Sup Constants,* Computer and Mathematics with Applications, 38 (1999), pp. 243–261.

[344] G. Strang and G. Fix, *An analysis of the finite element method,* Prentice Hall, Englewood Cliffs, 1973.

[345] T. Strouboulis and J. T. Oden, *A posteriori error estimation of finite element approximations in fluid mechanics,* Comput. Meth. Appl. Mech. Engrg. 78 (1990), pp. 201–242.

[346] S. Sun and M. F. Wheeler, *$L^2(H^1)$-norm a posteriori error estimation for Discontinuous Galerkin approximations of reactive transport problems,* Journal of Scientific Computing 22/23 (2005), pp.501–530.

[347] J. L. Synge, *The hypercircle method,* Studies in numerical analysis (papers in honour of Cornelius Lanczos on the occasion of his 80th birthday), Academic Press, London, 1974, pp. 201–217.

[348] R. Temam, *Navier–Stokes equations. Theory and numerical analysis,* Studies in Mathematics and its Applications, vol. 2, North-Holland, Amsterdam, 1979.

[349] ——–, *Problèmes mathématique en plasticité*, Bordas, Paris, 1983.

[350] S. Tomar and S. Repin, *Efficient computable error bounds for dicontinuous Galerkin approximations of elliptic problems*, Radon Institute of Computational and Applied Mathematics, Preprint 2007-39.

[351] K. Tsuruta and K. Ohmori, *A posteriori error estimation for Volterra integro-differential equations,* Mem. Numer. Math. 1976, pp. 33–47.

[352] S. Turek, *Efficient solvers for incompressible flow problems: An algorythmic and computational approach*, Springer-Verlag, Berlin, 1999.

[353] J. P. Vanyo, *Rotating fluids in engineering and science,* Dover Publ. Inc., New York, 2001.

[354] R. Verfürth, *A posteriori error estimators for the Stokes equations,* Numer. Math. 55 (1989), pp. 309–326.

[355] ——–, *A posteriori error estimates for nonlinear problems. Finite element discretisations of elliptic equations,* Math. Comput. 62 (1994), pp. 445–475.

[356] ——–, *A review of a posteriori error estimation and adaptive mesh-refinement techniques*, Wiley and Sons, Teubner-Verlag, New-York, 1996.

[357] ——–, *A posteriori error estimates for nonlinear problems. $L^r(0,T;L^p(\Omega))$-error estimates for finite element discretizations of parabolic equations,* Math. Comput. 67 (1998), pp. 1335–1360.

[358] ———, *A posteriori error estimators for convection-diffusion problems,* Numer. Math. 80 (1998), pp. 641–663.

[359] ———, *A posteriori error estimates for finite element discretizations of the heat equation, Calcolo* 40 (2003), pp. 195–212.

[360] ———, *Robust a posteriori error estimates for sationary convection-diffusion equations,* SIAM J. Numer. Anal. 43 (2005), pp. 1766–1782.

[361] A. Vesser, *Efficient and reliable a posteriori error estimators for elliptic obstacle problems,* SIAM J. Numer. Anal. 39 (2001), pp. 146–167.

[362] M. Vishik , *La methode des projections orthogonales pour les equations differentielles conjuguees a ellesmemes,* Doklady Acad. Sci. URSS (N.S.) 56 (1947), pp. 115–118.

[363] M. Vogelius and I. Babuška, *On a dimensional reduction method I. The optimal selection of basis functions,* Math. Comput. 37 (1981), pp. 31–46.

[364] ———, *On a dimensional reduction method III. A posteriori error estimation and an adaptive approach,* Math. Comput. 37 (1981), pp. 361–384.

[365] M. Vohralik, *Equivalence between mixed finite element and multi-point finite volume methods,* C. R. Acad. Sci, Paris, Ser. I 339 (2004), pp. 525–528.

[366] L. B. Wahlbin, *Superconvergence in Galerkin Finite Element Methods*, Lecture Notes in Mathematics, vol. 1605, Springer-Verlag, Berlin, 1995.

[367] J. Wang, *Superconvergence analysis of finite element solutions by the least-squares surface fitting on irregular meshes for smooth problems,* J. Math. Study 33 (2000), pp. 229–243.

[368] J. Wang and X. Ye, *Superconvergence analysis for the NavierStokes equations,* Applied Numerical Mathematics 41 (2002), pp. 515527.

[369] H. Weil, *The method of orthogonal projections in potential theory,* Duke Math. J. 7 (1940), pp. 411–444.

[370] J. R. Whiteman and G. Goodsell, *A survey of gradient superconvergence for finite approximations to second order elliptic problems on triangular and tetrahedral meshes,* The Mathematics of Finite Elements and Applications VII, edited by J. R. Whiteman, Academic Press, 1991, pp. 55–74.

[371] N.-E. Wiberg, F. Abdulwahab, and S. Ziukas, *Enhanced superconvergent patch recovery incorporating equilibrium and boundary conditions,* Int. J. Num. Meth. Engng. 37 (1994), pp. 3417–3440.

[372] B. I. Wohlmuth and R. H. W. Hoppe, *A comparison of a posteriori error estimators for mixed finite element discretizations by Raviart–Thomas elements,* Math. Comput. 68 (228) (1999), pp. 1347–1378.

[373] T. Yamamoto, *A note on a posteriori error bound of Zabrejko and Nguen for Zincenko's iteration,* Numer. Funct. Anal. Optim. 9 (1987), pp. 987–994.

[374] K. Yosida, *Functional Analysis,* Springer-Verlag, New Tork, 1996.

[375] S. Zaremba, *Sur un probleme toujours possible comprenant, a titre de cas particuliers, le probleme de Dirichlet et celui de Neumann,* J. math. pures et appl. 6 (1927), pp. 127–163.

[376] E. Zeidler, *Nonlinear functional analysis and its applications. I. Fixed-point theorems,* Springer-Verlag, New York, 1986.

[377] Z. Zhang and A. Naga, *A new finite element gradient recovery method: superconvergence property,* SIAM J. Sci. Comput. 26 (2005), pp. 1192–1213.

[378] J. Z. Zhu, *A posteriori error estimation – the relationship between different procedures,* Comput. Methods Appl. Mech. Engrg. 150 (1997), pp. 411–422.

[379] O. C. Zienkiewicz, *Achievements and some unsolved problems of the finite element method,* Int. J. Numer. Meth. Engrg. 47 (2000), pp. 9–28.

[380] O. C. Zienkiewicz, B. Boroomand, and J. Z. Zhu, *Recovery procedures in error estimation and adaptivity: Adaptivity in linear problems,* Advances in Adaptive Computational Methods in Mechanics, edited by P. Ladevéze and J. T. Oden, Elsevier, New York, 1998, pp. 3–23.

[381] O. C. Zienkiewicz and K. Morgan, *Finite elements and approximation*, Wiley and Sons, New York, 1983.

[382] O. C. Zienkiewicz and J. Z. Zhu, *A simple error estimator and adaptive procedure for practical engineering analysis,* Internat. J. Numer. Meth. Engrg. 24 (1987), pp. 337–357.

[383] ———, *Adaptive techniques in the finite element method,* Commun. Appl. Numer. Methods 4 (1988), pp. 197–204.

[384] ———, *The superconvergent patch recovery and a posteriori error estimates. Part 1: The recovery technique,* Int. J. Numer. Meth. Engrg. 33 (1992), pp. 1331–1364.

[385] M. Zlámal, *Some superconvergence results in the finite element method. Mathematical aspects of finite element methods,* Lecture Notes in Math., Vol. 606, Springer-Verlag, Berlin, 1977, pp. 353–362.

[386] ———, *Superconvergence and reduced integration in the finite element method,* Math. Comput. 32 (1978), pp. 663–685.

Notation

$:=$	equals by definition
$\forall$	for all
$\Rightarrow$	implies
$\mathbb{R}^d$	space of real d-vectors
$\mathbb{M}^{d\times d}$	space of real $d \times d$ matrixes
$\mathbb{M}_s^{d\times d}$	space of symmetric real $d \times d$ matrixes
Ω, ω	open sets in $\mathbb{R}^d$
$\Gamma, \Gamma_1, \Gamma_2$	boundary of Ω and its parts
n	exterior unit normal
$\mathrm{meas}_d\{\omega\}$	Lebesgue measure of a set $\omega \in \mathbb{R}^d$
$\mathrm{diam}\, S$	diameter of the set S
$a \cdot b$	scalar product of vectors
$a \otimes b$	tensor product of vectors
$a \times b$	vector product of vectors
$\sigma : \varepsilon$	scalar product of tensors
$\mathrm{tr}\sigma$	trace of σ
σ^{D}	deviator of σ; $\sigma^{\mathrm{D}} := \sigma - \frac{1}{d}\mathbb{I}\,\mathrm{tr}\,\sigma$
$\{g\}_\omega$	mean value of g on ω; $\{g\} := \frac{1}{\lvert\omega\rvert}\int_\omega g\,dx$
$\widetilde{g}_\omega$	$\widetilde{g}_\omega := g - \{g\}_\omega$
$\mathrm{supp}\, g$	support set of g
$[g]_\Gamma$	jump of g on Γ
$(g)_-$	negative part of g
$(g)_+$	positive part of g
$\mathrm{Ker}\, T$	kernel of the operator T
$v_{,i}$	partial derivative with respect to i-th coordinate; $v_i := \frac{\partial v}{\partial x_i}$
∇	gradient of a scalar-valued function; $\nabla\phi := (\phi_{,1}, \ldots, \phi_{,d})$
div	divergence of a vector-valued function; $\mathrm{div}\, v := \sum_{i=1,d} v_{i,i}$
Div	divergence of a tensor-valued function; $(\mathrm{Div}\,\tau)_j := \sum_{i=1,d} \tau_{ij,i}$
curl	rotor of a vector-valued function; $\mathrm{curl}\, v := (v_{3,2} - v_{2,3}; v_{1,3} - v_{3,1}; v_{2,1} - v_{1,2})$
Δ	Laplace operator; $\Delta v = \mathrm{div}\,\nabla v$
$\overline{\mathfrak{M}}$	error majorant
$\underline{\mathfrak{M}}$	error minorant
e	error
$\mathcal{E}$	error indicator
I_{eff}	efficiency index of an error estimate
$\mathcal{B}(x, r)$	ball of the radius r centered at x

Υ	set of uncertain data
$\boldsymbol{S}$	set of possible solutions
P	projection operator
$L^p(\omega)$	space of functions integrable in ω with power p
$\widetilde{L}^2(\Omega)$	set of functions in $L^2(\Omega)$ with integral mean 0
$\Vert\cdot\Vert_{p,\omega}$	$L^p(\omega)$ norm
$\Vert\cdot\Vert$, $\Vert\cdot\Vert_\Omega$	$L^2(\Omega)$ norm
$\Vert\cdot\Vert_{\infty,\Omega}$	supremum norm
$W^{l,p}(\Omega)$	Sobolev space of functions having generalized derivatives up to order l integrable with power p
$\Vert\cdot\Vert_{l,p,\Omega}$	$W^{l,p}(\Omega)$ norm
$\vert\Vert\cdot\vert\Vert$	energy norm
$\vert\Vert[\cdot]\vert\Vert$	"broken" energy norm
$\Vert\cdot\Vert_{[\phi]}$	weighted norm with weight ϕ
$C_{F\Omega}$	constant in the Friedrichs inequality for the domain Ω
$C_{P\Omega}$	constant in the Poincaré inequality for the domain Ω
$\Sigma(\Omega)$	space $L^2(\Omega, \mathbb{M}^{d\times d})$ of tensor-valued functions with square summable components
$\Sigma_s(\Omega)$	space $L^2(\Omega, \mathbb{M}_s^{d\times d})$ of symmetric tensor-valued functions with square summable components
$\mathrm{S}(\Omega)$	subspace of $L^2(\Omega, \mathbb{R}^d)$ formed by divergence-free functions
$\overset{\circ}{\mathrm{S}}{}^1(\Omega)$	the closure of smooth solenoidal functions with compact supports in Ω with respect to the norm of $H^1(\Omega, \mathbb{R}^d)$
$H^{-1}(\Omega)$	space dual to $\overset{\circ}{H}{}^1(\Omega)$
$H(\Omega, \mathrm{div})$	$:= \{v \in L^2(\Omega) \mid \mathrm{div}\, v \in L^2(\Omega)\}$
$\Vert\cdot\Vert_{\mathrm{div}}$	norm of $H(\Omega, \mathrm{div})$
$H(\Omega, \mathrm{Div})$	$:= \{\tau \in L^2(\Omega, \mathbb{M}^{d\times d}) \mid \mathrm{Div}\, \tau \in L^2(\Omega, \mathbb{R}^d)\}$
$\Vert\cdot\Vert_{\mathrm{Div}}$	norm of $H(\Omega, \mathrm{Div})$
$\mathcal{L}(X, Y)$	space of linear bounded operators acting from X to Y
$\langle v^*, v\rangle$	duality pairing between $v^* \in V^*$ and $v \in V$
F^*	polar functional (Young–Fenchel conjugate of F), $F^*(v^*) := \sup_v\{\langle v^*, v\rangle - F(v)\}$
F^{**}	bipolar functional $F^{**}(v) := \sup_{v^*}\{\langle v^*, v\rangle - F^*(v^*)\}$
$D(v, v^*)$	compound functional, $D(v, v^*) := F(v) + F^*(v^*) - \langle v^*, v\rangle$
$\partial F(v)$	subdifferential of F at v

Index